Plant Collectors in Angola

Regnum Vegetabile
Volume 161

REGNUM VEGETABILE is the book series of the International Association for Plant Taxonomy and is devoted to systematic and evolutionary biology with emphasis on plants, algae, and fungi. Preference is given to works of a broad scope that are of general importance for taxonomists.

All manuscripts intended for publication in REGNUM VEGETABILE are submitted to the Editor-in-chief. Authors interested in publishing in the series are requested to send an outline of their book, including a brief description of the content, to the Editor-in-chief. Proposals for new book projects will be scrutinized by the editorial advisory board.

Plant Collectors in Angola

Botany, Exploration, and History in South-Tropical Africa

Estrela Figueiredo and Gideon F. Smith

The University of Chicago Press
Chicago and London

The University of Chicago Press, Chicago 60637
The University of Chicago Press, Ltd., London

Published 2024
Printed in China

33 32 31 30 29 28 27 26 25 24   1 2 3 4 5

ISBN-13: 978-0-226-83206-7 (cloth)
ISBN-13: 978-0-226-83208-1 (paper)
ISBN-13: 978-0-226-83207-4 (e-book)
DOI: https://doi.org/10.7208/chicago/9780226832074.001.0001

Library of Congress Control Number: 2023018057

♾ This paper meets the requirements of ANSI/NISO Z39.48-1992 (Permanence of Paper).

This book is dedicated to

José Alberto de Oliveira Anchieta (1832–1897), maverick Portuguese naturalist who settled in Angola and represents all of those who sacrificed comfort and riches for the love of science.

(Image: Artist unknown. Reproduced from *O Occidente* 30 November 1897. Public domain.)

Maria Fernanda Duarte Pinto Basto da Costa Ferreira (1938–), Angolan technician who represents those working behind the scenes, without whom many collections made in Angola would remain unprocessed and undetermined.

(Image: In 2003, at Herb. LISC. Photographer and private collection of Estrela Figueiredo.)

CONTENTS

FOREWORD

It is fitting that *Plant collectors in Angola* should be published in *Regnum Vegetabile*. The second volume in this series, which began almost 70 years ago, was the first of a seven-part guide to plant-collectors worldwide and the location of their herbarium specimens. While that guide is enormously useful in helping botanists track down elusive specimens, it only provides the most basic biographical detail for each collector: birth and death dates along with country of activity and collecting dates, when known. And, of course, lists of the herbaria that house the specimens made. It is therefore very satisfying to see what formerly were basic records for collectors active in Angola, such as "Gossweiler, John (João, 1873–1951)" and "Welwitsch, Friedrich Martin Josef (1806–72)," transformed into full-bodied biographical sketches that flesh out their activities and personalities, while also shedding some light on what motivated them to explore the rich flora of southwestern Africa. As anyone who has dealt with Angola's flora knows, the names Gossweiler and Welwitsch are effectively synonymous with Angolan botany because their collections are the ones most likely to be encountered. These two "collectors extraordinaire" were the most prolific individual collectors active in Angola, amassing 14,600 and 10,000 numbers, respectively. In addition to expanded biographical accounts of Gossweiler and Welwitsch, the authors of *Plant collectors in Angola* provide us with information on over 350 other individuals who prepared herbarium specimens in Angola, a good number of them previously recorded simply as a name and some of them previously overlooked entirely.

Compiling useful information on plant collectors and their activities is a labour-intensive and time-consuming task, and it requires tremendous attention to detail. The seed for the present volume on Angola evidently was planted some time ago and only now has it matured into the work presented here. In 2008, Figueiredo and Smith published a checklist of the vascular flora of Angola (*Plants of Angola / Plantas de Angola*), which included a basic list of collectors active in Angola. Subsequently, they published *Common names of Angolan plants* (2012; 2017, 2nd ed.). Both compilations involved substantial research including the examination of herbarium specimens, many deposited in Portuguese institutions as well as southern African ones. I know from experience that anyone who has spent time focused on studying the herbarium material from a defined region quickly learns the idiosyncrasies of the different collectors who were active there; what they collected; where they collected; the names of their collecting companions; how they labelled and sequenced their specimens; and where the specimens (and duplicates) were deposited. Sharing this information is critically important if we want to expedite and facilitate revisionary, floristic, and conservation research.

Estrela Figueiredo and Gideon Smith are especially well-qualified to have written this account of plant collecting in Angola. Figueiredo has extensive experience in Portuguese-speaking countries in West Africa as well as in South Africa, while Smith has had a lengthy and productive career in South Africa that has combined research, administration, and participation in many international botanical initiatives. In anticipation of *Plant collectors in Angola*, Figueiredo and Smith authored or co-authored a score of articles detailing the botanical activities of plant collectors in Angola. Their collaborative efforts, however, are not confined solely to Angolan plants. They also have jointly published on the systematics and nomenclature of southern African succulents, including field and horticultural guides.

Our knowledge of plant life in tropical areas first developed, in part, as the by-product of a clash of cultures: western European nations competing with one another to exploit the resources of "newly discovered" continents while subjugating the inhabitants of the same. The story is not always pretty and certain parts of the introductory materials and some of the biographical sketches presented by Figueiredo and Smith should make us uncomfortable. The slave trade especially casts a pall over the earliest botanical exploration of Angola. Additionally, more than one of the collectors treated in this volume was in service to or otherwise involved with Leopold II, who created the Congo Free State (now the D.R. Congo); another heinous episode in human history. In more recent times, Angola's war of independence from Portugal (1961–1974) followed by its protracted civil war (1975–2002) effectively halted botanical exploration for several generations. A number of the biographies in this volume bear witness to the consequences of these disruptions: not only did these armed conflicts negatively impact individual lives, families, and careers, but they also led to the loss of established institutional herbarium collections.

Happily, botanical exploration is much more than the product of imperialistic ambitions. There are an infinite number of reasons why one might choose to collect plants, some of them as simple as curiosity. One can take delight in the foibles and tribulations of some Angolan expeditions and stand in awe at the amazing drive of others. One has to be amused by Joaquim Monteiro and his wife on their last trip to Bembe, who were kitted out with everything they needed including an "inflatable rubber bathtub". One has to have admiration for the pluck of two women, Mary Pocock and Dorothea Frances Bleek, who traversed the continent on foot and in *machila* (hammock) by themselves with native porters, and yet managed to photograph, paint, write in their journals, and collect a significant number of herbarium specimens. One cannot help but be amazed by the determination of Gilbert Reynolds in his single-minded pursuit of aloes, struggling to tow his caravan (camper) where motorized vehicles were never intended to travel, yet persisting and discovering new species.

When one scans all the biographical entries in *Plant collectors in Angola*, the relative paucity of Angolans who have made herbarium collections in their own country is striking: the biographies are overwhelmingly those of western European men. Most of their collections, however, would not exist if the people of Angola had not accepted subsidiary albeit essential roles as translators, guides, drivers, cooks, porters, and servants. Some Angolans successfully transcended these supporting roles and stand out as important plant collectors in their own right. The collecting activities of Angolan-born Joaquim Teixiera rival those of the Swiss-born Gossweiler or the Austrian-born Welwitsch. The Tchokwe *soba* (chief) Matende Sanjinje's work for the *Museu do Dundo* in Lunda, appears to have begun simply but it clearly took on increasing complexity over time. Several more contemporary Angolan botanists, especially herbarium curators and curatorial staff, appear to have benefited from support from the Southern African Botanical Diversity Network (SABONET), but the profiles in *Plant collectors in Angola* include only those active up until 2000, which is just before the cessation of armed conflict in the country. One hopes that the demographic composition of plant collectors in Angola has and will change to include more African-born botanists, but much depends upon improving economic conditions and a different, hopefully improved relationship with western European and American (including Brazilian) botanists and funding agencies.

I have no doubt that *Plant collectors in Angola* will become an essential reference tool for everyone studying African plant specimens. It will join Gunn and Codd's *Botanical exploration of southern Africa* (1981) and Polhill and Polhill's *East African plant collectors* (2015), exemplary models of "historical-botanical studies" focused on continental Africa, on the bookshelves of not only botanists but also Africanists. One hopes that *Plant collectors in Angola* will be a prelude to a new floristic effort. It certainly will make future research on the flora of Angola and southern continental Africa easier and more scholarly. In any case, this effort by Figueiredo and Smith combined with their earlier checklist and guide to common names of Angolan plants places them among a select few botanists whose publications have had a lasting impact on the literature essential for understanding the flora of Angola.

Laurence J. Dorr
Curator and Research Scientist,
U.S. National Herbarium, Smithsonian Institution,
Washington, D.C., U.S.A.

PREFACE

When we first started profiling now largely forgotten biodiversity specialists – especially plant collectors – who often operated well away from the countries where they were born, we soon came under the impression of not only the often idiosyncratic nature of the individuals, but most profoundly also their dedication and resourcefulness, and the often ingenious compulsions and desires that drove them all. Even those who earned their living from the act of collecting specimens, often on behalf of thankless foreign governments or wealthy patrons, had to have a commitment to what we today know as biodiversity science that few can fathom. Collecting biodiversity specimens can hardly feed material dreams, even if being paid, or perhaps more accurately underpaid, to do so, as the rewards reaped, including discovering species new to science, are most usually a long time coming. In the 21st century it is often difficult to fathom the challenging conditions under which early explorers had to operate, and what they had to endure, at least in the tropics, including being exposed to largely unknown diseases without the benefit of easy access to modern – any, really – medical science.

This book is not only about the particulars of the lives and activities of committed individuals – essentially about plant collectors in Angola – and their remarkable achievements while often enduring hardships well beyond modern grasp. To better understand and contextualise their work, information about the often less-than-desirable socio-economic and -political circumstances under which they operated must be taken into account to better appreciate how it came about that today scores of herbaria, museums, and other institutions with preserved plant collections under their custodianship, especially in Europe but also elsewhere, hold material from Angola (Figueiredo & Smith, 2021a). Regardless, for many of the collectors, their activities in the field were self-expression that came from an inextinguishable desire to amass collections that will feed science and an educated, curious public waiting to be impressed by the next significant find, like the *Hoodia* of Andrew Curror, a "noble flowering specimen" in the words of William J. Hooker who described it (Hooker, 1844), or the *Welwitschia* of Friedrich Welwitsch, "out of all question the most wonderful plant ever brought to [Britain] – and the very ugliest" (J.D. Hooker as cited in Swinscow, 1972: 269). For the collectors it was often a simple way of nourishing their imagination and expressing their freedom to roam foreign lands, albeit without modern modes of transport, and to discover places he or she might never otherwise have had the opportunity and privilege to visit.

It is likely that many collectors knew that they would never see the fruit of their endeavours. Yet, their commitment and work ethic remained unwavering, and several were prepared to study their own, and others', specimens day in and day out once the opportunity arose. To tackle the writing of a Flora, for example, of a little-known country from which very few specimens had been collected requires long, laborious hours of herbarium and laboratory work, with no guarantee that the work will ever be finished. (Some Floras took many decades to write, and some remain incomplete, that of Angola, the *Conspectus florae Angolensis*, being a case in point.) In the view of many collectors, meticulously collecting specimens and recording information for a collecting label has the imagined power to change the world – is this the specimen that will prove a theory? Or is it plant material that will yield a chemical long sought after? Perhaps it belongs to a species that fills the long-existing knowledge gap and that is required to confidently speculate about infrageneric relationships. Fortunately, biodiversity specimens are meant to be preserved, treasured, and shared, and generations later, the material collected by pioneering collectors is still being studied by a fresh crop of plant scientists.

With a land surface area of 1,246,700 km^2 [481,354 sq. mi.], Angola is the largest country in Africa south of the equator – it is the 7th-largest African country; it is rich in plant diversity and has a flora of more than 7000 species. With two plant diversity catalogues comprehensively detailing scientific and common names (Figueiredo & Smith, 2008, 2012, 2017) having become available, floristic studies in Angola have truly come of age, as few other political regions globally can lay claim to having two such documents at their disposal (Freitas, 2017: 9).

Up to now, a full, contextualised record of the botanical exploration of Angola has been lacking. Fortunately, the enormous volume of data generated through the studies that gave rise to the plant diversity and common names catalogues finally paved the way for the completion of the present volume: a complete and illustrated history of the botanical exploration of Angola from the earliest times until 2000, with biographical accounts of the plant collectors who contributed to amassing specimens for herbaria and so concomitantly recording the flora of the country. We did not apply a minimum cut-off figure – even collectors who made only a single collection that ended up in a herbarium are included in this book. Therefore, every effort was made to record all the collectors who were active in Angola. However, as the databasing and digitisation of herbarium collections progress it is possible that more names will come to light. The number of collectors treated in this book is already over 10% higher than the figure given in a preliminary list that was provided in a broad overview of collecting in Angola, which emphasised aspects such as gender and race issues (Figueiredo & Smith, 2021a).

Very few plant collectors who were active in Angola have been well anthologised and documented. The best-known of these are Friedrich Welwitsch and John Gossweiler, who were active in Angola from the 1850s to the early 1860s and the first half of the 20th century, respectively. However, the vast majority of the collectors were sorely in need of biographical treatment and analysis when this project was initiated. The profiles presented in this book necessarily have to be concise. After all, more than 350 collectors are recorded here. Yet behind most of the summaries presented, stories that need to be told abound. For this reason, among other things, as comprehensive as possible a bibliography is included.

INTRODUCTION

"What, they inquired, can these men want in the interior, if they are not seeking trade?" (Capelo & Ivens, 1882: 121)

Historical background

As early as the 15th century, Portuguese ships were travelling further and further south along Africa's western Atlantic coast as part of efforts to find a sea route to India, and beyond, to access and ideally monopolise the trade in spices that Europe desired (Richards & Place, 1960: xvi–xx; Wiencek, 1981). In 1488, Bartholomeu Dias rounded the Cape, which he named the "Cape of Storms", but which was later rechristened the "Cape of Good Hope" by Portugal's King John, after Dias had returned to Lisbon.

Portugal did not establish a victualing station at the Cape and it was only in 1652 that the Dutch, through the Dutch East India Company, founded what eventually developed into Cape Town, expanded into the Cape Colony, and later still, today's Republic of South Africa. The region held little attraction for the Portuguese who were after minerals, slaves, and opportunities to trade European wares for local goods. Portuguese seafarers did, however, leave a mark at the southern tip of Africa, for example through the post office tree in Mosselbaai, with the bay itself first being named *Aguada de São Bras* by Vasco da Gama in 1497 and, later, *Golfo dos Vaqueiros* by João da Nova in 1501, before the Dutch navigator Paulus van Caerden named it Mosselbaai in July 1601 (Raper & al., 2014: 343). Portuguese seafarers also contributed to the names of marine life they encountered off the African coast. For example, through Dutch, the Afrikaans word *pikkewyn* is derived from the Portuguese *pinguim*, and the common Afrikaans fish name *daeraad* was derived from the Portuguese *dourado*, which, when translated directly, means "gold coloured"

(Scholtz, 1941: 3–7). Interestingly, the break of day (*daeraad*) is, of course, also often golden – a fortunate coincidence.

In 1571, Bartholomeu Dias's grandson Paulo Dias de Novais was granted colonising rights in Angola and a few years later he established the town of São Paulo da Assunção de Loanda, now the city of Luanda. A second town, São Filipe de Benguela (now the city of Benguela), was established in 1617. The occupation of Angola by the Portuguese was slow at first, as Brazil was then the focus of their emerging economic interests and ventures in overseas territories. Brazil required the importation of labour to work in gold mines and in plantations, including of sugar cane, and Portugal became a major player in the slave trade. For example, between 1700 and 1810, about 1.8 million slaves were taken to Brazil. Although the slave trade in Brazil was abolished in 1831, it was only in 1850 that slave markets closed, and it took decades before all the slaves were emancipated.

In fact, the trade in slaves continued long after it was abolished, and slavery in Africa, in one form or another, persisted for even longer. A series of decrees were issued by Portugal with the aim of abolishing trade in human cargo, but they allowed for slavery to be continued. Likewise, later decrees to abolish slavery allowed it to continue in other forms. Portugal abolished the external slave trade in 1836, and in 1854 there was a partial abolition of slavery with government slaves becoming *libertos* (freed). In 1856, the children of all slaves became *libertos* but they had to work for their former owner for 20 years. In 1858, it was decreed that all slavery should end in 20 years. As a subterfuge, money paid for a slave was considered a redeeming fee, and the slaves then called *redimidos* (redeemed). In 1869, the immediate abolition of all slavery in all colonies was instituted. However, the *libertos* had to work for their former owner until 1878, otherwise they would be conscripted for forced labour (Leitão, 2015). In 1875, Monteiro (1875, vol. 1: 75) observed that at the time there were several sugar cane and cotton plantations in Angola where slaves, then referred to as "freed", worked. It was only in 1878 that finally the state of *libertos* was abolished. Nevertheless, until as late as 1962, the people then classified as "native" could be conscripted to do forced labour and were legally treated as minors who stood, under the tutorship of their foreman, in what was in fact a form of semi-slavery (Leitão, 2015).

In 1808, fleeing the armies of Napoleon during the Napoleonic war, the royal family of Portugal arrived in Brazil and remained there until 1821 when João VI returned to Portugal. The presence of the Portuguese royals in Brazil established and enhanced a sense of national pride in the country, and by 1822 Brazil gained its independence, with Pedro I, the son of João VI, declaring himself as emperor. With Portugal now having had to give up one of its most important international assets, it increasingly turned its attention to Angola and Mozambique on the western and eastern coasts of Africa, respectively. However, it was only in the late 19th century, with the western European "Scramble for Africa", that several regions in Angola were occupied. For a long time, the occupation was restricted to coastal settlements and trading posts in the interior. A deliberate outcome, and perhaps the most lasting effect of the travels of numerous explorers during the emerging imperialistic period, was the impact that "discovering" new lands and their biological and mineral riches had on stimulating European colonialist expansion in many parts of the world, which continued for several hundred years.

With a long history, having been founded in the 16th century, a unique society and social structures developed in Luanda, with these being different from those found in the rest of the country. Since early days in Luanda, people were not segregated by race but rather by class and culture (Birmingham, 2015). In the 19th century, Luanda had an elite of assimilated black people (*assimilados*) who had Portuguese as a mother tongue, and a religion and culture that coincided with that of Portugal. Some of the *assimilados* were appointed to local administrative positions in the provinces. Others were important and extremely wealthy slave-traders, the equivalent of today's magnates. By the mid-19th century, when Luanda was a major slave-trading city, several of these wealthy entrepreneurs were women. Ana Joaquina da Silva, a black woman known as the "Baroness of Luanda", was one of them. Her business ventures extended to Europe and South and Central America. She lived in a palace, owned 1000 slaves and did her shopping in Rio de Janeiro. According to Tams (1845), in a single shopping spree she had spent the equivalent of 20 million *reis*. That was 80 times the monthly salary earned by Friedrich Welwitsch (q.v.) at about that time, in 1853. However, at the end of the 19th century, the *assimilados* began to lose their status in relation to white people.

At Ambaca (Cuanza Norte), where the Portuguese had a presence since the 17th century, an elite of Portuguese-speaking, literate black people, the *ambaquistas*, also flourished. They, for example, held important positions as interpreters and scribes. Their trade routes were followed by many explorers who also used their services as interpreters.

The indigenous peoples were considered non-assimilated (*não-assimilados*), and from 1926 they were classified as "natives", as opposed to "citizens"; they had to pay poll tax, carry a pass, and could be conscripted to do forced labour, for example as porters. Karl Jordan observed women and children conscripted by the local administration to do road maintenance carrying the soil on their heads (Jordan, 1936: 61). The forced labour (for government and plantations) was still in place in 1947 when it was denounced by the politician

Henrique Galvão in the Portuguese parliament, which later resulted in his arrest. To have become "citizens", the "natives" had to prove to an inspector that they spoke fluent Portuguese, were monogamous, ate with a knife and fork, and wore European clothes (Birmingham, 2015).

In the second half of the 19th century, the majority of white people who travelled to Angola from Portugal were convicts (*degredados*), and from 1880 convicted Portuguese were sent exclusively to Angola. By the 1930s, there were some 2000 *degredados* living in the country. Few women were included among the convicts. The lack of white women in Angola resulted in a widespread mixed population, not unlike the "coloured" people that originated in Cape Town in South Africa.

The expeditions

In the 19th century, during the "Scramble for Africa", several expeditions were undertaken in Angola by German and Portuguese explorers funded by geographical societies (Figueiredo & al., 2020). The publications that resulted from these explorations included observations on the flora, fauna, landscape, culture, and society, as well as information on the way expeditions were

Fig. 1. Mary Pocock in a *machila* at Cuelei River, Angola in 1925. Photographer unknown. Selmar Schonland Herbarium, Albany Museum, Grahamstown. Reproduced with permission.

undertaken. Large caravans of dozens to hundreds of people were generally involved, including foremen, guides, interpreters, the personal servants of the explorers, *machila* (or *tipóia*, a hammock slung from poles; Figure 1) carriers, oxen keepers, porters and accompanying women (who transported the porters' possessions), and children. For example, in 1879, the caravan accompanying Maximilian Buchner (q.v.) consisted of c. 180 people: not only those at his service but additionally about 40 *ambaquistas*, relatives of his guide who joined the group for slave-trade purposes (Heintze, 2010).

In the next century, as motorised vehicles became available, porters were no longer needed, unless the expeditions were made on foot. The journey that Mary Pocock (q.v.) undertook in 1925 was made on foot and in *machila* (Figure 1) and it was likely the last expedition of that kind. In 1937, the second *Missão Botânica* (q.v.) included a large entourage of local men, albeit not porters. As late as the 1960s, expeditions still relied heavily on local labour. The second *Campanha de Angola* 1959–1960 (q.v.) had, in addition to the four collectors, five servants who were likely assigned to each collector as personal valets/cooks, as was done during the *Missão Botânica* of 1937.

In the 19th century, explorers travelled on mules (Figure 2), in *machilas*, on oxen (Figure 3), and even on the shoulders of the porters (Figure 4). In a *machila*, the traveller could settle "comfortably into the half-drowsy state

Fig. 2. Mary Pocock (in front) with Mrs Procter going to Cuelei River for a picnic on 19 September 1925. Photographer unknown. Selmar Schonland Herbarium, Albany Museum, Grahamstown. Reproduced with permission.

which the swaying motion of the hammock produces" (Johnston, 1895: 2). In the 20th century, the expeditions benefitted from rail travel that became available along the routes of some of the earlier itineraries. In 1909, when Henry Pearson (q.v.) visited Angola, 107 km of railway track was available from Moçâmedes. However, to ascend to the Huíla plateau, the trip was still made by ox-wagon or through a steeper but quicker ascent on horseback or by *machila*.

Avoiding exposure to water in streams and rivers seems to have been a major concern, likely because of a fear to contract water-borne diseases such as schistosomiasis, also known as bilharzia. In 1873, husband and wife Joachim and Rose Monteiro (q.v.) crossed streams on a Madeira-cane chair carried by porters employed for that purpose. Hermenegildo Capelo (q.v.) and Roberto Ivens (q.v.) improvised a bridge consisting of a line of men leaning forward so that the explorers could crawl over their backs or they crossed rivers on the shoulders of carriers (Capelo & Ivens, 1881: 79). Much later, in the 20th century, crossing rivers with vehicles was more complicated, as experienced in Hans Hess's (q.v.) expedition in 1951/1952 (Figure 5). It could even be a mission in itself, as when Gilbert Reynolds (q.v.) crossed the Cunene River with a car and a caravan (camper) on "the worst pont he had found in Africa" (Figure 6) (Reynolds, 1960).

The 19th century expeditions carried cargo that included all the equipment and comforts needed by the explorers, loads of merchandise (the equivalent of

Fig. 3. Maximilian Buchner arriving on an ox to meet Hermenegildo Capelo and Roberto Ivens at Bango, Malange, Angola in 1879. Artist unknown. Reproduced from Capelo & Ivens (1881).

money) that would be traded for local commodities or paid as tributes to chiefs or to allow river passage or as payment to porters, and, of course, provisions for the whole caravan. In 1873, Joachim and Rose Monteiro travelled with thirty porters carrying a tent, provisions, bedding, clothes, the aforementioned Madeira-cane chair, pots and pans, soap, drying papers and boxes for collections, and most remarkably an inflatable rubber bathtub and irons for ironing clothes (Monteiro, 1875). In 1879, Buchner travelled with 4000 kg of cargo (Heintze, 2010). Much later, in 1925, Mary Pocock entered Angola with 35 men, six *machila*-bearers, a cook, a waiter, four men to carry the meal for the men, and 17 porters to carry tent, stretchers, bedding, food, stores, clothes, camping gear, books, collecting equipment "and so on" (Dold & Kelly, 2018: 58). In 1937, but still in 19th century style, the *Missão Botânica* also carried an enormous amount of luggage ("too much stuff") that filled three trucks, a van, and a car, and included evening dresses and pearl necklaces (Exell, 1937b: 17–18). This convinced Mildred Exell (q.v.) that "the Portuguese have no idea at all of simplifying life when in camp" (Exell, 1937b: 14) Admittedly, this expedition was not only scientific but also a type of official tour with a preoccupation of impressing the local inhabitants, and formally meeting the local authorities and influential members of the society, with considerable time spent on ceremonial meetings and formal meals and functions, to the frustration of the more pragmatic members of the group. The luxuries were required for occasions such as when they were hosted by the Governor, with

Fig. 4. Hermenegildo Capelo and Roberto Ivens crossing a river. Artist unknown. Reproduced from Capelo & Ivens (1881).

much pomp and formal ceremonious meals where they were waited on by servants dressed in white uniforms and white gloves, with rose petals spread on the table. The black men of the expedition were "provided (at considerable expense) with white trousers and coats with big brass buttons down the front as well as khaki shorts and shirts. It seems that this splendour is in case we have visitors" (Exell, 1937b: 14).

Sourcing provisions while travelling was a major problem so large quantities of tinned foods and other preserved foods had to be carried from the outset. Harry Johnston (q.v.) noted that "native food [was] almost non-existent" at Quicembo (Zaire Province) when he visited in 1882 (Johnston, 1895). In 1937, at Dala (Lunda Sul), Mildred Exell noted that "food [is] difficult to get here – no meat, eggs etc." (Exell, 1937b: 26). Strangely, even in the white settlements in the interior, meat was scarce and the settlers relied on salted cod brought from Europe. In 1937, the meals offered to the team of the *Missão Botânica* were often the "usual 'bacalhau' [salted cod]" (Exell, 1937b: 20). Other meals were prepared by the expedition servants and were "excellent meals – our cook is an expert" (Exell, 1937b: 25).

In general, the visitors praised the hospitality they received along the way from European settlers. In 1882, Johnston was entertained by an English trader with a meal consisting of "mock turtle soup, salmon cutlets, lobster, curried

Fig. 5. Hans Hess's jeep, a team member, and helpers crossing a river in Angola in 1951/1952. Photographer unknown. ETH-Bibliothek, University Archives, Akz.-2006-37. Reproduced with permission.

rabbit, roast beef and boiled mutton with preserved potatoes, game patty, asparagus, plum pudding, peaches and strawberries, tea and biscuits" (Johnston, 1895: 3–4). On his way to Humpata (Huíla) he also received "charming hospitability" and a well-cooked dinner served by servants in white uniforms at the house of a Portuguese *degredado* (Johnston, 1923). During the 1937 *Missão Botânica*, lunches were mostly an "elaborate business" (Exell, 1937b: 15).

All the 19th century explorers complained in their writings about their contracted men. Frequently they used violence to dominate them. Explorers depended entirely on their entourage, not only for the obvious tasks of carrying their goods, preparing meals, etc., but also for guidance in the field, for translation, and for compliance with protocol when contacting the locals. Usually, at the start of an expedition there was a clash of wills and a measuring of force, until eventually a balance was struck. Buchner, who remonstrated about the "villainy, malice, wickedness and infidelity" of his men, later, in retrospect, admitted that what the men saw in him was an immensely rich and extraordinarily crazy individual who was interested in things that had no value in themselves. He acknowledged that German peasants might have treated a weird foreigner even worse (Buchner, as cited by Heintze, 2010). In some cases, the staff was highly praised and esteemed and some even accompanied the explorers on their return to Europe.

Fig. 6. The car and caravan of Gilbert Reynolds crossing the Cunene River, Angola in 1959. Photographer unknown. South African National Biodiversity Institute. Reproduced with permission.

As noted, porters were needed on the 19th and early 20th century expeditions to carry the huge expeditionary cargoes (Figure 7). According to Capelo and Ivens (1881) a porter's load could weigh more than 30 kg, "the trifle of seventy pounds" (Capelo & Ivens, 1882: 29). However, Hermann von Wissmann (q.v.) stated that the weight carried by a porter could reach 60 kg (Heintze, 2010). The contracting of foremen and porters was always problematic as, understandably, many men were not willing to travel through foreign (often enemy) lands and territories for obscure purposes. Rather thoughtlessly, Johnston (1895: 2) complained that porters were difficult to find as men preferred other jobs and "little cared for the more fatiguing task of carrying a white man in a hammock" (Figure 8). Porters were often slaves or conscripts for forced labour. In 1925, Mary Pocock had to engage porters within each district she travelled through and their employment was regulated by the district's administrative officials. It is possible that some of her porters were conscripted for forced labour.

Fig. 7. The procession of Hermenegildo Capelo and Roberto Ivens. Artist unknown. Reproduced from Capelo & Ivens (1881).

Men and women absconding from a caravan happened frequently, often with loss of cargo and guns. Many that tried to escape were killed by their guards. For the explorers, absconding was a constant worry. Once, Capelo and Ivens retired for the night and the next morning they were told "Fugiu tudo, senhores!" (Everyone escaped, gentlemen!) (Capelo & Ivens, 1886: 97). Dozens of men and their cargo were gone. Therefore, "it was not without great emotion" that they conducted a roll call in the morning and found that not a man or a load was missing (Capelo & Ivens, 1886: 147). Even when present, porters frequently refused to continue the march and had to be cajoled with further rations or pay.

Interpreters were essential members of the expedition; without whose services the explorers could hardly communicate with the local population. Buchner's expedition in 1879 was much impaired by his failure to engage a good interpreter (Heintze, 2010). Several interpreters became well-known. For example, Germano was an *ambaquista* who was originally from Mozambique, had been taken to Portugal as a slave, and after being freed became a trader in Angola. He travelled on expeditions for many years and in 1875 joined the expedition of Paul Pogge (q.v.), who considered him to be a "highly civilised black man" (Heintze, 2010). Afterwards, in 1878, Germano accompanied Otto

Fig. 8. Joachim and Rose Monteiro travelling by *machila* in Angola in the early 1870s. Drawn on wood by Edward Fielding from sketches by Rose Monteiro, and from photographs. Reproduced from Monteiro (1875).

Schütt (q.v.) and, in 1883, Hermann von Wissmann. Lourenço Bezerra, also known as Lufuma, was another *ambaquista* and ivory trader, who eventually settled in the Lunda Kingdom, and in 1875 provided much information to Pogge. Lourenço was also a source of information for Schütt when they met in Mona Quimbundo in 1878. His brother Joannes Bezerra, also known as Caxavala, accompanied Schütt in 1878 and Pogge and Wissmann in 1881, and served as informant for the Portuguese explorer Henriques Dias de Carvalho (Heintze, 2010). The interpreters could engage in conversations or hear long allocutions required by protocol, after which they would report phlegmatically that so far nothing had been said. ("'For now,' he replied very phlegmatically, 'he hasn't said anything yet'", Capelo & Ivens, 1886: 372).

Likewise, securing the services of good guides (Figure 9) was crucial for the success of any expedition (see, for example, Richards & Place, 1960: vi–vii on the exploration of East Africa). Buchner, during one of his violent

Fig. 9. Cateco, one of the guides of Hermenegildo Capelo and Roberto Ivens. "His gait was almost as regular as that of an automaton, his step firm and assured as a hunter's should be, and with his long gun over his shoulder, he climbed mountains or strode down into valleys, crossed ravines or forded brooks with the same equanimity, […]" (Capelo & Ivens, 1882). Artist unknown. Reproduced from Capelo & Ivens (1881).

outbursts, scared off his two guides who abandoned the expedition, and when they returned said that they preferred to return their payment and leave. From then on, when it came to guides, Buchner never dared to scold (Buchner cited in Heintze, 2010).

The incomprehension between 19th century explorers and local people was mutual. As noted by Heintze (2010), on one side there was pride, contempt, and incomprehension, and in the other side suspicion, fear, and incomprehension. The explorers felt vastly superior to the local people, and few refrained from being judgmental of what they encountered in the light of their western civilisation. Their writings are as much about the explorations as about how to "civilise" the Africans. Adding to all the grievances, most explorers, from Capelo and Ivens (1881: 61) to Exell (1937b), protested about the "interminável batuque" (endless drumming) that went on throughout the nights. Furthermore, and into the 20th century, communicating with the Portuguese authorities and traders was problematic for the explorers who did not speak the language. The peculiar way of life of the settlers in Angola, with their mixed families, was also noted by visitors such as Mary Pocock, in 1925, who was particularly prejudiced against the "Portuguese race" (Dold & Kelly, 2018: 73).

For the indigenous people, the visiting white men were unfathomable and regarded with suspicion, the purposes of the expeditions could not be

Fig. 10. A member of Hans Hess's expedition in conversation with local people in Angola in 1951/1952. Photographer unknown. ETH-Bibliothek, University Archives, Akz.-2006-37. Reproduced with permission.

understood, and were mostly suspected of being nothing more than a cover for seeking trade opportunities. Many local people that the explorers encountered had never even seen a white man before. Explorers sometimes had to undress to show that it was not only their faces that were white and so satisfy the curiosity of the people (Capelo & Ivens, 1881). According to Jorge Paiva (1933–; pers. comm., 2019) this was still happening to unwary plant collectors who ventured into remote areas as recently as the 1960s. Female travellers raised even more interest among the locals. In the early 1870s, Rose Monteiro was the first white woman to visit a village in the interior of Boma (D.R. Congo). Crowds gathered to cheer at her arrival, and men wanted to meet and congratulate Joachim Monteiro, her "owner" (Monteiro, 1875: 89). Throughout the centuries of foreign exploration in Angola, and very likely elsewhere in Africa, the indigenous populations consistently remained curious about the visitors (Figure 10).

In the 20th century, the peculiar societal and interracial relationships in Angola were sometimes noted by the visitors, but not as much as could be expected. Karl Jordan (q.v.), who travelled with a "half-cast" and a black man, noted that when wanting to take a meal at a restaurant, only the black man was not allowed in (Jordan, 1936). Mildred Exell (1937b: 23), without further details, noted that the contracted black men were discontent about the way they were treated by some members of the expedition and that they would "do better and respect one if they [were] treated as human beings at least". She also felt that all the shouting and cursing seemed unnecessary, and she witnessed a "vigorous application of 'palmatória' in [a] guard-room. Air rent with yells and groans of unfortunate [blacks]" (Exell, 1937b: 37). A *palmatória* was a wooden paddle used to hit the palm of the hand. It was very common in the 19th and early 20th centuries and used to punish children.

Overall, the always present tropical diseases were the main danger faced by the explorers. Diseases caused the early death of many collectors such as Christen Smith (q.v.) at 30, Andrew Curror (q.v.) at 33, Pogge at 44, and José Anchieta (q.v.) at 64, as well as numerous members of the entourages. Capelo & Ivens (1881) provided a list of names of the nearly 70 men, women, and children they had lost: about 30 had died of illness, accidents, exhaustion or hunger, the others had mostly absconded.

The collectors

As in other parts of the world that piqued the interest of European nations eager to explore foreign shores, early plant collectors in Angola were essentially of two types: some were trained botanists, while others – the majority – only made incidental, curiosity-driven, and largely voluntary collections that found their way into the preserved collections of, typically, the herbaria of private individuals or those of botanical institutions in the United Kingdom and continental Europe.

Most long-distance ocean-faring vessels required that a ship's surgeon be appointed as part of the crew. These medical practitioners were expected to see to the well-being of the crew, and in the case of slave ships, also the human cargo (see for example Domico, 2018: 54–57 on the role of ship surgeons on convict ships headed to Australia, at the time a penal colony). Given their academic background, these surgeons were also often interested in natural history and the acquisition of preserved and living specimens of plants and animals where the ships made landfall along the coasts of the continents that were being navigated (Kean, 2019). Although many surgeons might have abhorred the slave trade, their attachment to vessels that explored comparatively unknown territories, such as along the African coast, enabled them to enrich and expand the private plant and animal collections of wealthy noblemen as well as those of what eventually became public institutions. For example, William Anderson, the surgeon on the third and final voyage (1776–1780) of Captain Cook, was both a medical doctor as well as an accomplished natural historian, as was Thomas H. Huxley, the doctor (assistant surgeon)-naturalist on the voyage of H.M.S. *Rattlesnake* to Australia's eastern seaboard and New Guinea from 1846 to 1850 (Desmond, 1999: 14). In fact, Sir Joseph Dalton Hooker, famed explorer and plant collector, was urged to qualify as a medical doctor in The Royal Navy in preparation for and apparently as a precondition for his appointment as assistant surgeon and botanist on H.M.S. *Erebus*, captained by Sir James Clark Ross (Desmond, 1999: 13). Some explorers, like Mungo Park (11 September 1771–1806), who travelled considerable distances overland in West Africa (excluding Angola), were also both medical practitioners and natural historians (Lawson, 1985: 2–6). However, at least in some instances exploration ships deliberately separated the duties of surgeon and scientist, manifesting in, for example, the French survey of 1800 to 1804 of, inter alia, part of the Australian coastline, which had 24 scientists on board, including several botanists (Desmond, 1999: 15). Some of the earliest collectors in Angola too were naval surgeons. These include John Kirckwood (q.v.), who collected at the beginning of the 18th century, and Andrew Curror (q.v.), a surgeon with H.M.S. *Waterwitch* of the West Africa Squadron, who made collections in the 19th century.

The first collector who ventured into the interior of Angola in the late 18th century, Joaquim Silva (q.v.), collected plants while participating in a slave-seeking expedition.

In many cases, colonial powers were predominantly interested in the medicinal and other uses and potential economic value of the plant material they collected in "their" colonies. After all, for centuries, the study of plants – hardly "botany" as we know it today – was considered to be a branch of medicine (Bellorini, 2016). Placing emphasis on accessing plants of economic importance was a global trend with more attention being paid to plant use and the collecting, introduction, and development of living material of potentially new medicinally important, agricultural, and horticultural crops, rather than in documenting floristic (and faunal, for that matter) diversity. In Angola, this was the case for Friedrich Welwitsch (q.v.), who was contracted to "get the most complete knowledge possible about the natural products of each province and use that knowledge to increase the wealth and well-being of its inhabitants and to strengthen their relations with the metropolis [Portugal]" (translated from Dolezal, 1974: 49), therefore the mandated and main purpose of his collecting activities was to focus on economic plants.

Overall, up to the year 2000, the number of Portuguese collectors in Angola equals the number of foreign collectors. With three collectors active in the 17th century and two in the 18th century, it was only by the mid-19th century that the number of collectors in Angola increased significantly, reaching a peak in the 1960s, after which it decreased sharply as a result of the diaspora of civil servants following Angolan independence and the civil war that was soon to follow. The 1930s were very active collecting years, particularly in terms of foreign collectors (18 out of the total of 28 who flourished in that decade).

In the 1950s and 1960s, the majority of the collectors were Portuguese, including Angolan-born Portuguese. Especially in the 1950s, collecting and research were still carried out under the scope and auspices of the colonial missions that emanated from Portugal, such as *Missão botânica de Angola e Moçambique*. Several collectors were employed by this *Missão*, including Francisco A. Mendonça (q.v.), António Rocha da Torre (q.v.), and Eduardo Mendes (q.v.). A few years later, in the 1960s, collectors were mostly employed by Angolan institutions, namely the *Instituto de Investigação Científica de Angola* (created in 1957) and the *Instituto de Investigação Agronómica de Angola* (in 1961, IIAA). It must be noted that it was only in 1963 that finally tertiary education at university-level became operational in Angola, under the name *Estudos Gerais Universitários*. This institution became the University of Luanda in 1968, and after independence, in 1976, University of Angola (now University Agostinho Neto).

In the late 1960s, Angola finally had the requisite infrastructure and the local staff in place to develop botanical studies. However, it soon deteriorated and eventually all but collapsed. Following the revolution that occurred in Portugal in April 1974, a great number of people left Angola, fearing the unrest and the inevitable upcoming independence of the country. Angola became independent in November 1975 and civil war immediately broke out. Most of the collectors listed here that worked for Angolan institutions, such as the IIAA, returned to Portugal c. 1975, leaving the institutions without technical staff. One of the last collecting expeditions of the period may have been a survey of the Moçâmedes region, undertaken by Gilberto Cardoso de Matos (q.v.) and Alberto Castanheira Diniz (q.v.) in May 1975, during which they had to travel through areas that were occupied by the military, with pass control barriers in place. They were even requisitioned to change their itinerary to transport an ill soldier to hospital. Collecting activity in Angola only resumed in the next century.

Out of the 358 collectors recorded, at least 17 were missionaries. Only one of these was Portuguese, José Maria Antunes (q.v.), a missionary of the Congregation of the Holy Spirit (Spiritans) at *Missão da Huíla*, a mission with a relevant role in botanical collecting in Angola. At least eight Spiritan missionaries were collectors. In various parts of the world, gardens established by missionaries additionally fulfilled a role in trialling and evaluating new crops (Kingsbury, 2009: 46).

The great majority of women collectors were from foreign countries. The first woman to make plant collections in Angola was Rose Monteiro, who was active in 1858 (she was 18 years old at the time). The first Portuguese woman was Maria Chaves (q.v.) in 1886 (but see the biographical text about her). It took over 70 years before, in 1959, the next (likely) Portuguese woman, Ermelinda de Oliveira (q.v.), was to be active. Among the missionaries, it is only in Protestant communities that there were women collectors, with three being recorded. Although there were Catholic nuns associated with the Spiritans in Angola, none is recorded as collector. At *Missão da Huíla*, a few nuns of the congregation Sisters of St. Joseph of Cluny ran a girl orphanage since 1886. In 1893 there were eight sisters and 130 girls. The girls were taught mostly domestic chores and reading and writing. The sisters (and the girls) provided “relevant services” to the priests and male students, such as doing the laundry of the whole community (Brásio, 1940: 67). There are no records of any activity of nuns in association with studies.

Very few collectors of African ancestry are recorded as having been active in Angola. From the time that the botanical exploration of the country first was chronicled, a large number of plants were not collected by the collector whose name is on the specimen label, but rather by other people such as local people or by helpers who were instructed to take the material to the collector. It appears that the first names of African collectors recorded were in association with *Museu do Dundo*, *Companhia de Diamantes de Angola* (Diamang) in Lunda, from the late 1940s to early 1970s. This was likely not unrelated to the fact that during that period the *Laboratório de Investigações Biológicas* of the museum was directed by António Barros Machado (q.v.), known for his left-wing political inclination and progressive views. Among these African collectors, Matende Sanjinje (q.v.), who was the chief of ancillary staff and main guide of the laboratory, is the best-known. The names of a few other African people appear on the specimen labels of material of the Diamang collection (q.v.) that date from that period, and in some cases, collectors are referred to as "Indigenous". Elsewhere, also during that period, there were professional collectors whose names were never recorded. For example, Agostinho Chipa (q.v.), an assistant technician at the IIAA, who collected and prepared many specimens, is only recorded from oral testimony. His name was never mentioned on labels.

In Angola, several collectors amassed high numbers of specimens. Of these, John Gossweiler (14,600 numbers; q.v.) and Friedrich Welwitsch (10,000 numbers, each number often grouping several collections, q.v.) are considered the most prolific in the country, as their numbers refer mostly to Angolan collections. Joaquim Brito Teixeira (q.v.) also holds a high quantity of 13,000 numbers; however, these include material collected by many co-workers during expeditions. Unlike these collectors, others who were active in Angola and have a higher number of collections were also active in other countries. For example, Luis Grandvaux Barbosa (q.v.), whose collection also includes specimens collected by others, was active in several countries, having amassed a total of 22,000 numbers, of which 15,000 are from Africa). The bulk of the collections of António Rocha da Torre (q.v.) of 19,000 numbers is from Mozambique.

During the 20th century, some Portuguese botanists, for example Flávio Ferreira Pinto de Resende (28 February 1907, Cinfães, Portugal – 1 January 1967, Lisbon, Portugal) visited Angola (Figueiredo & al., 2019b: fig. 2; Smith & al., 2020). However, they are not recorded in this work if they did not collect in the country.

The localities

"The Portuguese delight in changing geographical names, which is very disconcerting for the naturalist who finds the old names on labels in museums and does not know their equivalent on recent maps" (Jordan, 1936: 51). Indeed, over the years many Angolan localities had more than one change in their toponyms and determining where the explorers were and where they collected plants can be difficult. Furthermore, in Angola, the same toponym can be applied to different locations in different districts or even provinces. For example, Munhino is a toponym applied to different locations in the provinces of Huíla and Namibe. There is Munhino as a region in Namibe, or as the town presently known as Capangombe in that region. Additionally, Munhino may also refer to a historical collecting locality near the town of Huíla, in the province of Huíla, which was visited by many collectors. Welwitsch, for example, collected at Munhino by the Lopolo River and at Morro de Munhino, located on the Huíla plateau. The Munhino mission is similarly a frequent collecting locality that is situated on the plateau. The same name is applied to the river in the vicinity of Capangombe that, after confluence with the Bero River, becomes the Giraul River. Furthermore, Munhino was also variously spelled Monino or Monyino.

Often it is not possible to determine the precise localities mentioned by collectors. For example, Bango, where Buchner met Capelo and Ivens (Figure 3), refers to an area shown in Capelo & Ivens (1881: vol. 2 opposite p. 41) as situated north of Malanje. Buchner was stationed in Malanje from 28 February to July 1881. Previously, in March 1879, he had travelled up to the Bango Hill to explore it and to meet Capelo and Ivens who were camped there and on their way to the interior. Capelo & Ivens (1881) referred to that meeting place as Bango. Nevertheless, Bango is a name used for several localities in Angola (Thompson & Page, 1986). There is also a collection by Buchner recorded from "Soba Bango". Sobato Bango was a chiefdom located west of Malange, in Golungo Alto region, Cuanza Norte.

Some changes in the boundaries of provinces and districts occurred in Angola. The actual province delimitations of Angola are relatively recent. In the 1860s, the area purportedly under Portuguese rule was divided into districts extending from the coast to c. longitude 18° (Bandeira & Leal, 1864). Thus, on labels of collections made in the 19th century, regional names correspond to districts, to kingdoms or to chiefdoms. In 1934, five provinces were created. The *Conspectus florae Angolensis*, of which the first fascicle of the first volume was published in 1937, followed the division into districts that dated from 1927 (Mendonça, 1937). This division, shown on a map published in some of the volumes, was to be used in subsequent issues of the *Conspectus*, with the exception of the last fascicle (Diniz, 1993) that followed the current delimitation of provinces. By 1951, Angola was considered to be an overseas province of Portugal; then its provinces reverted to being called districts.

Angolan districts and provinces also changed in number and boundaries (Figure 11). Particularly troublesome for analysis of collection localities are the changes in Bié, the south of which became the province of Cuando

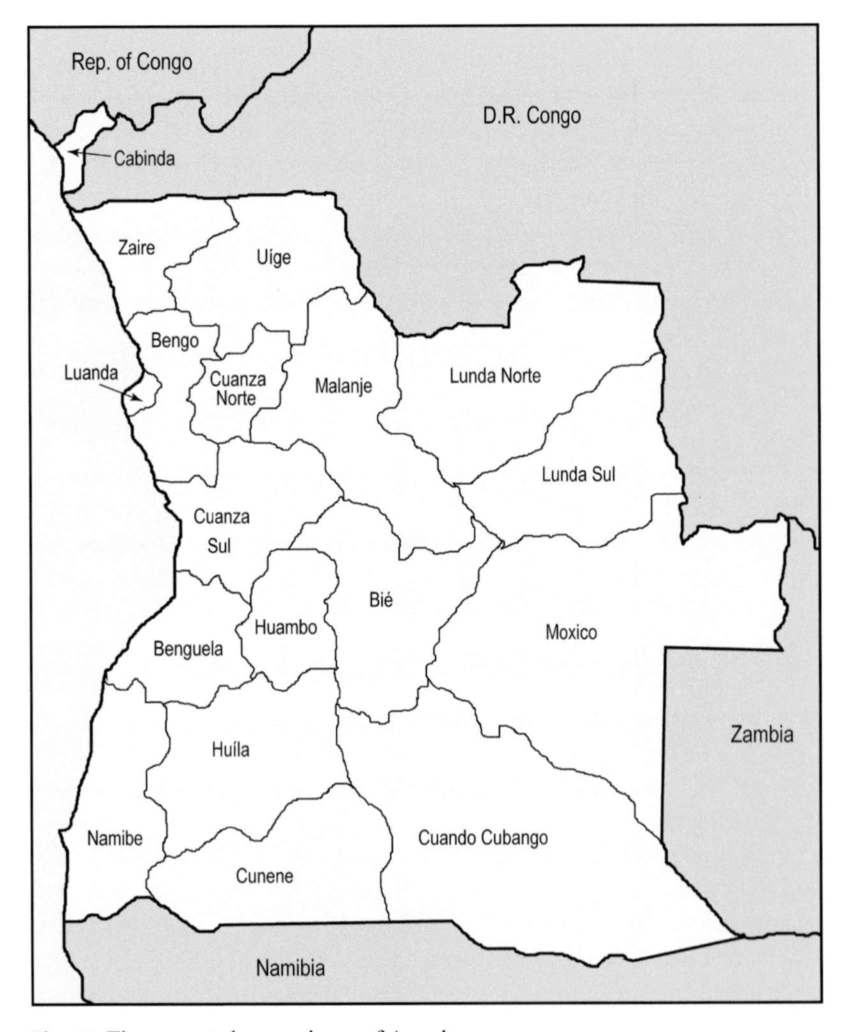

Fig. 11. The present-day provinces of Angola.

Cubango in 1961, and in Huíla, the southern area of which became the province of Cunene in 1970. Most of Luanda became Bengo in 1980. Lunda became Lunda Norte and Lunda Sul. It must be noted that even if the name of the collecting locality is the same as that of a present-day province, this does not signify that the locality falls within that province. For example, Lunda in the 19th century refers to the Kingdom of Lunda, whose territory is now in both Angola and the D.R. Congo; Congo refers to the kingdom that is now included in Cabinda (Angola), the D.R. Congo, and the Rep. of the Congo.

At present, the name of six Angolan provinces is also the name of the main town of that province. Originally some of these town names were longer and distinct from the name of the region (for example São Paulo da Assunção de Loanda, now Luanda, and São Filipe de Benguela, now Benguela), but once they were abbreviated, town and province names became the same. The implication is that on certain labels it is not possible to know if the collecting locality is the town or the region in a broad sense. The province of Namibe was previously called Moçâmedes. The latter name was used for the town until 1976 when both province and town became known as Namibe. That is, until 2016, when the town name reverted to Moçâmedes. Localities on labels and in literature therefore have to be considered in the context of the date.

Many collecting localities bear the name of the *soba* (chief) of the region, as noted by Montenegro (1957–1964) in a compilation of Angolan toponyms. Mendonça (1937) further noted that these locality names changed whenever a new chief came to power and for that reason it was often difficult or impossible to locate them. On itinerary maps, such as that included in Wissmann & al. (1891), numerous localities are named "Soba [...]".

A locality that was frequented by several 19th century explorers was "Kimbundu". The present-day town of Mona Quimbundo, by the Luvo River, was known as Feira [market] Kimbundu in the 19th century. At the time, this was the most easterly trading station of the Portuguese and a locality where several routes crossed. Mona Quimbundo means "son of Quimbundo", referring to the chief of the region. On his way to Lunda, Pogge met this chief.

"Musumb", a term that appears on many labels, is not the name of a specific locality, but rather means "capital". The Lunda Kingdom had a capital, a Musumb, which was the place where the king took up residence. The location of the Musumb changed place whenever a new king was on the throne. Bearing this in mind is clearly important when collecting localities have to be determined. For example, the Musumb where Buchner arrived in December 1879 was one day's travel distant from the Musumb Pogge visited in December of the previous year.

Some place name changes in Angola

Arthington Falls (after Robert Arthington, 1823–1900) – Quedas do Mebridege
Bela Vista (Huambo) – Cachiungo
Duque de Bragança – Calandula
Namibe town (from late 1970s to 2016) – Moçâmedes
Novo Redondo – Sumbe
Roçadas (also the name of the fort, after an army officer) – Xangongo
Sá da Bandeira – Lubango
Salazar – Dalatando – Ndalatando
Santo António do Zaire – Soyo
São Salvador do Congo – Mbanza Congo
Silva Porto – Cuíto
Teixeira de Sousa – Luau
Vila Arriaga – Bibala
Vila General Freire – Nambuangongo
Vila Henrique de Carvalho – Saurimo
Vila Luso – Luena
Vila Paiva Couceiro – Quipungo
Vila Pereira d'Eça – Ondjiva
Vila Serpa Pinto – Menongue
Vila Teixeira da Silva – Bailundo

Herbaria

Historians are gradually beginning to explore the consequences of the civil war of the last quarter of the 20th century for Angola. Natural historians however have yet to fully grasp the effects that this conflict wrought on enriching the domestically held preserved natural history collections of plants and animals. For example, South Africa has at least 72 herbaria, housing more than 3.2 million plant specimens; Angola has only four herbaria that collectively held 125,000 specimens in 1999 (Smith & Willis, 1999: 140). Herbaria remain indispensable resources for taxonomic research. Most of these South African herbaria were established between 1970 and 1990, a period during which the Angolan civil war was at its height and collecting for herbaria likely at its lowest. In the first decade of this period, 17 new herbaria were founded in South Africa, and 13 more in the ensuing decade. In contrast, in the decade before 1970 only nine new herbaria were established in South Africa, and eight in the decade after 1990. Very few were established at any time before or after this 50-year period (Smith & Willis, 1999).

It also has been shown that from 1975 to 1985, the first ten years of the Angolan civil war, more specimens were collected and lodged in herbaria of the South African National Biodiversity Institute than in any other decade (Victor & al., 2015). At the time, these herbaria were separately managed by the National Botanic Gardens (Herb. NBG, incorporating SAM) and the Botanical Research Institute (Herb. GRA; Herb. PRE, incorporating TM, and including NH, STE, and WIND).

The relatively brief history of Angolan herbaria is nonetheless rather complicated with herbarium closures, the splitting of collections, and the transfer of herbaria between towns and institutions. The damage to collections that resulted from the civil war remains unassessed and the condition (and even existence) of some collections is still unknown.

The first herbarium created in Angola dates from 1941 and became known as Herb. LUA. It was initially in Luanda but was later integrated into the *Instituto de Investigação Agronómica de Angola* at Chianga, c. 13 km from Huambo (Huíla). By the 1990s it held about 40,000 specimens (Holmgren &

Fig. 12. The Herbarium LISC in Lisbon in 2003 with (right to left) Maria João Tendeiro, Maria Fernanda Pinto Basto, and Luis Catarino. Photographer and private collection of Estrela Figueiredo.

al., 1990; Smith & Willis, 1999; Sampaio Martins & Martins, 2002; Figueira & Lages, 2019). During the civil war, with the initiative of Elizabeth Matos (q.v.), the herbarium was transferred from Huambo to Luanda, to be kept at the herbarium of the former *Centro Nacional de Investigação Científica*, Herb. LUAI. After several decades it was returned to Huambo.

The next herbarium to be created was established by the *Companhia de Diamantes de Angola* (DIAMANG) at its *Museu do Dundo* in Lunda in 1947. The herbarium (Herb. DIA) became inactive and was reported as no longer extant (Thiers, 2021). However, according to Luis Ceríaco (pers. comm., March 2019), Herb. DIA may still exist, at least in part, in boxes in storage at the Museum.

With the creation of the *Instituto de Investigação Científica de Angola* in 1958, a new herbarium appeared in Lubango (Huíla), which was assigned the code LUAI. After the independence of Angola, part of this herbarium was transferred to Luanda and incorporated in the *Centro Nacional de Investigação Científica* at the University of Luanda. It would later become part of the *Faculdade de Ciências Agrárias* and it is estimated to hold c. 35,000 specimens (Smith & Willis, 1999; Sampaio Martins & Martins, 2002; Figueira & Lages, 2019).

The collections that remained at Lubango became the basis of the herbarium of the *Instituto Superior de Ciências da Educação* (Herb. LUBA), which has c. 15,900 specimens (Smith & Willis, 1999; Sampaio Martins & Martins, 2002; Figueira & Lages, 2019) not 50,000 as recorded by Thiers (2021).

Finally, a herbarium was created in the 1960s at the *Faculdade de Agronomia e Silvicultura* of the then *Estudos Gerais Universitários de Angola* (later University of Luanda), operating at Chianga, Huambo. This would become Herb. LUAU. After the independence of the country, the faculty was transferred to Luanda to become the *Faculdade de Ciências Agrárias*. It is not known what happened to this herbarium that had about 5000 specimens (Smith & Willis, 1999; Sampaio Martins & Martins, 2002).

Numerous herbaria in Europe and the U.S.A, as well as elsewhere, for example in other African countries, have collections from Angola. This is just one of the consequences of the long period during which Angola was a colony, with collections made in Angola infrequently remaining in the country. As a result, the types of names published for Angolan plants are deposited in a range of herbaria in formerly imperialistic nations (see Figueiredo & Smith, 2010). Even after herbaria had been created in Angola, before the country became independent, the best specimens from the collections made by the staff at Herb. LUA were selected for deposition at Herb. LISC in Lisbon (M.F. Pinto Basto, pers. comm., 2019). In Portugal, collections from Angola are at three universities: University of Coimbra (Herb. COI), University of Porto (Herb. PO), and University of Lisbon (Herb. LISU+LISC, incorporating LISJC).

The herbarium of the former *Jardim Colonial*, Herb. LISJC in Lisbon, was incorporated into the herbarium of the former *Centro de Botânica* of the *Instituto de Investigação Científica Tropical*, Herb. LISC (Figure 12). With the extinction of this institute, the herbarium moved to the *Museu Nacional de História Natural e da Ciência* of the University of Lisbon, to there conjoin the museum's herbarium (Herb. LISU). All the specimens from Angola at Herb. LISC were databased but the data made publicly available through websites do not represent the total dataset available, as c. 10% of the information has not been disclosed.

ABOUT THIS BOOK

In this book we discuss, as far as we could determine, all the collectors who amassed botanical specimens (vascular plants, bryophytes, and algae; collectors of fungi are excluded, although some botanical collectors also collected fungi) in Angola for deposition and permanent safe keeping in herbaria. The collectors range from some who collected tens of thousands of specimens, to those who collected only a small number of specimens.

In the "Collectors" chapter, the collectors are listed alphabetically according to surname. Note that where a person's surname consists of more than one [sur-]name, the alphabetical sequence is determined by the final [sur-]name in the set of such names. Therefore, some collectors who almost invariably appear on herbarium labels with a single surname are here listed under that surname, although they are often known under, and generally referred to, by two surnames. For example, in sched., "Teixeira" and "Barbosa" refer to "Brito Teixeira" and "Grandvaux Barbosa", respectively; in the literature they are often referenced in the latter way, i.e., by two surnames. Where using this sequencing methodology will cause confusion between collectors and to avoid the use of initials, a combination of two surnames is used. For example, Gilberto Cardoso de Matos, who is often mentioned in the text and cited in the literature references, is referenced as (and listed under) the surnames he is widely known by, i.e., "Cardoso de Matos". The two brothers António Rocha da Torre and Albano Rocha da Torre (who incidentally have the same set of initials) are listed, respectively, under "Torre" (the name by which the first-named brother is known and that is used in sched.) and "Rocha da Torre" (the second-named brother, an infrequent collector). Furthermore, in cases where a collector has a combination of surnames and that collector is not known by the last surname alone, the combination of surnames is adopted here. For example, "Gomes e Sousa" is used for António de Figueiredo Gomes e Sousa, as he is not known as "Sousa"; "Pinto Basto" is used for

Maria Fernanda Duarte Pinto Basto da Costa Ferreira, as she is not known as "Ferreira"; and "Crawford Cabral" is used for João Crawford de Meneses Cabral, who is not known as "Cabral" nor as "Meneses de Cabral". The purpose of this methodology is to facilitate the identification of the individuals and to improve the readability of the text. This is therefore not an attempt to establish standardised names for citing the collectors. During the period covered in this work, Portuguese names usually consisted of one or two Christian names (also referred to as "given names" or "baptismal names") and up to six surnames derived from those of both parents, normally with the father's surname(s) being the last one(s). Names that are listed in the text by more than one surname are also cited in that way in the list of references.

Entries for collectors cited in the literature or in herbarium databases but for whom there is doubt whether they collected in Angola are given in square brackets in the "Collectors" chapter.

As far as possible, all the collectors are discussed under standard headings, where such information was available, and could be accessed, verified, and catalogued. The headings are:

Bio.: consisting of concise biographical information, including dates and places of birth and death.
Note: where information particular to the collector is recorded. This heading is also used when no biographical information could be traced.
Angola: dates, places, and collecting numbers where known. If joint collections were made, co-collectors are also mentioned.
Herb.: refers to those herbaria where collections made by the collector are known to be deposited. The herbaria listed are deliberately not only those where Angolan material is known to exist. Rather, the herbaria listed include those that do, or likely, also hold collections that the collector made in other countries or regions. Furthermore, because the majority of the herbaria cited are not databased it is not possible to know if they hold any collections from Angola made by a particular collector. Therefore, to facilitate further searches and investigations, the intention is to provide a list of herbaria that could hold duplicates of Angolan material.
Ref.: literature and electronic references, as well as personal communications, used for compiling information. This often includes references to major works, such as books written by the collectors, which provide brief biographical and associated information not available elsewhere.

We made use of as large as possible a set of published and unpublished resources while writing this book. Information about collectors is scattered among a vast range of documents and we cite all the references we managed to trace and locate.

A comprehensive list of references is provided. This gives readers and scholars access to the sources of information used in writing this book. Some items will not be easily accessible, except in well-resourced specialist libraries, but many of the books and scientific and popular articles can be found in institutional libraries. Today, an astonishing number of items can be obtained by searching the Internet using a thoughtfully structured set of keywords; this applies to books, or at least large parts of books, and scientific and popular papers. In the "Literature cited", journal titles are abbreviated to conform with BPH Online (https://huntbot.org/bph), unless use of an abbreviation will result in confusion, or the journal or periodical does not usually carry botanical information. In those instances, abbreviations follow BPH rules for abbreviations.

For clarity, where places of birth and death are given in the text pertaining to the biographical information of collectors, present geographical names are used. However, where necessary to foster accuracy, reference is made to historical place names.

In many instances we consulted with former and current colleagues, friends, and family members of collectors. Information so obtained is cited directly in the treatments as "pers. comm." (i.e., personal communications), with the date added on which the information was provided.

Herbaria are abbreviated according to the standard herbarium codes used in Index Herbariorum (Thiers, 2021). Some terms are abbreviated: University as Univ.; Herbarium as Herb.; birth as b.; death as d.; and flourished (meaning the overall period that the person was active, at least in Angola) as fl.

The terms "collections", "numbers", and "specimens" as used in the text have different meanings: a collection may consist of several specimens with the same collection number. When known, we give the number of collections or specimens collected in Angola by the individual as first collector. In some herbaria, such as Herb. LISC, several specimens (duplicates) with the same collection number were deposited in the main collection. For example, collections made by Gossweiler consist of several specimens, with *Gossweiler 6931* (LISC) comprising 21 sheets.

Some specimens are cited using their barcodes to facilitate searching for them in websites.

In a few instances, information on dates or places of birth or death, which was retrieved from genealogy websites, is given in square brackets followed by a question mark, if the information could not be verified.

COLLECTORS

Abranches Bizarro, Clemente Joaquim (c. 1805–1845)
Bio.: b. Portugal, c. 1805; d. Luanda, Angola, 9 July 1845. Medical surgeon who obtained his degree at the Univ. of Lisbon in 1828 and practised in Lisbon for some years and published on medical subjects until being deported to Angola as a consequence of voicing criticism against the state. An abolitionist, he became Curator of the Board of Superintendence for Emancipated Negroes, which was created after the anti-slavery Anglo-Portuguese Treaty of 1842. Tams (1845) recorded meeting him in Benguela sometime between October 1841 and March 1842, as the only European living there who had an interest in nature and knowledge of plant products; Abranches "had made a collection of specimens of aloes". **Angola:** No herbarium specimens have been located either at the herbarium of the Univ. of Coimbra (Herb. COI) or at the herbarium of the Univ. of Lisbon (Herb. LISU). **Ref.:** Tams, 1845; Torres, 1904; Bossard, 1993; Romeiras, 1999; Figueiredo & al., 2008; Santos & al., 2016.

Abreu, J. Guilherme de (fl. 1937–1938)
Bio.: Agricultural technician with *Repartição Técnica dos Serviços de Agricultura. Comércio, Colonização e Florestal de Angola* and head of the *Repartição das Dunas*, in Moçâmedes, Namibe. **Angola:** 1937–1938; Namibe. **Herb.:** BM, COI, LISC (4 specimens), MO, PRE. **Ref.:** Gossweiler, 1939; Lanjouw & Stafleu, 1954; Bossard, 1993; Romeiras, 1999; Figueiredo & al., 2008.

Ackermann, G. Wilhelm
(c. 1837–1862)
Bio.: b. Breslau, Germany (now Wrocław, Poland), c. 1837; d. Luanda (?), Angola, 19 April 1862. Horticulturalist. He studied at the *École d'horticulture de Gendbrugge* in Belgium and was employed by the renowned nurseryman Louis Benoît van Houtte (1810–1876). In May 1860, Ackermann travelled to Africa embarking in Ghent, Belgium. He made a 10-month stopover on the island of São Tomé in the Gulf of Guinea, where he was hosted by the Governor. There he met the collectors Gustav Mann (1836–1916) and Welwitsch (q.v.). On 11 October 1861, Ackermann sailed from São Tomé for Luanda, Angola. He carried a handwritten letter from the King of Portugal D. Pedro V (1837–1861) that served as introduction to the local Governor and guaranteed his support. Furthermore, the influential merchant and slave trader Francisco António Flores whom Ackermann befriended could facilitate the exploration of the interior. Six months later, on 16 April 1862, likely in Luanda or in Ambriz (in Bengo), Ackermann suddenly fell ill with fever; he died three days later. The Flores family administered the last rites. Ackermann is not to be confused with George Ackermann, who collected plants in South America during 1826–1831 and is commemorated in some succulent plant names. **Angola:** 1861–1862; likely collected living material for Van Houtte in Belgium but it is not known if any herbarium specimens were prepared. **Ref.:** Koch, 1862; Exell, 1944; Lanjouw & Stafleu, 1954; Dolezal, 1974; Ferreira (R.), 2015; Figueiredo & al., 2020.

Afonso, G.F.
(fl. 1950s–1960s)
Angola: 1950s–1960s; collected with Araújo (q.v.). **Herb.:** LUA. **Ref.:** Bossard, 1993; Romeiras, 1999; Figueiredo & al., 2008.

Aguiar, Fernando Queiroz de Barros
(1931–)
Bio.: b. Santa Maria Maior, Viana do Castelo, Portugal, 16 January 1931. Agronomist. In 1958, he graduated from the *Instituto Superior de Agronomia* (ISA) in Lisbon. In the same year, he moved to Angola, working initially at Cela (Cuanza Sul) where he was eventually in charge of the *Centro de Estudos*. In 1962, he joined the *Instituto de Investigação Agronómica* in Huambo. He worked there for 12 years on several projects in soil surveying and cartography, including, with Castanheira Diniz (q.v.), in the project *Zonagem Agro-Ecológica de Angola* of which he was one of the leaders for eight years. In 1974, he left Angola and returned to Portugal. He settled in Évora and taught

at the *Escola de Regentes Agrícolas* before joining the Univ. of Évora. Later, he was attached to the *Instituto Nacional de Investigação Agrária* in Lisbon and lectured at ISA. In 1984, he returned to Angola where he lectured at the Univ. Agostinho Neto and, for four years, he was a consultant for and leader of a PNUD/FAO project in Huambo that aimed to revive soil cartography studies in the country. **Angola:** 1967; Cuanza Sul. **Herb.:** LISC (1 specimen), LUA. **Ref.:** Aguiar, 1984, 2021.

Alcochete, António
(1963–)
Bio.: b. Angola, 1963. Botanist. **Angola:** 1991; Cunene, Huíla, Namibe; collected with Gerrard Reis (q.v.), (E.) Matos (q.v.) and Newman (q.v.). **Herb.:** K, L. **Ref.:** Goyder & Gonçalves, 2019.

Alice, C.
(fl. before 1999)
Angola: No information recorded. **Herb.:** LUBA. **Ref.:** Smith & Willis, 1999; Figueiredo & al., 2008.

Almeida, José Joaquim de
(c. 1863–1933)
Bio.: b. c. 1863; d. Lisbon, Portugal, 5 May 1933. Agronomist. After graduating as an agronomist in 1887, he accepted a position as a lecturer at the *Instituto Lauro Sodré* in Belém, Brazil, from 1889 to 1901. He then moved to Angola to work as an agronomist at Lunda from 1901 to 1903 and Luanda from 1904 to 1906. Returning to Portugal, he became professor at the *Instituto Superior de Agronomia* in Lisbon and later director of the first *Jardim Colonial* that was created in Lisbon in 1906, in the gardens of the palace of the Count of Farrobo. In 1911, he was commissioned to audit the *Repartição de Agricultura* of Mozambique and stayed in the colony as director of the institution until 1919, when he returned to Lisbon. **Angola:** 1903; Bengo, Bié, Cabinda, Cuanza Norte, Cuanza Sul, Huambo, Huíla, Malanje, Uíge; over 450 specimens. **Herb.:** COI, LISC (463 specimens). **Ref.:** Anonymous, 1933; Gossweiler, 1939; Lanjouw & Stafleu, 1954; Bossard, 1993; Liberato, 1994; Romeiras, 1999; Figueiredo & al., 2008.

Almeida, Luís Correia de
(fl. 1960s)
Angola: 1960s. **Herb.:** LUA. Collected with (Brito) Teixeira (q.v.). **Ref.:** Bossard, 1993; Romeiras, 1999; Figueiredo & al., 2008.

Almeida, Marcelino de
(fl. 1952)
Angola: 1952; Luanda. **Herb.:** LISC. **Ref.:** Bossard, 1993; Romeiras, 1999; Figueiredo & al., 2008.

Almeida, Pompeu Ferreira de
(fl. 1953–1964)
Bio.: Forester who published on tropical flora. **Angola:** 1953–1964; Cuando Cubango, Moxico. **Herb.:** LISC (125 specimens), LUA, PRE. **Ref.:** Almeida, 1956, 1961; Bossard, 1993; Romeiras, 1999; Figueiredo & al., 2008.

Almeida, Viriato de
(fl. 1953–1954)
Angola: 1953–1954; Benguela (Cubal, Dende, Ganda). **Herb.:** COI, LISC (24 specimens), LUA, PRE. **Ref.:** Bossard, 1993; Romeiras, 1999; Figueiredo & al., 2008.

Alves, J.C.
(fl. 1950s–1960s)
Angola: 1950s–1960s; Moxico. **Herb.:** LISC (3 specimens), LUA. **Ref.:** Bossard, 1993; Romeiras, 1999; Figueiredo & al., 2008.

Amado de Melo Ramalho, Paulo
(1864–1926)
Bio.: b. Formoselha, Montemor-o-Velho, Coimbra, Portugal, 1 March 1864; d. 1926. Army officer. After leaving school, he joined the army in 1880 and was sent to Angola in 1886. He married Sophia Moller, daughter of Adolpho Moller, the head gardener of the *Jardim Botânico* of the Univ. of Coimbra. From 1889 onwards, Amado was stationed in Angola, in action in Bié, Humbe, and Cuango. In 1894, he was in Humpata, when, due to an injury, he had to retire from active military duty. He held several administrative positions in Angola and Mozambique, where he was stationed in service during World War I. He retired in 1923 and died a few years later. His collections were likely done on request of his father-in-law or by his wife Sophia Moller. He is alternatively listed as "A. de Mello Ramalho" (e.g., Glen & Germishuizen, 2010: 351). **Angola:** 1894; Huíla (Humpata). **Herb.:** B, COI. **Ref.:** Gossweiler, 1939; Vasconcelos, 1942; Lanjouw & Stafleu, 1954; Bossard, 1993; Liberato, 1994; Romeiras, 1999; Figueiredo & al., 2008.

Anchieta [Portes Pereira de Sampaio], José Alberto de Oliveira (1832–1897)

Bio.: b. Lisbon, Portugal, 9 October 1832; d. Chicambi, Caconda, Angola, 13 September 1897. Naturalist. He had an erratic academic career. He was the son of an infantry general and when nine years old was sent to the military school in Lisbon. However, he left two years later as he was bored with military habits and the need to salute everybody. By then he was already known for his desire to learn. In 1849, he started studying at the *Escola Politécnica* in Lisbon. He led a bohemian life, playing the violin and avidly reading on matters that interested him, while missing exams and dropping from courses for which he was enrolled. In 1852, he moved to Coimbra and enrolled in the faculties of Philosophy and Mathematics of the university as a first-year student. As before, he could not conform to the rigors of academic obligations and instead pursued his own interests. He failed to pass but managed to further his reputation of being a carefree eccentric, as later recalled by friends. At the end of the academic year, he was summoned back to Lisbon by his father and returned to the *Escola Politécnica*. Shortly thereafter, in c. 1854, he left the *Escola* to travel to the Cape Verde Islands. While there, busy making collections, an epidemic of cholera broke out in 1855. Already with some knowledge of medicine from his readings, Anchieta travelled to the island of Santo Antão to attend to the sick. The epidemic resulted in the lockdown of the island. Having lost friends that he had to bury and with the prospect of illness and hunger he set out on foot to the interior of the island, a strenuous journey across the mountains, eventually reaching the other coast where he was picked up by a passing steamship. He returned to Lisbon and afterwards travelled to London to study medicine, remaining there for one year. He attended public courses and frequented anatomy theatres, self-learning. From London he moved to Paris around 1857 to continue his studies, but the lack of means of subsistence made him return to Lisbon without being awarded a degree. He arrived in Lisbon in September 1857 with the same good-humoured philosophical indifference to both the nuisances and the comforts in life and proceeded to enrol again at the *Escola Politécnica.* After another failed exam, he left the *Escola* in 1858. He travelled to Angola for the first time likely in 1863. His luggage was minimal: a box of pins for insect collections, a small bundle of white clothes, jacket and some paper collars, and his violin or guitar. According to a friend (Pato, 1894) he had broken the violin on someone's head during a skirmish and replaced it with a guitar afterwards. His books and violin were his most cherished possessions and his only solace when facing life's sorrows. He went to Angola for the love of science and intended to lead an "almost primitive life": mostly outdoors, hatless under the sun, and eating manioc and bananas (Rebelo, 1881; Pato, 1894). During this first trip, he visited Cabinda and the Congo and Cuilo rivers in 1864, sending collections to Lisbon in that same year. In 1865, in addition to those localities, he visited Malembo and Luango in Cabinda and also Luanda. He collected mostly birds, reptiles, and insects. In 1865, he returned to Lisbon, and in December 1865, he again enrolled at the *Escola*. On 14 July 1866, he married Maria Amália de Almeida (d. 1883), and two days later, on 16 July, the couple left for Benguela where they settled. For the next two years they travelled in Angola with the purpose of collecting zoological specimens for the museum of the *Escola Politécnica*. They travelled to Moçâmedes and to the interior as far as Lake Ivantala (northwest of Lubango, Huíla) where they camped for a few days. Amália was unable to endure the tough conditions during the voyages and the tropical diseases so she returned to Lisbon in 1868, while Anchieta remained in Angola for the rest of his life, continuing his explorations for a further 29 years. The meagre funding provided by the Portuguese government and the lack of institutional support were constant challenges. Anchieta only received a small stipend from the government: it was less than half of what Welwitsch (q.v.) had earned over a decade earlier. He nevertheless persisted and developed his exploratory endeavours while driven by his insatiable curiosity and love of science; he was "a martyr for science" according to Bocage (cited in Andrade, 1985: 29). At the same time, he practised medicine; he did not charge for these services but accepted contributions, sometimes in the form of biological specimens. In 1868, he was called to Moçâmedes to treat the local doctor Lapa e Faro (q.v.). Afterwards, he was asked to provide medical services to the Moçâmedes community in Lapa e Faro's absence. Anchieta was based at Caconda (Huíla) when in January 1878 he was visited by Capelo (q.v.), Ivens (q.v.), and Serpa Pinto (q.v.). According to Serpa Pinto (1881b), Anchieta's house was in the ruins of a church and had the walls covered in bookshelves with books, mathematical instruments, cameras, telescopes, microscopes, retorts (flasks), (stuffed) birds of all colours, glassware, crockery, bread, bottles with colourful liquids, surgical kits, piles of plants, medicine, cartridges, and clothes. His house was almost a field station. He fathered two children from African women. The son, José (born c. 1873), was educated by him and instructed in the preparation of zoological specimens. Anchieta had high hopes for his son, but after a promising youth eventually the junior Anchieta became an alcoholic. Little is known of the daughter, Maria Carlota. In 1897, finally the diseases caught up with Anchieta; he died at the age of 64 during a final expedition to Cusse, in Chicambi, on the way back to his home in Caconda. An intelligent, persistent, and modest man (Capelo & Ivens, 1881), righteous and poor in the midst of the vice and immorality that surrounded him (Serpa Pinto, 1881b), a candid and good soul (Andrade, 1985), Anchieta was well-known for his eccentricity, unorthodoxy, and disregard for material comforts and wealth. He was recalled by an African woman as a "Tchinluahúco" (lunatic), a weird white person who always kept his hair too long, never took off his hat, even inside the house, did not utter a word some days, and did not eat like the other whites. When someone "wanted to die" (was about to die), Tchinluahúco was called because he saved everyone, and when he died, blacks were more distressed than whites (Andrade, 1985: 10). Anchieta is commemorated in dozens of scientific names. Although contracted to obtain zoological specimens and best-known for those, he was a general naturalist taking an interest also in other subjects such as geology and botany. He not only collected specimens, but also studied them as shown by the identifications he provided in his lists. According to Bocage, his zoological collections included 4386 specimens of birds, specimens of over 100 species of mammals, and 170 species of reptiles. He also collected fishes and insects and collected plants in areas that had not been explored by Welwitsch. According to Romariz (1952), the number of

Fig. 13. José de Anchieta. Photographer unknown. Reproduced from Capelo & Ivens (1881).

Anchieta plant collections at the herbarium of the *Museu Nacional de História Natural e da Ciência* (Herb. LISU) total 619 with numerous duplicates. These were received on more than one occasion. Lists of the plants sent by Anchieta are reproduced in Andrade (1985) and refer to c. 420 numbers; the last set of plants was dispatched in July 1889. At least one set of plants (list numbered 34 to 83) was sent to Júlio Henriques at the Univ. of Coimbra (Herb. COI). Collections were made mostly in and around Huíla and at "Quindumbo", an unplaced locality that Anchieta described as situated in the Quiaca Kingdom at c. 1400 m above sea level and at eight days travel from the source of the Catumbela River; it is probably in Bié Province. **Angola:** 1867–1889 (plants: 1872–1888); Bengo (Dande River mouth), Benguela, Cuanza Norte (Cazengo, Dondo), Huíla (Caconda, Chela, Quilengues), Namibe (Bibala, Capangombe, Moçâmedes), Cuanza River (plants: Huíla [Caconda, Cuando River, Cusse], Bié ["Quindumbo"]). **Herb.:** B, BM, BR, COI, K, LISU (822 specimens), MO. **Ref.:** Anonymous, 1852; Rebelo, 1881; Serpa Pinto, 1881b; Pato, 1894; Bocage, 1897; Dias, 1939; Gossweiler, 1939; Romariz, 1952; Lanjouw & Stafleu, 1954; Andrade, 1985; Patterson, 1988; Bossard, 1993; Liberato, 1994; Romeiras, 1999; Figueiredo & al., 2008. Figure 13.

Andrada, Eduardo de Azevedo Gomes Campos de (1913–2004)

Bio.: b. S. Domingos da Rana, Parede, Cascais, Portugal, 28 January 1913; d. Cascais, Portugal, 2004. Forester. He graduated from the *Instituto Superior de Agronomia* in Lisbon in 1939. He worked in Angola and Mozambique from 1939 to 1951. Afterwards he moved to Madeira where he remained for 23 years. Later he returned to Cascais, Portugal, where he died in 2004. **Angola:** 1940–1945; Benguela, Bié, Huambo, Lunda Sul, Moxico. **Herb.:** BM, COI, EA, K, LISC (132 specimens), LISU, LL, LMA, LMU, LUA, SRGH. **Ref.:** Lanjouw & Stafleu, 1954; Bossard, 1993; Romeiras, 1999; Figueiredo & al., 2008; Glen & Germishuizen, 2010.

Andrade, Abel Marques de (c. 1933–)

Bio.: b. Portugal, c. 1933. Portuguese botanical collector with the *Serviço de Agricultura de Angola* and the *Instituto de Investigação Agronómica de Angola*. He moved to Portugal and now resides in the Algarve. **Angola:** 1950–1964; Cuanza Sul (Amboim, Gabela), Huambo, Huíla; collected also with (Brito) Teixeira (q.v.). **Herb.:** COI, LISC (5 specimens), LISU, LUA, PRE. **Ref.:** Bossard, 1993; Figueiredo & al., 2008; M.F. Pinto Basto, pers. comm., April 2019, and G. Cardoso de Matos, pers. comm., January 2021.

Antunes, José Maria
(1856–1928)

Bio.: b. Santarém, Portugal, 22 May 1856; d. Paris, France, 16 December 1928. Missionary with the Congregation of the Holy Spirit (Spiritans). He entered the local Spiritan seminary in 1867, the year the seminary was created. The seminary operated in Santarém for three years. Then, after a year of interruption, it was moved to Braga, Portugal. Antunes became a priest in 1878. In July 1881, he was living in Braga and taught at a local private school, when, with the approval of the religious authorities, he penned a proposal to create a Catholic mission in Angola that would include schools for boys and for girls from both the indigenous populations and the Boers who had recently arrived from South Africa via Namibia. The proposal was soon approved by the government and arrangements for setting up the mission were initiated with the help of Duparquet (q.v.) of whom Antunes had been a student. On 5 October 1881, Duparquet, Antunes, the priest Carlos Wunemburger, and three assistants left for Angola, and in November, they were already reporting from Moçâmedes. That same month, Antunes and Duparquet left for Huíla where they established the *Missão da Huíla*, which Antunes headed. He founded all the missions in the province but one (Sendi mission). In August 1904, 23 years after his arrival in Angola, Antunes was appointed Provincial Superior

Fig. 14. José Maria Antunes. Photographer unknown. Missionários do Espírito Santo. Reproduced with permission.

for Portugal and had to return to that country. In 1925, he moved to Paris where he died at the age of 72. He is commemorated in several plant names, such as *Xylopia antunesii* Engl. & Diels and *Artabotrys antunesii* Engl. (both Annonaceae) of which he collected the types. Antunes made numerous collections, many of which are type specimens, in Huíla. Several duplicates that originated from *Missão da Huíla* were later labelled as having been collected by "Antunes vel Dekindt" because the collector is not specified on the labels. The label data on most duplicates are limited and the collection numbers do not follow a chronological order, but some specimens (e.g., at Herb. P) have labels with detailed information. The highest collection number recorded is 3229. The Antunes and Dekindt collection in Angola was donated to Herb. LUA in 1956. **Angola:** 1881–1903. **Herb.:** B, BR, COI (c. 100 specimens), E, K, LISC (1206 specimens, Antunes vel Dekindt), LISU (226 specimens), LUA, M, MPU, P, PC, PRE, US (grass type fragments). **Ref.:** Gossweiler, 1939; Brásio, 1940; Estermann, 1941; Lanjouw & Stafleu, 1954; Teixeira, 1957; Bossard, 1993; Liberato, 1994; Romeiras, 1999; Figueiredo & al., 2008; Vieira, 2012–2017. Figure 14.

Fig. 15. *Missão da Huíla.* The teachers of the seminary with the Bishop of Angola in 1891, including (in front) Eugène Dekindt (extreme left) and José Maria Antunes (second from the left). Photographer unknown. Missionários do Espírito Santo. Reproduced with permission.

Missão da Huíla

The *Missão Portuguesa do Espírito Santo da Huíla* or *Missão do Sagrado Coração de Jesus,* better known as *Missão da Huíla,* was a Catholic mission that resulted from a proposal that Duparquet (q.v.) made in December 1880 to the Portuguese government. The proposal was approved, and on 5 October 1881, a party of six, including Antunes (q.v.) and Duparquet, left Lisbon, for Luanda, Angola. In Luanda, the Governor of Angola offered them an area of 2000 ha for the establishment of the mission. They then proceeded to Moçâmedes. After some negotiations that resulted in the Boer settlers having to concede some of their land by the rivers Palanca and Mucha on Huíla, the land was demarcated, and construction soon started. A further 200 ha by the Munhino River that included its source, were later acquired by the mission. The missionaries translocated to the mission in September 1882, and the mission continued to develop, including establishing an orphanage for children rescued from slavery (by 1894 numbering 124 boys and 203 girls), a professional school, a high school, and a hospital. It also included a seminary that in 1891 had 48 students; it was transferred to Luanda in 1907. After the independence of the country, during a turbulent period the mission was confiscated. It was eventually returned to the church and is again fully operational. Several plant collectors passed through the mission. These included: Duparquet, Antunes, Bonnefoux (q.v.), who succeeded Antunes as superior of the mission, Dekindt (q.v.), and Estermann (q.v.), who succeeded Bonnefoux. **Ref.:** Brásio, 1940. Figure 15.

Araújo, Virgílio de Portugal Brito (1919–1983)

Bio.: b. Aldeia das Dez, Oliveira do Hospital, Portugal, 9 February 1919; d. Évora, Portugal, 1983. Biologist. He went to Angola when he was 20 years old, in 1939, and worked as a technician in apiculture. In 1965, he graduated in biology from the Univ. of Rio Claro in Brazil. In 1975, he left Angola and immigrated to Brazil, to work as a researcher at the *Instituto Nacional de Pesquisas da Amazónia*. He returned to Portugal in 1979 and for a while lectured at the Univ. of Évora. He died in Évora at the age of 64. **Angola:** 1956–1961; Benguela, Luanda, Moxico; collected also with Afonso (q.v.) and Conceição (q.v.) in Moxico. **Herb.:** COI, LISC (124 specimens), LUA, LUAI, PRE. **Ref.:** Kerr, 1984; Bossard, 1993; Romeiras, 1999; Figueiredo & al., 2008.

Arnold, Ernest
(fl. 1965)
Angola: 1965; Huambo; collected also with (Brito) Teixeira (q.v.). **Herb.:** K, LISC (1 specimen). **Ref.:** Bossard, 1993; Figueiredo & al., 2008.

Asher, J.
(fl. 1960)
Angola: 1960; Huambo. **Herb.:** K. **Ref.:** Figueiredo & al., 2008.

B

Bacelar, B.
(fl. 1925)
Angola: 1925; Lunda Norte. **Herb.:** LISC (1 specimen).

Balfour-Browne, John ("Jack") William Alexander Francis
(1907–2001)
Bio.: b. Larne, Northern Ireland, 15 May 1907; d. [Kirriemuir, Angus, Scotland?], 18 [11?] June 2001. Entomologist. Son of the well-known entomologist William Alexander Francis Balfour-Browne (1874–1967). Although the younger Balfour-Browne became deaf at the age of ten, he overcame the impediment by developing an outstanding ability to lip-read, eventually in several languages. He was educated at Rugby School and attended Oxford Univ. (1925–1927) and Cambridge Univ. (1928–1931). After working for a while as an entomologist in Madeira, Portugal, he joined the Natural History Museum, London, in 1934 as an unofficial worker. He specialised in aquatic Coleoptera and became a member of the staff in 1947. In 1948 [1944?], he married Frances Lottie [Lotte?] Stephens (1904–1980), a mycologist who worked at the Museum from 1929 to 1967. Jack Balfour-Browne also retired in 1967 but continued working as a volunteer. He collected in South Africa and southwestern Africa in 1954. The collections from Angola have been wrongly attributed to his father (e.g., Bossard, 1993; Romeiras, 1999; Figueiredo & al., 2008). This error originates from a reference cited by Bossard (1993) as "Vickery, R. (1989) *British Museum Ethnobotanical survey sheets: Angola.* London: British Museum – Natural History". These survey sheets were not published; rather, they consisted of cards, the present whereabouts of which are unknown according to the author Roy Vickery (M. Carine, pers. comm., 2019). Furthermore, Balfour-Browne is not to be confused with his wife,

who was also a collector; she used the name F.L. Balfour-Browne. **Angola:** 1954; Huíla (Humpata). **Herb.:** BM. **Ref.:** Bossard, 1993; Romeiras, 1999; Figueiredo & al., 2008; Darby, 2019.

Bamps, Paul Rodolphe Joseph (1932–2019)

Bio.: b. Louvain, Belgium, 6 February 1932; d. Brussels, Belgium, 28 February 2019. Agronomist and botanist. He graduated from the Catholic Univ. of Louvain in 1955 as *Ingénieur agronome des régions tropicales*. From 1957 to 1961, he was with the *Division de Botanique* of the *Institut national pour l'étude agronomique au Congo* based in the D.R. Congo. In 1961, he returned to Belgium. He was first attached to the *Institut Belge pour l'Encouragement de la Recherche Scientifique Outre-Mer* (IBERSOM) and later to the *Institut royal des Sciences naturelles de Belgique* (IRSNB), being in service at the *Jardin Botanique National de Belgique*, Meise, where he remained for 36 years. He became curator of the African Herbarium (Herb. BR) and head of department, retiring in 1997. During his career, he collected c. 10,750 specimens and published numerous papers, including a list of collectors that have specimens from Angola at Herb. BR. In 1973, he

Fig. 16. Paul Bamps in the 1990s. Photographer unknown. Published by Lachenaud & Fabri (2020). Reproduced with the permission of the Botanic Garden Meise, Belgium.

undertook a 3-month expedition to Angola during which he collected over 600 numbers. He is commemorated in many plant names, including the genus *Bampsia* Lisowski & Mielcarek (Linderniaceae). **Angola:** March–May 1973; Benguela (Cubal, Lobito), Bié, Cuando Cubango (Menongue), Cuanza Norte, Cuanza Sul, Huambo (Lunge, Mungo), Huíla (Tchivinguiro), Luanda, Malanje, Namibe (Tômbua); collected with Cardoso de Matos (q.v.), Maia Figueira (q.v.), Raimundo (q.v.), Sampaio Martins (q.v.), and (Manuel) Silva (q.v.); numbers 4000–4653. **Herb.:** BR (main, 642 numbers), K, LISC (450 specimens), LUA, P, PRE, UPS, WAG. **Ref.:** Bamps, 1973, 1975a, 1975b; Codd & Gunn, 1985; Bossard, 1993; Romeiras, 1999; Figueiredo & al., 2008; Glen & Germishuizen, 2010; Polhill & Polhill, 2015; Lachenaud & Fabri, 2020; E. Robbrecht, pers. comm., October 2018. Figure 16.

Baptista de Sousa, João Bernardo Nobre (fl. 1960–1970)

Bio.: Technician. He was with the *Instituto de Investigação Agronómica de Angola* at Chianga, Huambo, as an auxiliary technician (*auxiliar técnico 3ª classe*) in 1964. After the independence of the country, he moved to Portugal and worked at the *Instituto Superior de Agronomia* in Lisbon. **Angola:** 1960–1970; Huambo; collected with several others such as Barbosa (q.v.) &

Fig. 17. João Baptista de Sousa in Angola in 1971 with *Aloe mendesii*. Photographer unknown. From Centro de Botânica/Instituto de Investigação Científica Tropical (IICT), Universidade de Lisboa. Reproduced with permission.

(F.) Moreno (q.v.); Cardoso de Matos (q.v.) & Raimundo (q.v.); (Manuel) Silva (q.v.); and (Brito) Teixeira (q.v.). **Herb.:** BM, COI, LISC, LUA, LUAI. **Ref.:** Teixeira & al., 1967; Bossard, 1993; Romeiras, 1999; Figueiredo & al., 2008; M.F. Pinto Basto, pers. comm., April 2019. Figure 17.

Baragwanath, S.
(fl. 1994)
Angola: 1994; South-west. **Herb.:** PRE. **Ref.:** Figueiredo & al., 2008.

Barbosa, Luis Augusto Grandvaux
(1914–1983)
Bio.: b. Lourenço Marques (now Maputo), Mozambique, 5 December 1914; d. Praia, Santiago, Cape Verde Islands, 28 May 1983. Agronomist. He attended a secondary school in Évora, Portugal, and later studied at the *Instituto Superior de Agronomia* in Lisbon where he graduated as an agronomist in 1942. In 1946, he joined the *Repartição Técnica de Agricultura* in Mozambique as a plant ecologist. From 1955 to 1959, he was with the botany department of the *Centro de Investigação Científica Algodoeira* in Mozambique. He moved to Angola in 1960 to work at the *Instituto de Investigação Científica de Angola* and was later appointed as head of the *Centro de Estudos de Sá da Bandeira* at Lubango. From 1963, he also lectured at the university in Angola. In 1971, he obtained a doctorate at the *Instituto Superior de Agronomia* in Lisbon and in 1973 an aggregation. He became well-known for his *Carta fitogeográfica de Angola*.

Fig. 18. Luis Grandvaux Barbosa. Photographer unknown. Private collection of Kitty Grandvaux Barbosa. Reproduced with permission.

He also collaborated with Wild (q.v.) on the *Vegetation map of the Flora Zambesiaca area* and co-authored the first comprehensive vegetation map of Mozambique (Pedro & Barbosa, 1955). Barbosa died in the Cape Verde archipelago while lecturing and doing field work. He had the intention of producing a phytogeographic map of the islands, a project that his untimely death prevented, but that was later undertaken by Cardoso de Matos (q.v.). A prolific collector, Barbosa amassed c. 15,000 numbers in Africa and 7000 in Portugal. He is commemorated in several plant names and in the name of the national botanical garden of Cape Verde, the *Jardim Botânico Nacional Grandvaux Barbosa,* located in Santiago. Transcripts of his collection records are at the Univ. of Lisbon and digitised at JSTOR Global Plants (JSTOR, 2021). **Angola:** 1960–1972; collected with several others such as (R.) Correia (q.v.), (C.A.) Henriques (q.v.), and (F.) Moreno (q.v.). **Herb.:** B, BM, BOL, BR, BRLU, COI, DIA†?, FHO, FT, K, LISC (1821 specimens), LISU, LMA, LMU, LUA, LUAI, LUBA, M, MO, NU, P, PRE, SRGH. **Ref.:** Lanjouw & Stafleu, 1954; Pedro & Barbosa, 1955; Barbosa, 1970; Gomes e Sousa, 1971b; Bossard, 1993; Liberato, 1994; Romeiras, 1999; Costa, 2004; Figueiredo & al., 2008; Moreira & al., 2009; K. Grandvaux Barbosa, pers. comm., November 2020; G. Cardoso de Matos, pers. comm., 2021. Figure 18.

Barros Machado, António
(1912–2002)
Bio.: b. Famalicão, Portugal, 1 October 1912; d. Lisbon, Portugal, 30 May 2002. Entomologist. He graduated from the Univ. of Madrid, Spain, and later the Univ. of Porto, Portugal, in 1936. For political reasons, he was not allowed to pursue an academic career at the Univ. of Coimbra nor at Univ. of Porto. From 1937 to 1947, he was a secondary school teacher, and during that period, with his brother Bernardino Machado, he became a pioneer of speleology in Portugal, conducting significant research that resulted in the publication in 1945 of the first inventory of Portuguese limestone caves. He focused on the biology of the caves, studying the arachnids while exploring over 100 caves. Accepting an invitation to be director of the *Laboratório de Investigações Biológicas* of the *Museu do Dundo*, *Companhia de Diamantes de Angola*, at Lunda, he moved to Angola, where he remained from 1947 to 1973. Barros Machado was politically active and at some point, joined the political organisation *Movimento de Unidade Democrática* (MUD). This movement opposed the dictatorship that ruled Portugal at the time and was eventually made illegal. Due to this affiliation, Barros Machado was arrested when he visited Portugal on leave from the museum and had to stand trial together with several others who belonged to the MUD. His extensive biological collections

Fig. 19. António Barros Machado. Photographer unknown. Published by Regala (2014). Reproduced with permission.

in Angola cover many groups and he published on several subjects, ranging from insects to mammals. Of particular relevance is his publication on the systematics of *Glossina* (Diptera), a genus that includes the tsetse fly. He was in charge of the publications on biology of the *Companhia de Diamantes de Angola*. Over 2000 taxa were described by him and others as a result of this publishing initiative. His plant collections are integrated in the Diamang Collections (q.v.). In Portugal, after 1973, he also studied biogenic rocks and tropical soil evolution. In 1990, he received a doctorate *honoris causa* from the Univ. of Porto; he is commemorated in numerous scientific names. **Angola:** 1948–1973; Lunda Norte, Lunda Sul, Moxico; collected also with Fontinha (q.v.) and Peles (q.v.). **Herb.:** BM, COI, DIA†?, K, LISC (727 specimens), LUA, P, PC, US. **Ref.:** Barros Machado & Machado, 1945; Barros Machado, 1954; Vegter, 1976; Macedo, 1988; Bossard, 1993; Anonymous, 1996–; Romeiras, 1999; Figueiredo & al., 2008; Regala, 2014; Ceríaco & al., 2020. Figure 19; see also Figures 74 and 76.

Barroso Mendonça, Estevão
(fl. 1960–1973)
Bio.: He worked with R.M. Santos (q.v.) and died sometime before 1985. **Angola:** 1960–1973; collected also with (R.) Correia (q.v.), Menezes (q.v.),

and (R.M.) Santos (q.v.), under the name Barroso; not to be confused with F.A. Mendonça (q.v.). **Herb.:** COI, LISC (53 specimens), LUAI, LUBA, PRE. **Ref.:** Santos, 1989; Romeiras, 1999; Figueiredo & al., 2008.

Bastian, Adolf
(1826–1905)
Bio.: b. Bremen, Germany, 26 June 1826; d. Port-of-Spain, Trinidad and Tobago, 3 February 1905. Ethnologist. Son of a merchant, he studied in Germany: law in Heidelberg, biology in Berlin, Jena, and Würzburg; and obtained a degree in medicine in Prague, present-day Czech Republic. In 1850, he started the first of his many overseas travels as a naval surgeon. A voyage to Australia lasted several years, from 1851 to 1859. During that time, he visited Australia and the Pacific, South and Central America, and Asia, and on the return journey, West Africa, including Angola. During this stay in Angola in 1857 he travelled from Ambriz to Mbanza Congo (Zaire Province). Apparently, he did not collect any plants then, but he published on Mbanza Congo, then the capital of the Kingdom of Kongo, under the title *Ein Besuch in San Salvador* (Bastian, 1859). He travelled extensively worldwide and amassed a large collection of ethnographical objects and in addition

Fig. 20. Adolf Bastian in 1892. By Benque & Kindermann, Hamburg. Public domain.

was a prolific author. In 1867, he obtained his habilitation in ethnology at the Univ. of Berlin. The next year he was appointed curator of the ethnological and prehistoric collections of the Royal Museum of Berlin. In 1873, he established and became the first director of the *Königliches Museum für Völkerkunde* (now the *Ethnologisches Museum*) in Berlin and was one of the founders of the *Deutsche Gesellschaft zur Erforschung Aequatorial-Afrikas*. This institution financed his expedition to Loango in West Africa in 1873. The expedition aimed at exploring and advancing further into the interior of the continent from the Loango coast. From July 1873 to 12 October 1873, he visited several local Dutch factories. "Factory" – in Portuguese, *feitoria* – was the term then used for a trading post. He also made a few excursions inland, to the Loango and Congo rivers up to Boma (D.R. Congo), but further inland advancement was not achieved. He collected a few specimens of medicinal plants. Although the collections were recorded by Urban (1916), it is not known if any have survived. Bastian became known as one of the founders of ethnology in Germany. **Angola:** July 1873–12 October 1873, possibly also in 1857; Bengo?, Zaire? (1857), Cabinda (1873). **Herb.:** B (c. 30 numbers of medicinal plants, from 1873). **Ref.:** Bastian, 1859; Tylor, 1905; Urban, 1916: 327; Plischke, 1953; Lanjouw & Stafleu, 1954; Heintze, 1999a: 108–124, 2007: 104–120; Figueiredo & al., 2020. Figure 20.

Baum, Hugo
(1867–1950)

Bio.: b. Forst, Niederlausitz, Germany, 17 January 1867; d. Rostock, Germany, 15 April 1950. Horticulturalist. He studied in Guben on the Neisse River (then in Brandenburg, Germany now partly in Germany and Poland) and trained as a gardener in Nettkow and in Proskau (Silesia, then in Germany now Poland). After his military service he became a gardener at the Berlin Botanical Garden. Ten years later, in 1901, he was appointed head gardener at the Botanical Garden of the Univ. of Rostock. He was promoted to *Gartenoberinspektor* (head inspector of the garden) in 1926 and retired seven years later in 1933. In 1899, he was invited to participate as botanist in the *Kunene-Sambesi-Expedition* under the leadership of Pieter van der Kellen. The expedition was arranged under the auspices of the *Kolonial-Wirtschaftliches Komitee* (Berlin), the *Companhia de Mossamedes* (Paris, France), and the South West-Africa Company (London). It aimed to evaluate the economic potential of southern Angola and to undertake zoological and botanical exploration of the region. The group set off in three ox-wagons from Moçâmedes (Namibe) on 11 August 1899, heading to the Cunene, Cubango, Cuíto, and Cuando rivers. Almost eight months later, on 4 April 1900, after having travelled via Umpupa, Humbe, and

Fig. 21. Hugo Baum in 1926. Photographer unknown. Property of the Baum family, reproduced by Peter A. Mansfeld at http://www.level6.de. Reproduced under licence CC BY-SA 3.0 de.

Caiundo, they had reached the region beyond the Cuíto River and started their return journey via Menongue and Lubango. They arrived back at Moçâmedes, the place from which they departed, on 26 June 1900. Back in Germany, Baum worked on his collections and on the report of the expedition that was published in a volume edited by Warburg (1903). In 1925, he undertook a

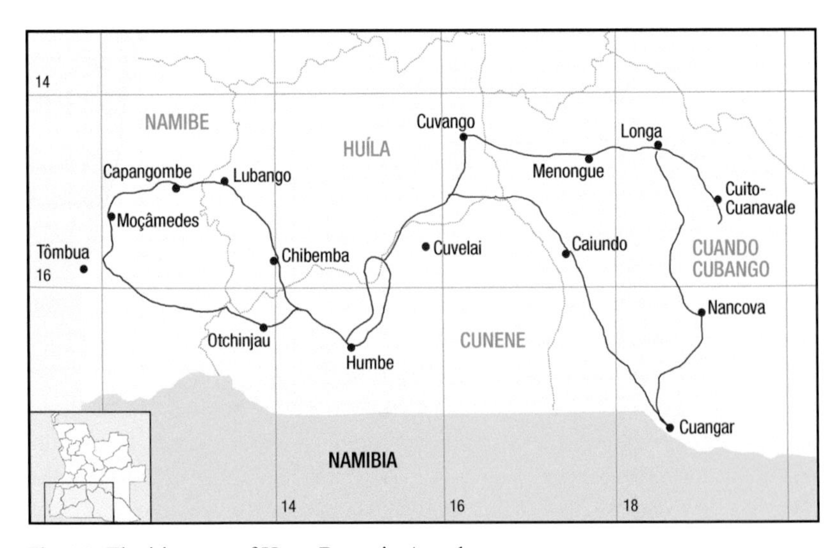

Fig. 22. The itinerary of Hugo Baum in Angola.

second and last expedition, to Mexico. A list of Baum specimens collected in Angola and a detailed itinerary of that expedition with a list of collecting localities including coordinates were provided by Figueiredo & al. (2009). Baum is commemorated in over 70 scientific names, including two genera: *Baumiella* Henn. (Pyrenomycetes) and *Baumia* Engl. & Gilg (Orobanchaceae), the latter being endemic to Angola. *Baumiella* H.Wolff (Apiaceae) is a nom. illeg. **Angola:** 11 August 1899–June 1900; Cuando Cubango, Cunene, Huíla, Namibe. **Herb.:** B (1016 numbers, partly destroyed), BM, BR, COI, E, G, GJO (lichens), HBG, K, LISC (1 specimen), M, NY, P, S, US, W, Z+ZT. **Ref.:** Baum, 1903; Warburg, 1903; Urban, 1916: 327; Gossweiler, 1939; Lanjouw & Stafleu, 1954; Bossard, 1993; Liberato, 1994; Heintze, 1999a: 125–128, 2007: 121–124; Romeiras, 1999; Figueiredo & Smith, 2008: 2, 3; Figueiredo & al., 2008, 2009, 2020; Mansfeld, 2012. Figures 21 and 22.

Beard, John Stanley
(1916–2011)
Bio.: b. Gerrards Cross, England, 15 February 1916; d. Perth, Western Australia, Australia, 17 February 2011. Ecologist. He was educated at Marlborough and Oxford, obtaining his Ph.D. from the Univ. of Oxford in 1945. He joined the British Colonial Forest Service, based in Trinidad, remaining there during World War II. After the War, he was appointed as a silviculturalist with the Natal Tanning Extract Company, South Africa, where he worked on crop improvement in the South African wattle industry. He remained in South Africa from 1947 to 1961 before moving to Australia where he was appointed as Foundation Director of the then new King's Park Botanic Garden, Perth. In 1970, he was appointed director of the Royal Botanic Gardens, Sydney, retiring to Perth a few years later to work on his Vegetation Survey of Western Australia. While based in KwaZulu-Natal in the 1950s, he developed an interest in the Proteaceae (see for example Beard, 1958, [1993]) and took an interest in the Botanic Gardens of Pietermaritzburg, now the KwaZulu National Botanical Garden. He is commemorated in several plant names. **Angola:** 1960; Benguela, Huambo, Huíla, Moxico. **Herb.:** A, COI, FR, K, LISC (11 specimens), MO, NU, PRE, TRIN. **Ref.:** Lanjouw & Stafleu, 1954; Beard, 1958, [1993]: back dustjacket text; Gunn & Codd, 1981; Dixon, 2006; Figueiredo & al., 2008.

Beatriz, Manuel Guerreiro
(fl. 1913–1930)
Bio.: Technician. He was a specialist on cotton, working from 1913 to 1919 with *Direcção de Agricultura e Comércio de Angola*. He collected a few

hundred specimens. **Angola:** 1929–1930; Bengo (Cacoba, Catete), Cuanza Sul, Luanda (Quiçama). **Herb.:** COI, LISC (1 specimen). **Ref.:** Gossweiler, 1939; Lanjouw & Stafleu, 1954; Bossard, 1993; Liberato, 1994 [as Manuel Beatriz Guerreiro]; Romeiras, 1999; Figueiredo & al., 2008.

Becquet, Augustin-Jean-Marie
(1899–1974)
Bio.: b. Ixelles, Belgium, 4 June 1899; d. Palma de Mallorca, Spain, 29 May 1974. Agronomist. He graduated from the Univ. of Leuven, Belgium, in 1921. He started to work in Africa in 1923. In August 1938, he arrived in Kinshasa (D.R. Congo). Due to the involvement of Belgium in World War II, he could not return to Europe for the next years and took leave in Angola where he made collections of plants and animals, especially in Namibe and the Lubango plateau. In 1946, he joined the *Institut National pour l'Étude Agronomique du Congo Belge* of which he later became director. **Angola:** April 1941; Huambo, Huíla (Lubango), Namibe; 190 collections; numbers 1025–1233 (fide Bamps, 1973). **Herb.:** BR, K, MO. **Ref.:** Lanjouw & Stafleu, 1954; Bamps, 1973; Jurion, 1977; Romeiras, 1999; Figueiredo & al., 2008.

Bell, Fiona Clare
(1967–)
Bio.: b. Grahamstown, Eastern Cape, South Africa, 4 March 1967. Horticulturalist. In 1997, she completed a National Diploma in Horticulture at the then Cape Technikon, now the Cape Peninsula Univ. of Technology. From 1994 to 2000, she worked in the Botanical Society Garden Centre at the Kirstenbosch National Botanical Garden, Cape Town, and in-between (1997–1998) at the San Michell Nursery, Atlantis, Western Cape Province. From 2001 to 2004, she conducted environmental rehabilitation work in Sekukhuneland, Limpopo Province, northeastern South Africa, while working for Jeremy Stubbs Landscaping. She was the Head of the Nursery of the South African National Biodiversity Institute's Pretoria National Botanical Garden from 2004 to 2005 and is currently self-employed as a landscaper in Cape Town. In 1996, she accompanied Clifford Thompson (q.v.) and his wife on a trip to Angola. Thompson collected material of a then unknown representative of the Apocynaceae, which Clare provided to Ernst J. van Jaarsveld, at the time attached to Kirstenbosch, who recognised it as novel. The holotype of the name *Tavaresia thompsoniorum* van Jaarsv. & R.Nagel, as '*thompsonii*', is *Clare Bell s.n.*, a specimen collected "between Iona and Oncucua [Oncócua] before [the] Abamalinda Mountain" on 14 July 1996 (NBG0200687-0). The herbarium in which the specimen was lodged was given as "NBI", which is

to be corrected to Herb. NBG. **Angola:** 1996; Cunene, Namibe. **Herb.:** NBG. **Ref.:** Van Jaarsveld & Nagel, 1999: 9; Figueiredo & al., 2008; F.C. Bell, pers. comm., October 2021.

Bequaert, Joseph Charles Corneille (1886–1982)

Bio.: b. Torhout, Belgium, 24 May 1886; d. Amherst, Massachusetts, U.S.A., 19 January 1982. Biologist. He graduated with a Ph.D. from the Univ. of Gand, Belgium in 1908. Afterwards he accepted an appointment as entomologist in the Belgian Sleeping Sickness Commission and later was in charge of a collecting expedition to Ruwenzori (D.R. Congo/Uganda) and Rutshuru (D.R. Congo). The expedition started off at the African coast in July 1913. He explored the coastal region, Mayombe, and Cabinda. After September, he travelled up the Congo River. The expedition took two years and resulted in 7500 collections, including dozens that represented new taxa. Because of World War I, his collection was initially kept at the herbarium of the British Museum (now Natural History Museum) in London (Herb. BM), and only moved to Belgium, to the herbarium of the Meise Botanic Garden (Herb. BR) after 1918. Bequaert immigrated to the U.S.A. in 1916 and five years later became a nationalised American citizen. From 1917 to 1922, he was "Research Associate in Congo Zoology" at the American Museum of Natural History in New York City and also worked as malacologist, entomologist, and botanist. In 1917, he participated as hymenopterist in an overland collecting expedition that started in Ithaca, New York, and ended in Berkeley, California, U.S.A.

Fig. 23. Joseph Bequaert. Photographer unknown. Reproduced under licence CC BY-SA 4.0.

He returned to Africa as part of the Harvard African Expedition of 1934. He later worked at Harvard Univ. (1923–1956), first as an assistant professor, then as "Curator of Recent Insects" in the Museum of Comparative Zoology. In 1951, he was appointed Alexander Agassiz Professor of Zoology and curator of insects at the Museum of Comparative Zoology. After retiring, he was appointed as professor at the Univ. of Houston, Texas (1956–1960) and the Univ. of Arizona, Tucson (1969–1971). **Angola:** August 1913; Cabinda (Cacongo, Landana, Chiloango River). **Herb.:** A, BM, BR, BUF, DIA†?, F, GH, K, LISC, MICH, NMW, NY, P, S, U, US, USBR. **Ref.**: Bradley, 1919; Lanjouw & Stafleu, 1954; Bamps, 1973; Carpenter, 1982; Romeiras, 1999; Pauly, 2001; Lawalrée, 2002; Figueiredo & al., 2008; Polhill & Polhill, 2015. Figure 23.

Berthelot
(fl. 1895–1908)
Note: The identity of this collector is unknown. The surname appears in the catalogue of the herbarium of the *Muséum national d'Histoire naturelle*, Paris (Herb. P), as Fr. Berthelot, indicating that it might refer to a friar. However, on specimen labels, the name is simply given as "Berthelot". The collections were originally kept in the herbarium of Henri[-y] de Vilmorin (q.v.), an horticulturalist who had a particular interest in acquiring seeds, no doubt as part of efforts to expand the range of material offered for sale in the Vilmorin's horticultural trade catalogues. Some specimens have notes by the collector commenting on the absence of seeds, which appears to indicate that the specimens were collected in response to a request by Vilmorin. Gossweiler (1939) suggested that Berthelot was a missionary. In the literature, there is reference to a Sister Maria Madalena Berthelot, who was Superior at Huíla in 1937 (Anonymous, 1937). This "Berthelot" is not to be confused with Sabin Berthelot (1794–1880), who did considerable work on the Canary Islands. **Angola:** 1895–1908; Cabinda, Huíla, Namibe. **Herb.:** P (c. 60 numbers). **Ref.:** Gossweiler, 1939; Lanjouw & Stafleu, 1954; Bossard, 1993; Romeiras, 1999; Figueiredo & al., 2008.

Bertrand, Henri P.I.
(1892–1978)
Bio.: b. 1892; d. Pyrenees, France, September 1978. Entomologist. He graduated with a B.Sc. from the Univ. of Bordeaux, France, in 1912, and with a D.Sc. from the Univ. of Paris in 1927. He became a specialist on Coleoptera. In 1944, he was research director at the French National Centre for Scientific Research (CNRS). From June to September 1957, he sojourned at the *Laboratório de*

Investigação Biológica of the *Museu do Dundo*, in Lunda, Angola, studying water beetles. During that time, he undertook extensive field work in the district. As a result, he published some papers on Angolan Coleoptera. He also collected some plants for the Diamang Collections (q.v.) that were deposited at the herbarium of the *Museu do Dundo* (Herb. DIA) with some duplicates elsewhere. In 1978, at the age of 86, he went missing while on a solitary entomological expedition in the French Pyrenees; his body was discovered one year later. **Angola:** 1957; Lunda. **Herb.:** DIA†?, LISC (5 specimens), PRE. **Ref.:** Bertrand, 1966; Bournaud, 1980; Romeiras, 1999; Figueiredo & al., 2008; d'Aguilar, 2013.

Bizarro, Clemente Joaquim Abranches – see **Abranches Bizarro**

Bonnefoux, Benoit-Marius
(1861–1937)
Bio.: b. Viverols, Auvergne, France, 8/9 September 1861; d. Huíla, Angola, 20 July 1937. Missionary with the Congregation of the Holy Spirit (Spiritans). He was ordained as a priest in 1884 and took orders in 1885 after which he was sent to the *Missão da Huíla* (see page 57) in Angola, where he arrived in November 1885. In 1888, he was in charge of the Orphanage, but

Fig. 24. Benoit-Marius Bonnefoux. Photographer unknown. Missionários do Espírito Santo. Reproduced with permission.

later, he succeeded Antunes (q.v.) as Superior of the Mission, being himself succeeded by Estermann (q.v.). From 1892 to 1904, he founded and directed the *Missão de Tchivinguiro.* In September 1904, he became the Superior Apostolic Prefect for the Cunene District. He died at *Missão da Huíla.* He was interested in many subjects, including in medicinal plants, on which he left unpublished manuscripts. He authored an Olunyaneka–Portuguese dictionary that was published posthumously in 1941. His precise sketches of the regions where he travelled were at the time very valuable for the cartography of Angola. **Angola:** 1885–1937; collected also with Villain (q.v.). **Herb.:** BM, P, UC, US. **Ref.:** Dias, 1937; Gossweiler, 1939; Brásio, 1940; Bonnefoux, 1941; Estermann, 1941; Lanjouw & Stafleu, 1954; Bossard, 1993; Romeiras, 1999; Figueiredo & al., 2008; Anonymous, s.d.(a). Figure 24.

Borges, Anabela
(fl. 1971–1974)
Bio.: Technician at *Instituto de Investigação Científica de Angola.* **Angola:** 1971–1974; Cunene, Huíla; collected over 400 numbers; also, with Constantino (q.v.) and Rey (q.v.). **Herb.:** BM, BR, COI, K, LISC (389 specimens), LUA, LUAI, LUBA, P, PRE. **Ref.:** Bossard, 1993; Liberato, 1994; Romeiras, 1999; Figueiredo & al., 2008.

Borges, José Baptista
(fl. 1940s)
Bio.: Agronomist who worked for *Junta dos Cereais.* **Angola:** 1940s; Cunene, Huambo, Huíla. **Herb.:** B, COI, K, LISC (1 specimen), LUA, US. **Ref.:** Bossard, 1993; Figueiredo & al., 2008.

Boss, Georg
(1903–1972)
Bio.: b. Biebrich (now Wiesbaden), Germany, 31 October 1903; d. Wiesbaden, Germany, 13 May 1972. Biologist. He studied natural science at the Univ. of Frankfurt, Germany, from 1922, graduating with a Ph.D. on the cytology of Ustilaginales in 1927. He worked as a teacher at the German high school in Swakopmund, Namibia, from 1931 to 1937. During that time, he undertook several trips to Angola where he made plant collections. After returning to Germany, he served as an officer in the German Luftwaffe during World War II. Afterwards he had various occupations, such as farmer, writer for the German Weather Service, organist, wine dealer, and medicinal plants grower. In February 1954, he returned to Angola with an expedition of the *Deutsche Forschungsgemeinschaft*, which included the ethnologist Hermann

Baumann (1902–1972; Braun, 1995) and a geography student, Manfred Topp. They travelled mostly in southwestern Angola. In Lubango (Huíla), they met Estermann (q.v.), who arranged for them to visit the prehistoric rock art site at Chitundo-Hulu, c. 12 km east of Capolopopo (Namibe). They also visited Caluquembe (Huíla) and Quibala (Cuanza Sul). Boss and Baumann had published a paper on the prehistoric stone graveyards of Quibala ten years earlier. According to Hertel & Schreiber (1988), Boss collected 600 numbers in Angola in 1954. Boss remained in Wiesbaden for the rest of his life. He published some articles on the Namibian flora and indigenous stone buildings in Angola and is commemorated in names such as *Blepharis bossii* Oberm. (Acanthaceae) and *Hydrodea bossiana* Dinter (Aizoaceae). **Angola:** 1932–1937 and 1954; Benguela, Cuanza Norte, Cuanza Sul (Cuanza River, Libolo), Cunene (Xangongo, then Roçadas), Huíla, Namibe (Moçâmedes). **Herb.:** B, BM, K, M (600 numbers), MO, P, PRE (incl. TRV). **Ref.:** Boss, 1927; Gossweiler, 1939; Lanjouw & Stafleu, 1954; Gunn & Codd, 1981; Bossard, 1993; Romeiras, 1999; Figueiredo & al., 2008, 2020; Heintze, 2010; Brandstetter & Hierholzer, 2017.

Boulton, Wolfrid Rudyerd ("Rud") (1901–1983)

Bio.: b. Beaver, Pennsylvania, U.S.A., 5 April 1901; d. Harare, Zimbabwe, 24 January 1983. Ornithologist and intelligence officer. In 1924, he graduated from the Univ. of Pittsburgh with a B.Sc. and was appointed as ornithology assistant at the American Museum of Natural History (AMNH) in New York City. In that capacity, he participated in the first AMNH Vernay Angola Expedition in 1925. In that year, he married his first wife, the ethnomusicologist Laura Craytor (1899–1980). In 1926, the couple moved to Pittsburgh where he assumed the position of assistant curator of birds at the Carnegie Museum of Natural History. In 1929, they participated in another AMNH expedition to Nyasaland (now Malawi), Uganda, and Kenya. Shortly afterwards, from August 1930 to May 1931, they joined the Carnegie Museum Pulitzer Angola Expedition, an endeavour of the publisher Ralph Pulitzer (1879–1939) to search for the Giant Sable Antelope. After returning to the U.S.A., the couple moved to Chicago, where Boulton became curator of birds at the Field Museum. They divorced in 1938. Boulton, who was considered a specialist on Angolan birds, left his ornithology career in 1942 and joined the U.S. Office of Strategic Services, an intelligence agency. He moved to Washington, D.C. and married his second wife, Inez Cunningham Stark (d. 1958), in that year. During World War II, he conducted intelligence work and afterwards he was with the C.I.A. He continued an association with the Field Museum and participated in their expedition to present-day Zimbabwe and Angola in 1957. In 1958, he retired officially, but perhaps not in practice, from the C.I.A. In that same year, he became a widower, and the following year, he married his third wife, Louise Rehm (d. 1974). The couple established a foundation, the Atlantica Foundation, which operated in Salisbury (now Harare, Zimbabwe), and they relocated to that country in 1960. The foundation became a base for several visiting researchers, especially from the U.S.A. During the ensuing years, the couple travelled extensively in the region. Boulton became a widower again in 1974 and remained in Zimbabwe for the rest of his life. No plant collections are known from his first visit to Angola as part of the 1925 AMNH Vernay Angola Expedition. Some plant collections were made during the 1930–1931 Pulitzer Angola Expedition. These were deposited at the herbarium of the Carnegie Museum of Natural History. **Angola**: 1930–1931. **Herb.:** CM. **Ref.:** Jacobs, 2015; Miriello, 2020.

Branquinho d'Oliveira, António Lopes
(1904–1983)

Bio.: b. Coimbra, Portugal, 1904; d. 1983. Phytopathologist. He obtained his doctorate at the Univ. of Cambridge, England. From 1942 to 1947, he was a professor at *Instituto Superior de Agronomia* in Lisbon and then director of the phytopathology department at the *Estação Agronómica Nacional*, Oeiras, until 1972. He was an expert on coffee rust and director of the *Centro de Investigação das Ferrugens do Cafeeiro*. **Angola:** 1956–1957; Cuanza Sul. **Herb.:** LISC (27 specimens). **Ref.**: Figueiredo & al., 2008; Anonymous, s.d.(b).

Brass, Leonard John
(1900–1971)

Bio.: b. Toowoomba, Queensland, Australia, 17 May 1900; d. Cairns, Queensland, Australia, 29 August 1971. Botanical collector and explorer. He trained as a botanist at the Queensland Herbarium in Brisbane, Australia, and was employed there for a while as herbarium assistant. After a period working on cattle stations, in 1925, he undertook his first expedition to New Guinea on behalf of the Arnold Arboretum, U.S.A. Over the next 34 years, he participated in nine collecting and exploring expeditions, amassing a large number of collections. From 1933 to 1959, under employment of the American Museum of Natural History, he was the botanist of six "Richard Archbold expeditions" to New Guinea and Australia. He also collected in the Solomon Islands for the Arnold Arboretum. During World War II, he served with the Canadian army and in 1944 he accepted an appointment as curator and resident botanist at the Archbold Biological Station in Florida, U.S.A. A few years later, in 1947, he obtained U.S. citizenship. He undertook two expeditions to Africa. The first one, to Nyasaland (now Malawi), took place in 1946 with a Vernay Angola Expedition. The second one was to tropical Africa and lasted from 1949 to 1950, and was on behalf of two commercial companies, Upjohn and S.B. Penick & Co., with the purpose of searching for alternative sources for the drug cortisone. On this expedition, Brass collected with Woodward (q.v.). They travelled in Angola and collected in several provinces. The labels of some of these collections have the header: "The Upjohn-Penick Expedition for Botanical Exploration". During his collecting career, Brass made over 33,000 collections and is commemorated in hundreds of plant names. In 1962, he was awarded an honorary doctorate by the Florida State Univ. in Tallahassee. He remained in Florida until 1966 when he retired and returned to Australia, settling in Cairns where he continued collecting. His plant collections are distributed among numerous herbaria; Angolan collections have been located at Herb. MO, NY, PC, and TENN. **Angola:** March–April 1950; Bié, Cuanza Norte, Cuanza Sul, Lunda Sul, Malanje, Moxico, Uíge, Zaire; collected with Woodward (q.v.). **Herb.:** A (main), AMES, B, BM, BO, BR, BRI, CANB, F, G, GH, K, L, MO, NMW, NSW, NY (main), P, PC, S, SING, TENN, U, UPS, US, W. **Ref.:** Lanjouw & Stafleu, 1954; Australian National Herbarium, 2021; JSTOR, 2021; Lohrer, 2021.

Bredo, Hans Joseph Anna Erich Richard (1903–1991)

Bio.: b. Belgium, 30 May 1903; d. Brussels, Belgium, 29 June 1991. Entomologist. He graduated in chemistry at the Univ. of Louvain, Belgium in 1929. In that year, he moved to the D.R. Congo to work as a government entomologist. He was in that position until 1944. During the period 1930s–1940, he collected in several countries in Central Africa and in Angola. In the 1930s, he surveyed the territory of the D.R. Congo for the permanent breeding grounds of the red locust. This formed part of an international effort to control locust plagues in which Henry Arnold Francis Lea (1907–1989) conducted a similar survey in Mozambique. From 1945 to 1952, Bredo directed the International Locust Control Program. He published on insects and is commemorated in several insect names. **Angola:** October 1940; Bié (Silva Porto), Moxico (Caianda, Luau); 32 collections; numbers 4670–4707 (fide Bamps, 1973). **Herb.:** BR, K, P. **Ref.:** Lanjouw & Stafleu, 1954; Bamps, 1973; Figueiredo & al., 2008; Beolens & al., 2014; Polhill & Polhill, 2015.

[Breteler, Franciscus ("Frans") Jozef]
(1932–)
Note: Dutch botanist, specialist on African flora, who made numerous collections. He was recorded in JSTOR Global Plants (JSTOR, 2021) as having collected in Angola. However, there are no Angolan specimens in the Dutch Herbaria database at Naturalis.

[*Brigada de Estudos Florestais do Maiombe*]
Note: This likely refers to M.E.F.A. (q.v.).

Brites, M.
(fl. 1967)
Angola: 1967; Huíla, Namibe; collected with (C.A.) Henriques (q.v.) and Menezes (q.v.). **Herb.:** COI, LUAI, LUBA, PRE. **Ref.:** Figueiredo & al., 2008.

Brown, Arthur H.
(fl. 1889)
Bio.: Brother of William H. Brown (q.v.). The two brothers joined an expedition to observe the solar eclipse of 22 December 1889 in West Africa, visiting Angola, where they made some collections. **Angola:** 1889; Bengo (Cunga), Luanda; the header on collection labels reads, in part, "United States Eclipse Expedition to Western Africa" and the collectors are "W.H. Brown & A.H. Brown". **Herb.:** US. **Ref.:** Brown, 1899; Lanjouw & Stafleu, 1954 (sub "Brown, W.H.").

Brown, William Harvey
(1862–1913)
Bio.: b. Des Moines, Iowa, U.S.A, 22 December 1862; d. Harare, Zimbabwe, 5 April 1913. Naturalist. He graduated from the Univ. of Kansas at Lawrence with a B.Sc. In 1886, he took part in an expedition to Montana to collect skins and skeletons of the American bison and afterwards joined the natural history department of the Smithsonian Institution in Washington, D.C. He was appointed naturalist of an expedition to observe the solar eclipse of 22 December 1889 in West Africa. During that expedition, he collected in Angola and Cape Town. His collections were eclectic, including mammals, birds, amphibians and reptiles, fishes, plants, and artefacts; they are deposited at the Smithsonian Institution. Brown arrived in Luanda on 6 December 1889. Through influence of the missionary (and also collector) Héli Chatelain (1859–1908), Brown and his group were authorised to occupy the quarters of one of the missions of Bishop William Taylor (1821–1902) of the Methodist

Episcopal Church in Luanda. Over the next days, based at the mission, they collected in the vicinities. It was then decided that the expedition would proceed sailing south to Cabo Ledo to make the eclipse observations there, while Brown was allowed to stay in Luanda for some weeks to continue his collecting activities. He was accompanied by four other men of the expedition team, one of whom was his brother Arthur H. Brown (q.v.), and two sailors, and ten contracted African men. The Governor of Angola put at their disposal a special train on the recently built railway. On 14 December, they embarked on this train that stopped whenever they wanted to shoot birds and take photographs. They passed Cacuaco and Quifangondo and finally reached Cunga, a small trading post near the Cuanza River. Chatelain arranged for accommodation for them, and they were based there for two weeks collecting in the vicinities. On 3 January, they left Cunga to return to Luanda, from where they sailed to Cape Town on 5 January. The expedition remained in Cape Town until 6 February and then returned home, leaving Brown and another team member who had decided to join the Pioneer Corps of the British South Africa Company as a trooper. The Corps marched to Mashonaland (a region in north Zimbabwe) where Brown settled. He continued collecting and sending specimens to the Smithsonian or to the Cape Town Museum. He was active in the First Matabele War in 1893. In 1899, he published a memoir of his adventures. Afterwards he settled near Harare and farmed. **Angola:** 1889; Bengo (Cunga), Luanda; the header on collection labels reads, in part, "United States Eclipse Expedition to Western Africa" and the collectors are "W.H. Brown & A.H. Brown". **Herb.:** P, US. **Ref.:** Brown, 1899; Anonymous, 1913; Lanjouw & Stafleu, 1954.

Brown(e), William
(fl. prob. 1706–1707)
Bio.: d. sometime before 1715. British naval surgeon. He collected plants for James Petiver (1658–1718). Petiver's collection was later bought by Hans Sloane (1660–1753) to integrate into his herbarium. Sloane's Herbarium, with over 70,000 specimens, was bequeathed to the British Museum, London, and became the basis of the herbarium of the Natural History Museum (Herb. BM). Among Petiver's collections in Sloane's Herbarium there are "numerous specimens from Angola" (Dandy, 1958) under "H.S. 154 and 155 Hortus siccus Africae continens". Brown(e) is one of the four pre-Linnaean collectors recorded for Angola, the others being Gladman (q.v.), Kirckwood (q.v.), and Mason (q.v.). **Angola:** before 1715, probably 1706–1707. **Herb.:** BM (Sloane Herb.), K. **Ref.:** Exell (A.W.), 1939, 1962; Dandy, 1958; Bossard, 1993; Liberato, 1994; Romeiras, 1999; Figueiredo & al., 2008.

Brühl, Ludwig Julius
(1870–unknown)
Bio.: b. Breslau, Germany, 17 August 1870. Ichthyologist. He studied zoology and medicine at the Univ. of Berlin, from 1889 to 1895, graduating as a medical doctor in 1898. In 1903, after an assistantship at the physiology institute in Berlin, he joined the *Institut für Meereskunde* of the Univ. of Berlin. He became curator in 1906 and specialised in ichthyology. He undertook several expeditions and collected material for the *Museum für Meereskunde* and for the Zoological and Botanical Museum of Berlin. These expeditions included an exploration aboard the steamboat *Helgoland* of Svalbard and Bear Island in 1898 and a separate Dead Sea Expedition from 1911 to 1912. He was commissioned by the holding company *Companhia do Fomento Geral of Angola* to study the maritime resources of Namibe, Angola, and in October 1922, he travelled to the country. He made collections of plants on the coastal deserts, including of *Welwitschia* Hook.f. (Welwitschiaceae), which were deposited at the herbarium of the Botanic Garden and Botanical Museum Berlin-Dahlem (Herb. B). Because he was of Jewish descent, he left Germany in 1934, immigrating to Tanga, in Tanganyika Territory (now Tanzania), where he died on an unknown date before 1953. **Angola:** October 1922–June 1923; Namibe (Moçâmedes). **Herb.:** B (main; 45 numbers, most likely destroyed; a specimen of *Welwitschia* is extant), PRE. **Ref.:** Urban, 1916: 333; Gossweiler, 1939; Lanjouw & Stafleu, 1954; Bossard, 1993; Romeiras, 1999; Figueiredo & al., 2008, 2020.

[Bruijn, J. de]
(1935–)
Note: He was with the technical staff at Herb. WAG from 1958 to 1990. He was recorded in JSTOR Global Plants (JSTOR, 2021) as collector in Angola. However, there are no Angolan specimens in the Dutch Herbaria database at Naturalis.

Buchner, Maximilian
(1846–1921)
Bio.: b. Munich, Germany, 26 April 1846; d. Munich, Germany, 7 May 1921. Medical doctor. In 1870, he graduated from the Univ. of Munich and served as a volunteer medical doctor in the Franco-Prussian War (1870–1871). After the war, from 1872, he was a naval surgeon for the *Norddeutscher Lloyd* in service on the route to either New York or Baltimore, U.S.A. He became a 1st class assistant doctor with the merchant navy in the following year. In 1874, he was arrested and given a one-year jail sentence because of getting involved

in a duel. In prison, he met Wissmann (q.v.), who had been arrested for the same reason. After being released, Buchner worked on a British vessel and travelled to New Zealand, Fiji, and Hawaii. His interest in travelling prompted him to contact the *Afrikanische Gesellschaft in Deutschland* to join one of their African expeditions. To prepare for the expeditions, he studied geology in Munich and astronomy in Berlin. He then undertook a four-year expedition to Angola with one aim being to follow Pogge's (q.v.) itinerary and meet the Mwat Yamv (in Portuguese *Muatiamvo*, *Muatiânvua*) in the Kingdom of Lunda. On 5 December 1878, Buchner arrived in Luanda, Angola, and continued along the usual route to the interior, up the Cuanza River to Dondo, where he met Mechow (q.v.) on 23 December. Buchner then proceeded on foot, and on 26 January 1879, again met Mechow, this time at Pungo Andongo. They arrived together at Malanje on 30 January 1879. Preparations for the expedition to the Mwat Yamv took several months. While Buchner was stationed at Pungo Andongo, he met Schütt (q.v.), who was on his way back from Lunda and gave him valuable advice and an itinerary map of his own journey. Buchner also met Capelo (q.v.) and Ivens (q.v.). On 19 March 1879, Capelo and Ivens were camped north of Malanje, near Bango, when Buchner arrived mounted on a *boi-cavalo* (an ox) accompanied by three or four Africans. News had reached Buchner of the arrival of Capelo and Ivens and he had wanted to meet them and, additionally, to explore the Bango Hill. They had lunch together and Buchner continued up the hill following a path that had been opened by Capelo and Ivens. Buchner collected several hundred numbers in Malanje, some of which originated from Bango. Finally, on 22 July 1879, Buchner's expedition departed with a party of 160 men. Although they aimed to follow in Pogge's footsteps, they followed a different route and rather headed north from a locality west of Mona Quimbundo. Buchner explored Lunda and the Cuango valley. On 11 December 1879, they reached the capital of the Kingdom of Lunda, the Musumb (in Katanga, D.R. Congo). They were not granted permission to continue further north and six months later, in June 1880, returned using a more northern route. After entering present-day Angolan territory, they proceeded in a northwesterly direction passing the Cuango River near Cassanje in December 1880. The expedition, by then reduced to eight men, arrived at Malanje on 28 February 1881. There, Buchner met Pogge, Wissmann, Mechow, and Teusz (q.v.). After some rest, he proceeded on foot towards the coast and arrived in Luanda in August 1881. On 1 September 1881, while in Luanda, he gave a presentation on the expedition. Henrique Dias de Carvalho (1843–1909), who would later also lead an expedition to the Mwat Yamv, was in the audience. Carvalho was the secretary of the society that organised the presentation. Buchner returned to Europe. On his way back, on 6 October 1881, he landed at Banana and undertook an excursion to the stations at Vivi and Isangila (D.R. Congo) on the Congo River. On 13 January 1882, he was back in Berlin. He returned to Africa to be an acting *Reichskommissar* in Cameroon from July 1884 to April 1885. There he was met by Büttner (q.v.). Back in Germany, Buchner became director of the Museum of Ethnology in Munich in 1889 and retired in 1907. He published several reports on the expedition to Lunda, which have been compiled by Heintze (1999b), but unlike the other German explorers, he did not publish a book. His meticulous photographs numbered at least 60, but most of them apparently have not been preserved. They include the first photograph, reproduced in Heintze (1999c), taken in the Kingdom of Lunda. Buchner is commemorated in numerous names, such as *Blepharis buchneri* Lindau (Acanthaceae). Buchner dispatched his collections to Berlin in at least three different consignments (Urban, 1892). The first consignment included numbers up to 167. Of these, the numbers up to 159 (with a few exceptions) are from Malanje and were collected up to July 1879. The second consignment consisted of numbers 168–500, and it was also shipped to Germany before Buchner returned to that country in 1881; these numbers were lost in the English Channel where the ship carrying them was wrecked. It is likely that Buchner collected at the Musumb (now in territory of the D.R. Congo) where he was stationed for six months. However, no records of material from that area were located. Those collections were likely included in the second consignment that was lost at sea. Buchner's third consignment of material consisted of 224 "species of phanerogams". Although there were 396 numbers at the herbarium of the Botanic Garden and Botanical Museum Berlin-Dahlem

Fig. 25. Maximilian Buchner. Artist unknown. From Wikimedia Commons. Public domain.

(Herb. B) (Urban, 1916), many specimens were destroyed during the World War II bombing of Berlin. Most of these do not have any duplicates. Some of the numbering of the specimens does not appear to be chronological. **Angola:** 5 December 1878–September/October 1881; Bengo, Cuanza Norte, Lunda Norte, Lunda Sul, Malanje. **Herb.:** B (main; 396 numbers according to Urban, 1916), BM, BR, COI, GH (few), K, L, P (few). **Ref.:** Buchner, 1881; Capelo & Ivens, 1881, 1882; Urban, 1892, 1916; Schnee, 1920: 248; Gossweiler, 1939; Lanjouw & Stafleu, 1954; Maull, 1955; Mendonça, 1962a; Bossard, 1993; Liberato, 1994; Heintze, 1999a, 1999b, 1999c, 2010; Romeiras, 1999; Figueiredo & al., 2008, 2020. Figures 25 and 26; see also Figure 3.

Itinerary of Buchner's expedition in Angola and the D.R. Congo
(Heintze, 1999b: 38–39; Figueiredo & al., 2020) (Figure 26, in part)
22 July 1879 – Left Malanje.
22 July to 19 August 1879 – In the Songo area.
26/27–29 July 1879 – In Sanza.
27 August 1879 – Crossed the Cuango River.
28 August to 10 September 1879 – In the Minungo area.
6 September 1879 – Crossed the Cucumbi River.
11 September to 7 October 1879 – In Chokwe (Kioko, Quioco) area.
14 September 1879 – Crossed the Cuilo River and proceeded in a NE direction to avoid Quimbundo.
16 September 1879 – Crossed the Luangue River.
19–22 September 1879 – Went along the Luvo River and crossed it.
27 September 1879 – Crossed the Luele River.
1 October 1879 – Crossed the Chicapa River.

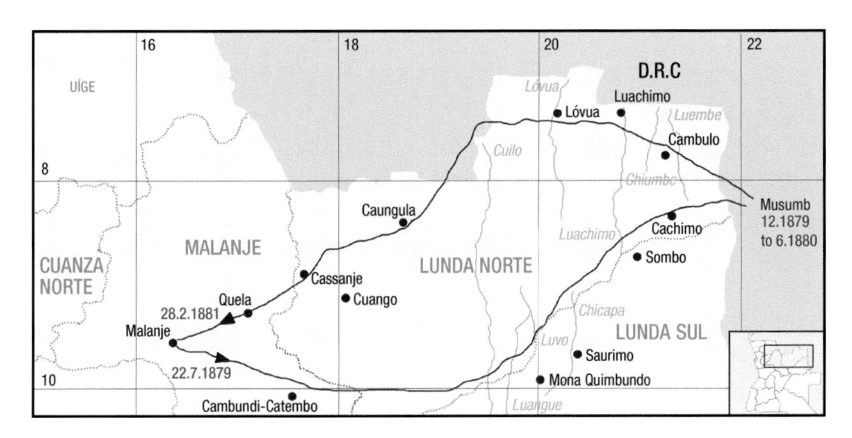

Fig. 26. The itinerary of Maximilian Buchner in Angola.

6/7 October 1879 – Crossed the Luachimo River.
22 October 1879 – Crossed the Chiumbe River.
c. 25 October 1879 – At the Luembe River.
11 December 1879 – Reached the Musumb.
June 1880 – Left the Musumb.
1 September 1880 – Between Cassai (Kassei) and Luembe.
December 1880 – Passed the Cuango near Cassanje.
28 February 1881 – Arrived at Malanje.
July 1881 – Left Malanje.
July and August 1881 – Marched towards Luanda via coffee plantations of Cazengo and Golungo.
End of August 1881 – Arrived at Luanda.
September/October 1881 – Left Luanda.
6–14 October 1881 – In Banana (D.R. Congo).
25–26 October 1881 – At Vivi (D.R. Congo).
31 October 1881 – Arrived in Isangila (D.R. Congo).
End of 1881 – Spent some weeks at Banana (D.R. Congo).

[Bullock, Arthur Allman]
(1906–1980)
Note: English botanist with Royal Botanic Gardens, Kew, England, since 1929 until retirement in 1968, with two interruptions: during World War II when he served in the Royal Air Force as a Flight Lieutenant and later, from 1949 to 1951, when he was stationed in Tanzania and Zambia with the Red Locust Control Service. He was recorded in JSTOR Global Plants (JSTOR, 2021) as a collector in Angola. However, the collections listed there as originating from Angola are, in fact, from Zambia, and no specimens from Angola were located. **Ref.:** Meikle, 1982; Polhill & Polhill, 2015.

Burger, Alfons M.
(fl. 1955)
Bio.: He lived with his family at Entre-Rios (now Dimuca) in Angola in the 1950s. He is likely the author of a note on oil production in Angola. **Angola:** 1955. **Herb.:** K, M. **Ref.:** Burger, 1955; Hellmich, 1957; Figueiredo & al., 2008.

Büttner, Oskar Alexander Richard
(1858–1927)
Bio.: b. Brandenburg/Havel, Germany, 28 September 1858; d. Berlin-Karlshorst, Germany, 11 September 1927. Naturalist and explorer. After graduating with a doctorate in Chemistry and Natural Sciences at the Univ. of Berlin in 1883,

he was contracted by the *Deutsch-Afrikanische Gesellschaft* to join the expedition to the Congo lead by Eduard Schulze (1852–1885), with Willy Wolff (b. 1852), Richard Kund, and Hans Tappenbeck. The expedition left Hamburg on 1 August 1884. While the ship stopped in Cameroon, Büttner met Buchner (q.v.), who was there as acting *Reichskommissar*. At another port-of-call, in present-day Gabon, he made some collections with Soyaux (q.v.), who was then running a local coffee plantation. On 13 November 1884, the expedition arrived at Banana (D.R. Congo). They started at Tondoa (Tunduwa or Underhill), an English mission station across from Vivi (both localities now in the D.R. Congo), and a few kilometres upriver from Nóqui (Zaire Province, Angola). On 12 December 1884, Schulze and Büttner departed for San Salvador (now Mbanza Congo, Angola), the capital of the Kingdom of Congo. They arrived six days later, on the 18th, explored the surrounding area and remained there until the end of February 1885. In January 1885, Büttner took a trip to Quedas do Mebridege, a three-day journey east of Mbanza Congo, where he collected some specimens. Shortly after, on 15 February 1885, Schulze died of malaria. The expedition team then split, with Büttner and Wolff travelling separately and following different routes. On 12 April 1885, Büttner attempted to reach the Cuango River with a few local porters, but only managed to proceed as far as Quisulo (Uíge Province, Angola) and eventually returned to Mbanza, Congo. He made a second attempt to reach the Cuango River on 27 June 1885, with a group of 80 porters from Loango. Finally, by July 1885, after passing Quisulo and Maquela do Zombo, he crossed the Cuango River near its confluence with the Cuilo River (at c. 05°50′S 16°20′E), leaving the territory of present-day Angola. He proceeded on the right bank of the Cuango and reached the residence of the chief of the Yaka, in the region of Kasongo-Lunda (Kwango, D.R. Congo), Mwene Mputo Casongo (in Portuguese, *Muene Puto Cassongo*), on 27 July 1885. He remained there until 12 August. The return

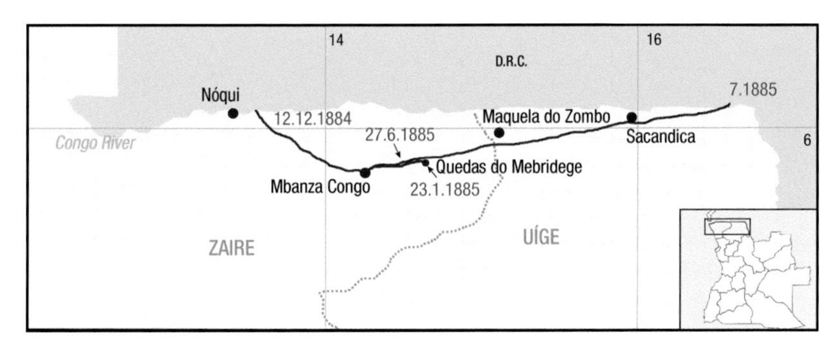

Fig. 27. The itinerary of Oskar Büttner in Angola.

journey followed the left bank of the Kwango River in a northerly direction, towards the Congo River, then downriver, on the left bank, to Stanley Pool (Malebo Pool). They arrived at Leopoldville (Kinshasa, D.R. Congo) on 20 September 1885. Büttner was stationed there for a while and conducted some excursions up- and downriver, and in Leopoldville, he also met Wissmann (q.v.). On 3 April 1886, Büttner returned to Europe. He wrote an account of the expedition and published several new species based on his collections (Büttner, 1890b, 1891). In 1890, he went to Togo to direct a research station in Bismarckburg, and the following year returned to Berlin to take up a teaching position in which he remained until September 1923. He is commemorated in numerous names, including *Aloe buettneri* A.Berger (Asphodelaceae), of which he had collected the types. Urban (1916) stated that 615 numbers collected by Büttner from 1884 to 1886 were received at the herbarium of the Botanic Garden and Botanical Museum Berlin-Dahlem (Herb. B); this corresponds to material from Gabon, Angola, and the D.R. Congo. As Büttner's numbering is not chronological, the date is important to establish the locality. Many of his specimens were destroyed during the World War II bombing of Berlin in 1943. According to Mendonça (1962a), Büttner's itinerary from November 1884 to September 1885 was restricted to present-day Angolan territory. However, it is clear from Büttner's accounts that he crossed the Cuango River to what is now the D.R. Congo in July 1885; therefore, his stay in Angola was shorter. **Angola:** December 1884 (after the 12th)–July 1885 (before the 27th); Uíge (Sacandica, Maquela do Zombo, Quibocolo, Cuango), Zaire (Madimba). **Herb.:** B (main), BM, G, GOET, H, LE, M, P, PC, UC, W. **Ref.:** Büttner, 1890a, 1890b, 1891; Urban, 1916; Schnee, 1920: 262; Gossweiler, 1939; Lanjouw & Stafleu, 1954; Mendonça, 1962a; Bossard, 1993; Liberato, 1994; Heintze, 1999a: 180–191; Romeiras, 1999; Figueiredo & al., 2008, 2020; Wagenitz, 2009; Heintze, 2010; Grace & al., 2011. Figure 27.

C

Cabral, João Crawford de Meneses – see **Crawford Cabral**

Cameira, Fernando José Marçal
(fl. 1951–1953)
Bio.: Portuguese forester. **Angola:** 1951–1953; Bengo, Cabinda, Cuanza Norte, Luanda, Moxico, Uíge. **Herb.:** COI, LISC (226 specimens), LUA, PRE. **Ref.:** Bossard, 1993; Romeiras, 1999; Figueiredo & al., 2008.

Campanhas de Angola

August 1955–March 1956: The *Campanhas de Angola 1955–1956* was an expedition of the project *Missão Botânica de Angola e Moçambique.* It took place from August 1955 to March 1956 in the southwestern region of Angola. It was led by F.A. Mendonça (q.v.), with Mendes (q.v.) as assistant, from 7 September 1955 to 6 November 1955, and by Torre (q.v.) from 13 December 1955 to March 1956. The expedition started in Luanda where Mendonça and Mendes disembarked on 22 May 1955. Mendes travelled by road with a driver, a cook and a servant, arriving in Lubango on 28 August to rejoin Mendonça. The next day they arrived in Moçâmedes. R.M. Santos (q.v.) joined them a few days later. Mendonça returned to Portugal in November 1955, and Torre joined the team in December 1955, replacing Mendonça as leader. The expedition team varied as other people joined the group at different times. For instance, at Lubango they were joined by Salbany (q.v.), who travelled with them for a while. A total of 2745 collections were made during the expedition. These collections were made under separate series depending on the collector. Mendes collected a total of 1737 numbers (1–1737). Torre collected 665 numbers (8200–8864). Mendonça collected 163 numbers (4528–4690), and Santos collected 184 numbers (1–184). The provinces covered were Bengo, Benguela, Cuanza Norte, Cuanza Sul, Cunene, Huambo, Huíla, Luanda, and Namibe. Field books, an itinerary map, and transcripts of the expedition are digitised and accessible at JSTOR Global Plants (JSTOR, 2021).

December 1959–May 1960: The second expedition of the *Campanhas de Angola* lasted from December 1959 to May 1960 and aimed at exploring a little-known area in the southeastern region of Angola from 14° to 16°S and 16° to 20°E. The main team included Mendes (q.v.) as leader, Crawford Cabral (q.v.) as assistant, (R.M.) Santos (q.v.) as head of *preparadores*, (F.) Moreno (q.v.) as helper, and five African servants. Mendes and Crawford Cabral arrived in Luanda on 4 December 1959. After one week of struggling with bureaucratic issues, and a false start that ended with a vehicle breaking down and a forced return to Luanda, the group finally succeeded in departing in mid-December. Mendes took a flight to Moçâmedes (Namibe), while Crawford Cabral proceeded by road. They met at Lubango (Huíla) on 21 December and continued to the east. On 4 January, they were in the town of Cuvango, in the extreme east of Huíla, and then entered the Cuando Cubango Province. Although they intended to explore the little-known area in the southeastern corner of the country, as far as the southern and eastern frontiers, they were unsuccessful because of logistical limitations, such as lack of fuel beyond Menongue, and the impossibility of driving along some tracks during the rainy season. After travelling between Cuvango, Menongue,

Cuito-Cuanavale, and Caiundo, and exploring the surroundings of these localities, they returned to Lubango (Huíla) about three months later. By 13 April, they were near Quipungo (Huíla), on their way back to Lubango. In total, they made 2265 collections, representing numbers 1800–4064 of the Mendes collection. Field books, an itinerary map, and transcripts of the expedition are digitised and accessible at JSTOR Global Plants (JSTOR, 2021). **Ref.:** Mendes, 1962. Figure 28.

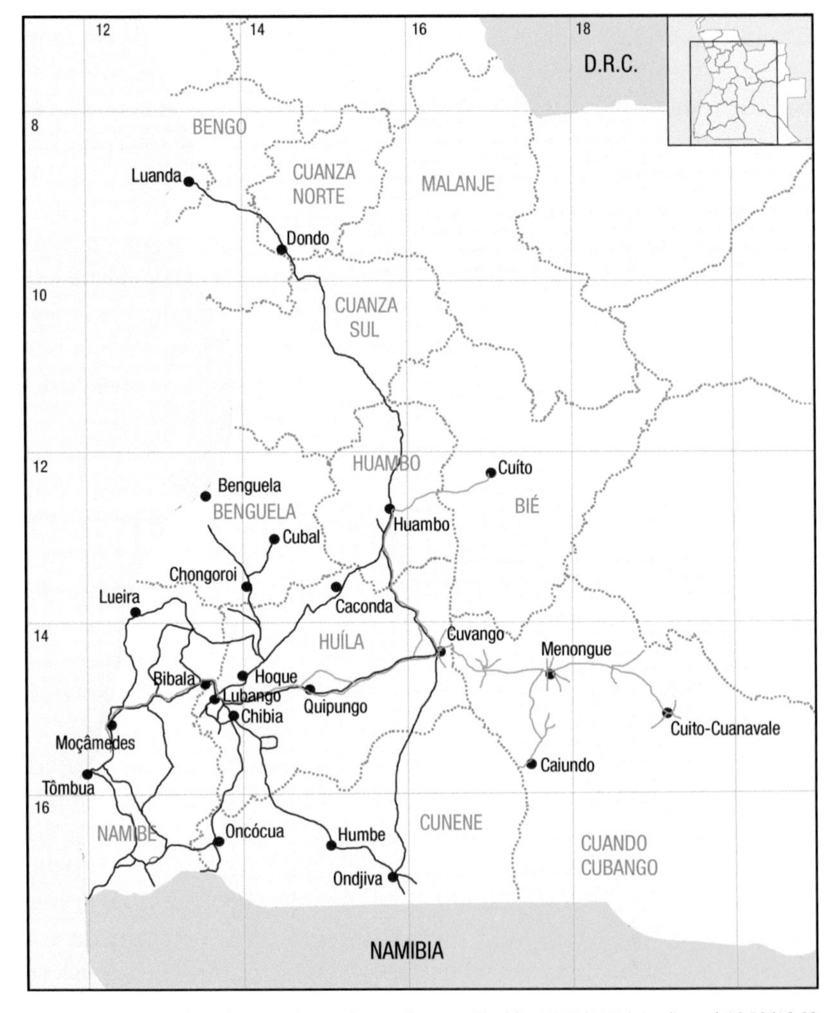

Fig. 28. *Campanhas de Angola*: main roads travelled in 1955/1956 (red) and 1959/1960 (orange).

Cannell, Ian Charles
(1937–)
Bio.: b. Maidstone, Kent, England, 12 October 1937. Civil engineer. He immigrated to Zimbabwe in 1947. He accompanied Leach (q.v.) on expeditions to Angola, Namibia, and Mozambique, and in Zimbabwe, and was commemorated by Leach in several names, such as *Aloe cannellii* L.C.Leach (Asphodelaceae) from Mozambique and *Euphorbia cannellii* L.C.Leach (Euphorbiaceae) from Angola. **Angola:** 1967, 1970, and 1973; collected also with Leach (q.v.). **Herb.:** BM, K, LISC (20 specimens), LUA, M, MO, PRE, SRGH. **Ref.:** Figueiredo & al., 2008; Glen & Germishuizen, 2010; Lavranos & Mottram, 2017a, 2017b.

Capelo, Hermenegildo Carlos de Brito
(1841–1917)
Bio.: b. Palmela, Portugal, 4 February 1841; d. Lisbon, Portugal, 4 May 1917. Naval officer and explorer. He joined the Portuguese Navy, enrolling as a student in the *Escola Naval* in 1855, and four years later, in 1859, at the age of 18, was commissioned as a navy officer. After a few voyages to Madeira, he sailed to Angola in 1860 in a fleet under the command of the then Prince D. Luis (later King D. Luis I) to support the military campaign against the Dembos. Capelo remained at the West African naval station for three years, after which he returned to Lisbon in 1863. The following year, he was promoted to the rank of second lieutenant and afterwards held several naval appointments, being deployed to Mozambique, Cape Verde, Guinea, and Macao, until 1876. He then progressed to the rank of first lieutenant and in 1877 was appointed to lead the Portuguese expedition to the interior of southern Africa (*Expedição portuguesa ao interior da África austral*, q.v.), with Ivens (q.v.) and Serpa Pinto (q.v.). Almost three years later, in March 1880, Capelo and Ivens were back in Lisbon. The following year, a narrative of the expedition was published in book form (Capelo & Ivens, 1881), and soon translated into English (Capelo & Ivens, 1882). On 12 March 1884, again with Ivens, Capelo left Lisbon on the second Portuguese expedition to the southern African interior. After they returned to Lisbon on 20 September 1885, this expedition was also the subject of a book authored by the two explorers (Capelo & Ivens, 1886). In 1887, Capelo left for Zanzibar to represent Portugal in negotiations with the Sultanate of the island. Capelo was later appointed as president of the *Comissão de Cartografia* and as vice-president of the *Instituto Ultramarino.* He was aide-de-camp to King D. Luis I and later to King D. Carlos I. A few weeks after the fall of the monarchy in Portugal, on 24 October 1910, he retired from the navy at the rank of rear admiral and

Fig. 29. Hermenegildo Capelo. Photographer unknown. Reproduced from Anonymous (1917).

died seven years later. His surname was spelled "Capello" before the 1911 orthographical reform of the Portuguese language. **Angola:** 1877–1885; Bengo, Benguela (Dombe, November 1877), Bié (Cuíto, numbers 137–180, March 1878), Cuando Cubango, Cuanza Norte, Cuanza Sul, Cunene, Huambo, Huíla (Caconda, numbers 1–86, January 1878), Luanda, Lunda Norte, Lunda Sul, Moxico, Namibe (Curoca, numbers 1–36 of series 2, March & April 1884), Uíge; 366 specimens, some with Ivens (q.v.). **Herb.:** B, COI, LISU (main). **Ref.:** Capelo & Ivens, 1881, 1882, 1886; Noronha, 1936; Gossweiler, 1939; Romariz, 1952; Lanjouw & Stafleu, 1954; Exell, 1960; Nowell, 1982; Santos, 1988; Bossard, 1993 (with incorrect name and birth/death dates); Liberato, 1994; Romeiras, 1999; Figueiredo & al., 2008; Lima, 2012. Figure 29.

Cardoso, Fernando Jorge (1940–2011)

Bio.: b. Eiras, Coimbra, Portugal, 7 November 1940; d. Coimbra, Portugal, 23 November 2011. Technician. He started working at the *Instituto Botânico* of the Univ. of Coimbra at the age of 15, in 1955, as a day labourer. During the first years of the Angolan War of Independence (1961–1974) he was conscripted to serve in the Portuguese army and was sent to Angola. Following a request from Jorge Paiva, he collected plant specimens while stationed there. After his return to the *Instituto Botânico*, he was successively appointed as collector, naturalist assistant, and technician, before retiring in 2003. **Angola:** 1962–1963; Bengo, Benguela, Cuanza Norte. **Herb.:** COI, E, EA, LISC (6 specimens), LMU, SRGH. **Ref.:** Romeiras, 1999; Figueiredo & al., 2008; J. Paiva, pers. comm., October 2018; Arlindo Cardoso, pers. comm., November 2018.

Cardoso, Helder
(fl. 1965)
Angola: 1965. **Herb.:** COI, LISC (2 specimens). **Ref.:** Figueiredo & al., 2008.

Cardoso, Henrique
(fl. 1933)
Angola: 1933; Cabinda, Huambo (Bailundo), Luanda, Malanje (Pungo Andongo). **Herb.:** LUA. **Ref.:** Bossard, 1993; Figueiredo & al., 2008.

Cardoso, Joaquim António
(unknown–1938)
Bio.: d. Luanda, Angola, 21 June 1938. Settler who lived in Angola for 42 years. He was a collector with the *Direcção da Estação de Policultura Planáltica do Bié* under the auspices of which he collected a few hundred plant specimens. He studied the uses of the fibres of *Sansevieria cylindrica* Bojer ex Hook. (Asparagaceae), for which he had the concession of exploration in Dande (Bengo). He is also listed as António Cardoso. **Angola:** 1929–1935; Bié. **Herb.:** COI, LISC (161 specimens), LISU. **Ref.:** Gossweiler, 1939; Bossard, 1993; Liberato, 1994; Romeiras, 1999; Figueiredo & al., 2008.

Cardoso de Matos, Gilberto ("Gil")
(1935–)
Bio.: b. Luanda, Angola, 12 November 1935. Agrarian engineer. He attended school in Luanda and went to Portugal to study at the then *Escola de Regentes Agrícolas* in Santarém, graduating in 1957. After working for a while for the *Junta dos Cereais* and *Missão de Inquéritos Agrícolas* in Angola, he became a technician with the *Instituto de Investigação Agronómica de Angola* in 1963. From July to November 1974, he was on leave in Portugal. Returning to Angola he could only stay for less than a year and joined the diaspora returning to Portugal in September 1975 with only a few belongings. The invaluable photograph and slide collection of his extensive field work was left behind and subsequently lost. In Portugal, he joined the *Instituto Nacional de Investigação Agrária*, in Lisbon. From 1976 to 1981, he worked at *Estação Agronómica Nacional*, in Oeiras and afterwards at *Estação Florestal*. In 1983,

he had begun to study the vegetation of Cape Verde Islands, and in 1986, he moved there. Until 1990 he worked in Cape Verde, developing a botany section for the government. He created the national botanical garden of Cape Verde, the *Jardim Botânico Nacional Grandvaux Barbosa*, on the island of Santiago. Afterwards he returned to Portugal. He retired in 1994 and remained involved in projects mostly in São Tomé and Príncipe for several years. Eventually he left Lisbon and moved to the Algarve where he still resides (February 2021). During his career, he collected a total of over 18,500 numbers, alone and with others in all the former Portuguese colonies in Africa (Angola, Cape Verde, Guinea-Bissau, Mozambique, and São Tomé and Príncipe). He is one of the main collectors on the islands of São Tomé and Príncipe (1994–1999, c. 3000 numbers) and Cape Verde (1983–1990; c. 3000 numbers). He collaborated with (Brito) Teixeira (q.v.) and Baptista de Sousa (q.v.) in producing vegetation maps for Quiçama (Angola), and with Castanheira Diniz (q.v.) in producing agro-ecological and vegetation maps for Cape Verde, São Tomé and Angola. **Angola:** 1963–1975; in most areas, except for Cabinda, Lunda,

Fig. 30. Gilberto Cardoso de Matos. Photographer unknown. Private collection of Gilberto Cardoso de Matos. Reproduced with permission.

Cazombo region (in Moxico), and N'riquinha region (in Cuando Cubango); over 12,000 numbers, mostly integrated into the collection of (Brito) Teixeira (q.v.); collected also with Bamps (q.v.), Maia Figueira (q.v.), Raimundo (q.v.), and Sampaio Martins (q.v.). **Herb.:** BR, BRLU, COI, K, LISC (163 specimens), LUA, MA, PRE. **Ref.:** Teixeira & al., 1967; Castanheira Diniz & Cardoso de Matos, 1986–1994; Bossard, 1993; Romeiras, 1999; Figueiredo & al., 2008; G. Cardoso de Matos, pers. comm., October 2018. Figure 30.

Carreira, Amílcar
(fl. 1940)
Angola: 1940; Cuanza Norte. **Herb.:** LISC (1 specimen), LUA. **Ref.:** Figueiredo & al., 2008.

Carrisso, Luiz Wittnich
(1886–1937)
Bio.: b. Figueira da Foz, Portugal, 14 February 1886; d. Namibe desert, Angola, 14 June 1937. Botanist and administrator. In 1908, he graduated with a degree in biology from the Univ. of Coimbra, Portugal, and three years later obtained his doctorate at the same university. The following year he joined the *Instituto Botânico* as a lecturer. In 1918, he became a professor and director of the institute. Two years later, in 1920, he took a sabbatical to work with Robert Chodat (1865–1934) in the botanical institute at the Univ. of Geneva, Switzerland, with whom he co-authored a paper. Back in Coimbra, Carrisso implemented extensive changes to revitalise the *Instituto Botânico*. For example, he created three separate sections: The laboratory, the botanical garden, and the herbarium (Herb. COI), and invited new members to join the staff, including Mendonça (q.v.), Aurélio Quintanilha (1892–1987), and the young Abílio Fernandes (1906–1994). He developed the library and the *Sociedade Broteriana*, initiating the second series of the journal of the society, the *Boletim*, in 1922, and established two new serials, *Memórias* (in 1930) and *Anuário* (in 1935). In 1927, with Mendonça, he undertook his first trip to Angola, the *Missão Botânica da Universidade de Coimbra a Angola* (*Missão Botânica I*, q.v.), which stimulated his interest in the flora of the country. In 1929, he again visited Angola, this time with a group of 22 lecturers and students on a teaching expedition. It was only five years later, in 1934, that a visit of A.W. Exell (q.v.) to Herb. COI to study collections from São Tomé and Príncipe, led to the development of the catalogue that was to become known as the *Conspectus florae Angolensis*. Exell suggested to Carrisso that the *Instituto Botânico* in Coimbra should collaborate with the British Museum on developing Carisso's ideas to produce a catalogue of

the Angolan flora. After contacts with the Keeper of Botany at the British Museum, such a collaboration was formally established. In 1935, Mendonça travelled to London to study the collections with Exell, and the two of them travelled to herbaria in Berlin and Brussels. Two years later, in February 1937, the first volume of the *Conspectus florae Angolensis* was published. In the same year, a second botanical expedition (*Missão Botânica II*, q.v.) took place. Also led by Carrisso, the expedition had a larger team that included Mendonça, Exell, F. Sousa (q.v.), Gossweiler (q.v.), the biologist Jara de Carvalho and the wives of Carrisso and Exell (i.e., M. Exell, q.v.). In addition to his academic career and the direction of the *Instituto Botânico*, Carrisso was also involved in university administration, for instance, as vice-chancellor from 1929 to 1931. He was extremely active in other fields, adding several roles to his fulltime job at the university, such as mayor of the city of Coimbra in 1935, director of the municipal museum of his hometown for 20 years, and serving in numerous councils. He was, in his own words, a man of action who derived great solace from work. He died untimely at the age of 51 of a heart attack while on the 1937 *Missão Botânica*, in the Namibe desert, Angola. He is commemorated in many plant names and became a revered figure at the *Instituto Botânico*, which he strengthened and directed for almost 20 years. **Angola:** 1927, 1937; Benguela, Cuanza Norte, Huambo, Huíla, Luanda, Lunda Norte, Lunda Sul, Malanje, Moxico, Namibe; collected with (F.A.) Mendonça (q.v.) and (F.) Sousa (q.v.). **Herb.:**

Fig. 31. Luiz Carrisso (second from the left) in 1937, with the other botanists being (from left to right) Francisco Mendonça, Mildred Exell, and John Gossweiler. Photographer unknown [probably Arthur Exell]. Private collection of Estrela Figueiredo.

BM, COI (main), K, LISC (321 specimens), M, MO, P. **Ref.:** Carrisso, 1928, 1930, 1932, 1937; Exell, 1938a, 1938b; Fernandes, 1939; Gossweiler, 1939; Lanjouw & Stafleu, 1954; Mendonça, 1962a; Bossard, 1993; Liberato, 1994; Romeiras, 1999; Figueiredo & al., 2008. Figure 31.

Carvalho, Eduardo Augusto Luna de – see **Luna de Carvalho**

Carvalho, José Amaral Tavares de
(fl. 1957–1958)
Bio.: Portuguese agronomist who was leader of the first two M.E.F.A. (q.v.) expeditions to Cabinda, Angola. **Angola:** 1957–1958; Cabinda. **Herb.:** COI, K, LISC (3 specimens), LUAI. **Ref.:** Liberato, 1994; Romeiras, 1999; Figueiredo & al., 2008.

Carvalho, Juliano de
(fl. 1915)
Note: This is likely Juliano António de Carvalho (d. 1965). He was a lieutenant in the Portuguese Navy and participated in the 1914–1915 military operation in southern Angola, during World War I. His navy battalion arrived at Gambos on 3 January 1915 and remained stationed there until May 1915. In April of the same year, Carvalho made a collection (PC0106970) that became the type of *Campylopus angolensis* Guim. & Dixon (Dicranaceae). **Angola:** Gambos. **Herb.:** PC. **Ref.:** Comando Geral da Armada, 2021; Lopes, 2021.

Castanheira Diniz, Alberto
(1923–2008)
Bio.: b. Portugal, 1923; d. Lisbon, Portugal, 14 December 2008. Agronomist. He went to Angola in 1953 to work for the chemical company *Companhia União Fabril*. He remained in the country and joined the Soil Laboratory of the *Instituto de Investigação Agronómica de Angola* in Huambo, where he worked until the independence of Angola in 1975. He specialised in agro-ecology and soil science and produced several works on agricultural potential with maps of soil and vegetation that became basic references for agro-ecology in Angola. He left the country at the time of its independence. Afterwards, in Portugal, he collaborated with Cardoso de Matos (q.v.) on the production of agro-ecological maps for Cape Verde. In 2006, the Soil Laboratory of the now *Instituto de Investigação Agrária* was named after him. **Angola:** 1960s; Benguela, Bié, Huíla, Lunda, Malanje, Namibe. **Herb.:** LUA. **Ref.:** Castanheira Diniz, 1973; Castanheira Diniz & Cardoso de Matos, 1986–1994; Bossard, 1993; Romeiras, 1999.

Castro, Mário de Antas Pereira de (fl. 1922–1950s)
Bio.: b. Porto, Portugal. Office-boy at the Univ. of Porto in the 1950s. He participated in the fieldwork of the *Missão Geológica de Angola* from 1922 to 1923, when he collected a few hundred plants in Benguela and Namibe. He is commemorated in *Geissaspis castroi* Baker f. (Fabaceae) and *Hibiscus castroi* Baker f. & Exell (Malvaceae), of which he collected the types. **Angola:** 1922–1923; Benguela, Huambo, Namibe. **Herb.:** BM (Bryo.), COI, K, LISC (2 specimens), M, PC, PO. **Ref.:** Gossweiler, 1939; Bossard, 1993; Liberato, 1994; Romeiras, 1999; Figueiredo & al., 2008; C. Vieira, pers. comm., July 2019.

Cavaco, Alberto Júdice Leote (1915–2001)
Bio.: b. Tavira, Portugal, 4 August 1915; d. Cascais, Portugal, 1 June 2001. Botanist. After concluding his first degree at the Univ. of Coimbra, Portugal, he worked as an assistant to António Rocha da Torre (q.v.) in Mozambique, during the expedition *Missão Botânica de Moçambique*, which was led by

Fig. 32. Alberto Cavaco. Photographer unknown. Private collection of Maria Helena Cavaco. Reproduced with permission.

Mendonça (q.v.). Cavaco then worked for a while at the *Laboratoire Arago* in Banyuls-sur-Mer, France, and afterwards obtained his doctorate (*Doctorat d'État*) at the Univ. of Montpellier, France. He became *Maître de Recherches Attachés* at the *Muséum national d'Histoire naturelle*, Paris, where he established and developed several scientific activities. In 1959, he published an account of the flora of Lunda, Angola, based on the collections of Gossweiler (q.v.), as he had also collected in that province. Towards the end of the 1970s, he collaborated with the Faculty of Sciences of the Univ. of Lisbon and was involved in the *Sociedade Portuguesa de Ciências Naturais*. He published over 200 new species and is commemorated in two plant names, the genus *Cavacoa* J.Léonard (Euphorbiaceae) and *Schizolaena cavacoana* Lowry & al. (Sarcolaenaceae). **Angola:** 1950–1965; Lunda; his collection numbers are over 1400. **Herb.:** LISC (2 specimens), LISU, P. **Ref.:** Cavaco, 1959; Exell, 1960; Bossard, 1993; Romeiras, 1999; Figueiredo & al., 2008, 2018a; Maria Helena Cavaco, pers. comm., 2018. Figure 32.

Chaves, Maria
(fl. 1886–1889)
Bio.: b. Porto, Portugal. Her husband worked for a Dutch company in Boma (D.R. Congo). According to Gossweiler (1939), she collected a few hundred specimens between the Cabinda border and Banana (D.R. Congo) in the estuary of the Congo River and probably not in Angolan territory. As noted by Gossweiler (1939), in the preface to volume 4(2) of *Flora of Tropical Africa* Thiselton-Dyer made a reference to a Major F. Chaves that may be a mistake. Gossweiler gives her name as Maria V. Garcia Chaves, while Henriques (1887) refers to Maria José Chaves. **Angola?:** 1886–1889. **Herb.:** COI. **Ref.:** Henriques, 1887; Gossweiler, 1939; Bossard, 1993; Liberato, 1994; Romeiras, 1999; Figueiredo & al., 2008.

Childs, Margaret
(1902–1986)
Bio.: b. Lausanne, Switzerland, 5 November 1902; d. January 1986. Née Marguerite Pfaeffli. She married the American ethnologist and missionary Gladwyn Murray Childs (1896–1975) in 1925. She lived in Angola from 1925 to the 1960s, while her husband was the principal of the Currie Institute at *Missão do Dondi* (Catchiungo, Huambo). She communicated many Umbundu vernacular names to Gossweiler. **Angola:** 1954; Bié (Chinguar), Huambo. **Herb.:** BM, COI, LISC (2 specimens). **Ref.:** Gossweiler, 1953; Bossard, 1993; Romeiras, 1999; Archives West, 2007; Figueiredo & al., 2008.

Chipa, Agostinho
(unknown–c. 2000)
Bio.: d. Angola, c. 2000. Technical assistant. He worked at *Instituto de Investigação Agronómica de Angola* in Huambo (Huíla) and participated in many expeditions, collecting and preparing material for (Brito) Teixeira (q.v.). His name does not appear on labels and his collections were integrated into the "Teixeira & al." (in sched.) series. **Angola:** 1960s–1970s. **Herb.:** LISC, LUA. **Ref.:** G. Cardoso de Matos and M.F. Pinto Basto, pers. comm., 2021.

Cleghorn, Willoughby Bruce
(1920–unknown)
Bio.: b. [24 November?] 1920. He lived in Rhodesia in the 1960s and published on pastures. He collected in Zimbabwe and Angola from 1964 to 1984. **Angola:** 1973; Luanda. **Herb.:** K, LISC (1 specimen). **Ref.:** Figueiredo & al., 2008.

Codd, Leslie Edward Wostall
(1908–1999)
Bio.: b. Vants Drift, Dundee Distr., KwaZulu-Natal, South Africa, 16 September 1908; d. Pretoria, South Africa, 2 March 1999. Geneticist, botanist, and administrator. He obtained an M.Sc. from the Natal Univ. College (now the Univ. of KwaZulu-Natal) after which he continued his genetics studies at the Univ. of Cambridge. This was followed by employment in Trinidad (1930), where he focussed on the breeding of cotton, and in Guyana (1931–1936), where he worked on rice breeding. He joined the South African department of agriculture in Pretoria in 1937 and remained employed in what was to become the Botanical Research Institute until he retired as director, a position he held from 1963 to 1973. He had a special interest in, inter alia, the genus *Kniphofia* Moench (Asphodelaceae) and the Lamiaceae, and wrote an early book on the trees of the Kruger National Park, an area that abuts the south-central Mozambican border with South Africa in the east and Zimbabwe in the north. He was co-author, with Mary D. Gunn, of the seminal work *Botanical exploration of southern Africa* (1981), which has served globally as a fundamental example of best-practice in historical-botanical studies. In 1952, he undertook an expedition to Zambia and Namibia and made some collections near Rivungo (Cuando Cubango), in Angola, across the border from Shangombo (Zambia). **Angola:** 1952; Cuando Cubango (Missão de Santa Cruz). **Herb.:** BM, BR, COI, G, K, KNP, M, MO, P, PRE, SRGH, Z+ZT. **Ref.:** Codd, 1951; Lanjouw & Stafleu, 1954; Gunn & Codd, 1981; Codd & Gunn, 1985; Anonymous, 1998; Romeiras, 1999; De Winter & Germishuizen, 2000; Figueiredo & al., 2008. Figure 33.

Fig. 33. Leslie Codd. Photographer unknown. South African National Biodiversity Institute. Reproduced with permission.

Coelho, Maria Angelina
(fl. 1961)
Angola: 1961; Bié. **Herb.:** LISC (1 specimen).

Colaço, E.J. Martinho
(fl. 1967)
Angola: 1967; Uíge. **Herb.:** LISC (5 specimens). **Ref.:** Figueiredo & al., 2008.

Compère, Pierre
(1934–)
Bio.: b. Aywaille, Belgium, 6 November 1934. Biologist and algologist. He graduated with a degree in botanical sciences from the Univ. of Liège. In 1959, he worked on a survey of the flora and vegetation of the lower Congo region. After independence of the Belgian Congo, he worked for the Belgian Institute for Overseas Scientific Research, mapping vegetation and soils of the Congo. In 1963, he initiated his studies in algology working for the National Botanic Garden of Belgium, Meise. He specialised in systematics and nomenclature of cyanobacteria. He retired in 2000. **Angola:** February 1960; Uíge (between Quibocolo and Maquela do Zombo); a few collections. **Herb.:** BM, BR (main), IUK, K, OXF, P. **Ref.:** Bamps, 1973; Figueiredo & al., 2008; Golubic & Wilmotte, 2014.

Conceição, A.J.P.
(fl. 1961)
Angola: 1961; Moxico; collected also with Araújo (q.v.). **Herb.:** LISC (3 specimens), LUA. **Ref.:** Bossard, 1993; Romeiras, 1999; Figueiredo & al., 2008.

Constantino, Alfredo Teixeira
(fl. 1970s)
Bio.: Agronomist who worked on pedology. **Angola:** 1970s. Collected with (A.) Borges (q.v.) and Rey (q.v.). **Herb.:** LISC. **Ref.:** Figueiredo & al., 2008.

Cookson, A.J.
(fl. 1959)
Angola: 1959; Moxico; collected with Drummond (q.v.). **Herb.:** COI, LISC.

Cooper, C.E.
(fl. 1997)
Angola: 1997; Luanda (Quiçama). **Herb.:** PRE. **Ref.:** Figueiredo & al., 2008.

Correia, Rui Indegário de Sousa
(1935–1999)
Bio.: b. Sá da Bandeira, Angola, 14 January 1935; d. Louis Trichardt, South Africa, 6 October 1999. Technician. After he left Angola c. 1975, he joined the Potchefstroom Univ. for Christian Higher Education (now the

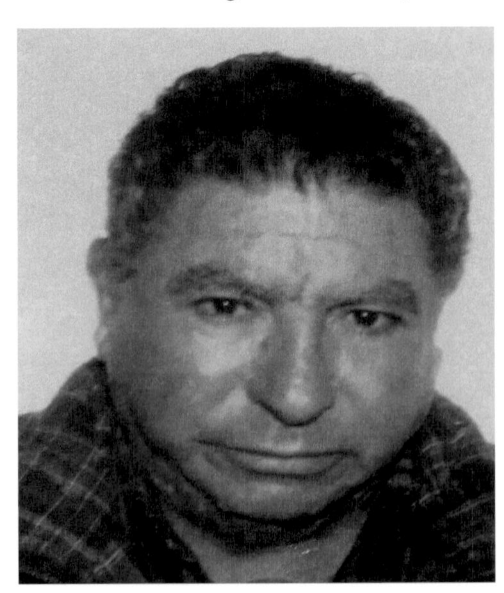

Fig. 34. Rui Correia. Photographer unknown. Private collection of Rui Correia filius. Reproduced with permission.

Potchefstroom campus of North-West Univ.), South Africa, and worked in reclamation, rehabilitation, and revegetation ecology. **Angola:** 1959–1966; Cunene, Huambo, Huíla. **Herb.:** BM, DIA†?, LISC (440 specimens), LUA, LUAI, LUBA, PRE, PUC; collected also with Barbosa (q.v.) and with Barroso Mendonça (q.v.). **Ref.:** Smith & Correia, 1988, 1989, 1992; Liberato, 1994; Romeiras, 1999; Figueiredo & al., 2008; R. Correia (filius), pers. comm., January 2019. Figure 34.

Costa, Esperança Maria Eduardo Francisco da (1961–)

Bio.: b. Luanda, Angola, 3 May 1961. Botanist. After attending school, she enrolled at the Faculty of Sciences of the Univ. of Luanda. She graduated with a B.Sc. in 1985, having done the research for her dissertation in Lisbon at the *Centro de Botânica* (LISC) of the *Instituto de Investigação Científica Tropical.* In the same year, she became a lecturer at the Univ. Agostinho Neto and the next year she ascended to head of department. In 1990, she started attending M.Sc. courses at the *Instituto Superior de Agronomia* in Lisbon and obtained her Ph.D. in 1997. Afterwards, she took various administrative positions in Angola, including vice-chancellor of the Univ. Agostinho Neto. She was recently the head of the university's *Centro de Estudos e Investigação Científica de Botânica* (*Centro de Botânica*), which was created in 2004 and includes the herbarium of the former *Centro Nacional de Investigação Científica* (Herb. LUAI). She co-authored the checklist of Poaceae of Angola published through

Fig. 35. Esperança Costa. Photographer unknown. South African National Biodiversity Institute. Reproduced with permission.

the SABONET project. **Angola:** No information recorded. **Herb.:** LUAI. **Ref.:** Anonymous, 2000; Costa & al., 2004. Figure 35.

Costa, Manuel Ribeiro da
(fl. 1968)
Angola: July 1968; Cuanza Norte (Colonato do Luinga, Camabatela). **Herb.:** COI, LISC (22 specimens). **Ref.:** Figueiredo & al., 2008.

Couto, Clara
(fl. 1972–1974)
Angola: 1972–1974; Cunene, Huíla, Namibe; over 370 numbers. **Herb.:** K, LISC (280 specimens), LUAI, MASS, PERTH. **Ref.:** Romeiras, 1999; Figueiredo & al., 2008.

Crawford [de Meneses] Cabral, João
(1929–)
Bio.: b. Funchal, Portugal, 1929. Zoologist. He was named João Crawford de Meneses Cabral, but is known by the surnames "Crawford Cabral", which he used in his publications, and later used hyphenated as "Crawford-Cabral". He graduated in biology from the Univ. of Lisbon in 1953. After training in plant identification with Mendes (q.v.) at the *Centro de Botânica* in Lisbon, he was contracted by *Missão Botânica de Angola e Moçambique* and participated in the 1959–1960 expedition to Cuando Cubango in southeastern Angola (*Campanhas de Angola*, q.v.). From 1960 to 1975, he was a researcher with the *Instituto de Investigação Científica de Angola* and was also in charge of the mammalogy section of the *Centro de Estudos de Sá da Bandeira* at Lubango from 1961 to 1975. After returning to Portugal, he became a researcher at the *Instituto de Investigação Científica Tropical* in Lisbon until he retired. His research focused on taxonomy and species distribution, and he published several papers on the mammal fauna of Angola. **Angola:** 1959–1960; Cuando Cubango; collected with Mendes (q.v.). **Herb.:** LISC (4 specimens), LUAI. **Ref.:** Crawford Cabral, 1967, 1970, Crawford Cabral & Mesquitela, 1989; Crawford-Cabral, 1998; Crawford-Cabral & Veríssimo, 2005; Figueiredo & al., 2008.

[Cruse, A.]
Note: Listed in Figueiredo & al. (2008) and likely an error from a database.

Cruz, Alberto Machado
(fl. 1956–1967)
Bio.: Teacher and social sciences researcher who was curator of the *Museu da Huíla* (now *Museu Regional da Huíla*) in Lubango. The museum was created in 1956 and focused on palaeontology, archaeology and ethnography. It housed ethnographic collections made by Cruz. **Angola:** Huíla. **Herb.:** LUAI, LUBA. **Ref.:** Cruz, 1967; Matos & al., 2021.

Curror, Andrew Beveridge
(1811–1844)
Bio.: b. Dunfermline, Scotland, 27 October 1811; d. off the coast of present-day Gabon (Gaboon River), 11 July 1844. Naval surgeon. He obtained a Diploma from the College of Surgeons in Edinburgh, Scotland, on 28 June 1831. He became an Assistant Surgeon (later rising to Surgeon) with the Royal Navy and on 13 July 1835 was appointed to serve on the British warship H.M.S. *Russell* with which he remained for three and a half years. On 19 February 1839, he obtained a certificate from the College of Surgeons in Edinburgh. Soon after, on 15 March 1839, he was appointed to H.M.S. *Waterwitch*, an icon of the *West Africa Squadron* that had been created to counter the Atlantic slave trade. Between 1840 and 1843, H.M.S. *Waterwitch* seized 40 vessels and freed 3791 slaves. During 1839, H.M.S. *Waterwitch* patrolled the region from Sierra Leone to the island of Príncipe (in the Gulf of Guinea) and from 13 January 1840 extended its cruising area to south of the equator via Ascension Island. It would remain in the Southern Hemisphere for over three years, patrolling mostly the area from Cabinda down to Baía dos Elefantes (Benguela, Angola). When possible, Curror made use of opportunities to explore the land and to collect plant and animal specimens. In Angola, Curror collected 200 specimens. He was on the *Waterwitch* between March 1839 and July 1843, with a four-month interruption from 15 January 1841 to 1 June 1841, during which he was deployed to H.M.S. *Fantome* as acting surgeon. On 4 May 1843, H.M.S. *Waterwitch* left the African coast for the last time, calling at St Helena, then proceeding to Ascension, before sailing north. It arrived back at Portsmouth on 29 June 1843. After a stay in Scotland, Curror applied for a further appointment as naval surgeon and was appointed to H.M.S. *Larne*, which sailed from England in March 1844. He died shortly afterwards, having served only eight months on the *Larne*. He is commemorated in several species names of which he collected the types, such as *Hoodia currorii* Decne. (Apocynaceae) and *Cyphostemma currorii* (Hook.f.) Desc. (Vitaceae), and the genus *Curroria* Planch. ex Benth. (Apocynaceae) is named after him. **Angola:** 1840–1843; Benguela (Baía dos Elefantes). **Herb.:** E, K. **Ref.:** Anonymous,

1835; Curror, 1843, 1844; Gossweiler, 1939; Lanjouw & Stafleu, 1954; Bossard, 1993; Liberato, 1994 (as "R. Curror"); Romeiras, 1999; Figueiredo & al., 2008; Glen & Germishuizen, 2010; Pearson, 2016; Figueiredo & Smith, 2019, 2020a.

The West Africa Squadron

After the Slave Trade Act of 1807 was passed by the British Parliament, a squadron was created in 1808 to counter the Atlantic slave trade. It became known as The West Africa Squadron. It consisted of a fleet that patrolled the west coast of Africa south to the region of Benguela in Angola, but later its reach was extended further south. The Squadron seized slave-trading vessels (captured ships were called "prizes"), and the crews received money for capturing ships that carried human cargo; ships seized were then dispatched for trial. Many personnel were required for these activities as men would be deployed to the seized ship as "prize crews" after a miscreant crew was arrested. The patrolling involved an intense movement of people between ships, and between ships and ports of call. Not only the "prize crews" and the seized crews had to be transported, but also freed slaves. Some of these slaves joined the Squadron as sailors.

Curtis, Anita Deidamia Grosvenor (1895–1980)

Bio.: b. Providence, Rhode Island, U.S.A., 11 March 1895; d. Beverly, Massachusetts, U.S.A., 15 February 1980. Née Grosvenor. With her husband Richard Cary Curtis (1894–1951), she undertook a hunting safari in Kenya, Mozambique and Angola in 1923. The couple disembarked in Lobito, Angola, and proceeded by train to Catchiungo (Huambo) and then by car to Capango (near Gamba, Bié). The march started there on 11 September, and they crossed the Cuanza River three days later. A campsite was established on 15 September; the march back to Capango started on 22 September. During the entire African expedition, she made c. 1000 collections, 16 of which (mostly from Angola) were later described as new taxa. Anita Curtis is commemorated in several names. **Angola:** 1923; Bié, Malanje; 311 numbers. **Herb.:** A, BM, GH (main), K, LISC (1 specimen), US. **Ref.:** Johnston, 1924; Curtis & Curtis, 1925; Gossweiler, 1939; Lanjouw & Stafleu, 1954 (as "Curtis, A.G."); Walker, 2004; Gunn & Codd, 1981; Bossard, 1993; Liberato, 1994; Romeiras, 1999; Figueiredo & al., 2008; Polhill & Polhill, 2015.

D

D'Orey, José Diogo Sampayo de Albuquerque
(1910–1992)
Bio.: b. Oeiras, Portugal, 28 January 1910; d. 30 November 1992. Agronomist. In 1931, he enrolled at the *Instituto Superior de Agronomia* in Lisbon, graduating in 1939 as an agronomist. In 1937, he joined the *Jardim Colonial* as a botanist in charge of cultures (*botânico chefe de culturas*). He became acting director of the *Jardim* in 1947 and in 1950, he became its director, a position he held until his retirement in 1975. From 1942 to 1945, he was assistant (*adjunto*) with the *Missão Botânica de Moçambique* and undertook two expeditions to that country, spending a total of one year in the field. From 1953 to 1954, he was the leader of *Brigada dos Estudos Florestais da Guiné*, and it was also under his directorship that the *Missão de Estudos Florestais a Angola* (M.E.F.A., q.v.) was undertaken. In 1960, he was the leader of the 4th expedition of this mission. He additionally lectured at the *Instituto Superior de Agronomia* and at the *Escola Superior Colonial* in Portugal. **Angola:** 1960, 1967; Cabinda (1960, under M.E.F.A.), Luanda (1967, a few collections). **Herb.:** BM, COI, K, LISC (11 specimens), LUAI. **Ref.:** Vegter, 1976 (sub "Mendonça"), 1983 (as "Orey"); D'Orey, 1982; Liberato, 1994; Romeiras, 1999; Figueiredo & al., 2008; Slewinski, 2008.

Dacrémont, Alfred Aloys
(1896–1952)
Bio.: b. Stekene, Belgium, 11 May 1896; d. Sint-Niklaas-Waas, Belgium, 20 December 1952. Not much is known about this collector, who has been described as an "oud koloniaal" [ex colonialist] (Jammart, 2021). There are over 400 numbers collected in the present-day D.R. Congo and Angola in the early 1930s, deposited at Herb. BR, which have labels with the collector given as "Dacrémont" or "A. Dacrémont". **Angola:** 1932; Zaire (Nóqui). **Herb.:** B, BM, BR (main), BRLU, C, K, LMA, P, PRE, US, WAG. **Ref.:** Lanjouw & Stafleu, 1954; Jammart, 2021.

Damann, João Baptiste
(fl. 1951–1964)
Bio.: Swiss missionary. He was one of the first eight La Salette missionaries in Angola. Initially stationed at *Missão da Ganda* (Benguela), later he was at *Missão Nossa Senhora de La Salette*, a mission that was created in 1947 in Quingenge (Huambo). He collected over 3000 numbers, which were deposited

Fig. 36. João Baptiste Damann (foreground) with Hans Hess (background) in Angola 1951/1952. Photographer unknown. ETH-Bibliothek, University Archives, Akz.-2006-37. Reproduced with permission.

at the herbarium of the *Instituto de Investigação Agronómica* in Huambo, Angola (Herb. LUA) and at the herbarium of the *Eidgenössische Technische Hochschule*, Zürich, Switzerland (Herb. ZT; ZT and the herbarium of the Institut für Systematische Botanik, Universität Zürich, Herb. Z, were amalgamated in 1991 and work together as Z+ZT). Hess (q.v.) met him in Angola and afterwards Damann visited Hess in Switzerland a few times. Damann died of schistosomiasis. Stopp (q.v.) deposited a collection made by Damann that contained specimens of a number of new taxa in the herbarium of the Johannes Gutenberg-Universität in Germany (Herb. MJG). Stopp (1964) commemorated Damann in *Ceropegia damannii* Stopp (Apocynaceae). **Angola:** 1951–1964; Benguela, Huambo. **Herb.:** K, LISC (1 specimen), LUA, MJG, US, Z+ZT. **Ref.:** Stopp, 1964: 122; Bossard, 1993; Romeiras, 1999; Figueiredo & al., 2008; A. Guggisberg, pers. comm., December 2018. Figure 36.

Dana, C.B.
(fl. c. 1880)
Angola: date and locality not recorded; specimens from Angola located at WIS. **Herb.:** WELC, WIS. **Ref.:** Lanjouw & Stafleu, 1954; Gunn & Codd, 1981.

Danckelman, Alexander Freiherr von (1855–1919)

Bio.: b. Gordemitz near Eilenburg, Germany, 24 November 1855; d. Schwerin, Germany, 30 December 1919. Geographer. From 1875 to 1878, he studied mathematics and natural sciences in Jena and Leipzig. Afterwards, and until 1881, he became a meteorological assistant at the Leipzig observatory. In that year, he joined the expedition of Henry Morton Stanley (1841–1904) to the Congo with the intention of establishing the foundation of the Congo Free State of Leopold II. Danckelman was in charge of the station at Vivi (now in the D.R. Congo). From 1882 to 1884, he travelled to the lower Congo and south to Namibe, Angola, laying the groundwork for a systematic meteorological exploration of southwestern Africa. After his return to Germany, he became general secretary of the *Gesellschaft für Erdkunde* in Berlin from 1885 to 1889 and was a member of the *Reichskolonialamt* from 1903 until 1911. He also collected in Cameroon. In 1884, he published an account of his visit to Angola. **Angola:** 1883; Huíla (Chela, Humpata), Namibe (Moçâmedes). **Herb.:** PC. **Ref.:** Danckelman, 1884; Schnee, 1920; Keil, 1957; Frahm & Eggers, 2001.

Daniel, José Maria (1943–2015)

Bio.: b. Lubango, Huíla, Angola, 12 September 1943; d. Lubango, Huíla, Angola, 25 January 2015. Technician. He was with the *Instituto de Investigação Científica de Angola* at the *Centro de Estudos do Lubango* from 1964 to 1967, then at Luanda (1968–1974). In 1987, he joined the *Instituto Nacional de Saúde Pública do Lubango* (*Laboratório do Centro de Estudo de Microbiologia Aplicada, Medicina Tradicional e Divisão Etnobotânica*). When this was terminated, he was transferred to the herbarium of the *Instituto Superior de Ciências da Educação* in Lubango (Herb. LUBA), which he curated and

Fig. 37. José Daniel. Photographer unknown. Provided by Francisco Maiato Gonçalves.

from where he retired in 2005. **Angola:** 1993–2008; probably collected in all provinces of Angola; c. 8000 specimens were recorded by him, 3000 have been located. **Herb.:** FT, LISC (17 specimens, post-2000), LUAI, LUBA (3000 specimens databased). **Ref.:** Figueiredo & al., 2008; F.M. Gonçalves, pers. comm., October 2018. Figure 37.

Davits, P.
(fl. 1953)
Angola: 1953; Bié; collected for Hess (q.v.). **Herb.:** Z+ZT.

Dawe, Morley Thomas
(1880–1943)
Bio.: b. Sticklepath, Southampton, England, 9 September 1880; d. Kyrenia, Cyprus, 14 July 1943. Horticulturalist. After training at the Royal Botanic Gardens, Kew, in 1902, he joined the botanical, forestry and scientific department of Uganda and became its head the following year. He also directed the Botanical Gardens of Entebbe. In 1910, he took the position of director of agriculture at the *Companhia de Moçambique* based in Beira (Sofala,

Fig. 38. Morley Dawe. Photographer unknown. Reproduced from Gomes e Sousa (1939).

Mozambique). After resigning from this position in early 1914, he worked in Brazil (1914) and Colombia (1915–1920), but later returned to Africa, this time to the west coast, with stays in Gambia (1920), Angola (1921–1922), and Sierra Leone (1922–1923). In Angola, he was contracted by the holding company *Companhia do Fomento Geral de Angola* to report on the agriculture possibilities of the country; he collected and sent plants to Kew from Mbanza Congo (Zaire Province) in August 1921, and from Cabinda in November 1921. Later, he became director of agriculture in Cyprus and director of agriculture and fisheries in Palestine. He retired due to health issues and moved to Cyprus. Soon after, World War II began and he served as Assistant Commissioner of Kyrenia, where he died before the end of the war. He is commemorated in *Aloe dawei* A.Berger (Asphodelaceae) and many other plant names. **Angola:** 1920–1922; Cabinda (Maiombe), Cuanza Norte (Lucala), Cuanza Sul (Libolo), Malanje (Calandula), Uíge (Maquela do Zombo), Zaire (Sumba). **Herb.:** B, BM, BR, ENT, FHO, K, MEDEL, NY, P, PRE, SL, US. **Ref.:** Spooner, 1925; Gossweiler, 1939; Sillitoe, 1944; Lanjouw & Stafleu, 1954; Bossard, 1993; Desmond, 1994; Liberato, 1994; Stafleu & Mennega, 1998: 111–112; Romeiras, 1999; Figueiredo & al., 2008; Glen & Germishuizen, 2010; Grace & al., 2011: 45; Polhill & Polhill, 2015. Figure 38.

Dechamps, Roger
(1930–1995)
Bio.: b. Brussels, Belgium, 17 June 1930; d. Brussels, Belgium, 6 January 1995. Technical agronomist. In 1950, he graduated at the Horticultural Institute of Vilvoorde, Belgium. From 1954 to 1960, he was a forest technician in Kasai (D.R. Congo). He returned to Belgium in 1960 and curated the wood collection of the Royal Museum for Central Africa in Tervuren, Belgium until his death in 1995. He was an “avid wood collector and had a keen interest in xylaria and wood identification” (Miller, 1999) and collected herbarium and wood samples in the D.R. Congo, Angola, Uganda, Cuba, and many European countries. In 1974, he undertook an expedition to Angola, with Murta (q.v.) and Manuel Silva (q.v.). During the c. six-week expedition, they visited eleven provinces, collecting over 600 numbers of herbarium specimens, as well as wood samples. **Angola:** February–March 1974; Benguela, Bié, Cuando Cubango, Cuanza Norte, Cuanza Sul, Cunene, Huambo, Huíla, Luanda, Malanje, Namibe; with Murta (q.v.) and (Manuel) Silva (q.v.); numbers 1001–1615 at BR. **Herb.:** BM, BR, K, LISC (527 specimens), LUA, P, WIS (incl. MAD); wood collections at Tervuren and MAD (now WIS), US. **Ref.:** Bossard, 1993; Miller, 1999; Romeiras, 1999; Figueiredo & al., 2008; Polhill & Polhill, 2015; Hans Beeckman, pers. comm., October 2018.

Dekindt, Eugène
(1865–1905)
Bio.: b. Caeskerke, Belgium, 21 July 1865; d. Lisbon, Portugal, 18 December 1905. Missionary with the Congregation of the Holy Spirit (Spiritans). He was educated in Alsace. In 1893, he was sent to the *Missão da Huíla* (see page 57) in Angola. At the time, the mission was headed by Antunes (q.v.). Dekindt taught theology at the mission seminary. During his stay, he became seriously ill and had to return to Europe. On 27 January 1904, he embarked for Lisbon, where he died the following year. He collected more than 1000 numbers mostly in the vicinity of the mission, along the Munhino, Caculovar, Nene, and Luala rivers, and at Tchivinguiro where the *Missão de Tchivinguiro*, run by Bonnefoux (q.v.), was established in 1892. Dekindt also developed some linguistics and ethnographic studies that were later continued by other missionaries. He is commemorated in several plant names. His collections are often labelled as "Antunes vel Dekindt" because the original labels lack some information. **Angola:** 1893–1904; Huíla. **Herb.:** B, BR, COI, E, K, LISC (273

Fig. 39. Eugène Dekindt. Photographer unknown. Missionários do Espírito Santo. Reproduced with permission.

specimens; see also under Antunes), LUA, MO, MPU, P (258 numbers), US (incl. grass type fragments). **Ref.:** Gossweiler, 1939; Brásio, 1940; Estermann, 1941; Lanjouw & Stafleu, 1954 (as "De Kindt"); Leeuwenberg, 1965; Bossard, 1993; Liberato, 1994; Romeiras, 1999; Vieira (G.), 2006; Figueiredo & al., 2008. Figure 39; see also Figure 15.

Dewèvre, Alfred-Prosper
(1866–1897)
Bio.: b. Brussels, Belgium, 20 March 1866; d. Kinshasa, D.R. Congo, 27 February 1897. Pharmacologist. He had a doctorate in natural sciences and an interest in economic plants. He undertook a two-year expedition to the Congo region in 1895, making some collections in Cabinda (Angola). He died of malaria during the expedition. **Angola:** August 1895; Cabinda (Landana); numbers 210–224 (fide Bamps, 1973). **Herb.:** B, BR, K, P. **Ref.:** Wildeman, 1948; Lanjouw & Stafleu, 1954; Liben, 1965; Bamps, 1973; Stafleu & Mennega, 1998: 253–255; Figueiredo & al., 2008.

Diamang Collections
Collections made by different collectors for the herbarium of the *Museu do Dundo* of the *Companhia de Diamantes de Angola* (Diamang) in Lunda Norte (Herb. DIA). The collections are attributed to individual collectors (see, e.g., Barros Machado, V. Martins, and Sanjinje), but follow a single numbering sequence. The numbers include year and month (e.g., "ANG.XII.54-92" meaning Angola number 92 collected in December 1954). Herbarium DIA is recorded as inactive in Index Herbariorum (Thiers, 2021). Although the zoological collections are still extant at the *Museu do Dundo* (Ceríaco & al., 2020), it is not known if the herbarium was preserved. Duplicates exist in other herbaria. The herbarium of the former *Centro de Botânica* (Herb. LISC), now incorporated in the Univ. of Lisbon, has c. 700 specimens recorded. Copies of the labels of three series of the Diamang plant collections (collection numbers of the 1st series: 1–221; 2nd series: 1–448; 3rd series 1–375) are deposited at the Univ. of Lisbon and digitised at JSTOR Global Plants (JSTOR, 2021).

Diniz, Alberto Castanheira – see **Castanheira Diniz**

Dold, Anthony ("Tony") Patrick
(1965–)
Bio.: b. Bulembu, Swaziland (now Eswatini), 8 June 1965. Botanist. He moved to South Africa in 1970. In 1991, he obtained the National Diploma in Forestry from Saasveld Forestry College, near George, Western Cape (now

a satellite campus of Nelson Mandela Univ., Gqeberha), and later M.Sc. *ad eundem gradum* from Rhodes Univ., Grahamstown (2002). He was Technical Assistant (1992–1993), Assistant Curator (1994–2002), and finally Curator (2003–present) of the Selmar Schonland Herbarium, Rhodes Univ. (Herb. GRA). He has a strong interest in ethnobotany. He collected widely, c. 6000 numbers, in Angola, Lesotho, Malawi, Mozambique, Namibia, South Africa, Swaziland, Tanzania, and Zimbabwe, and is commemorated in *Haworthia cooperi* Baker var. *doldii* M.B.Bayer (Asphodelaceae) and *Orbea doldii* Plowes (Apocynaceae). **Angola:** 1998; Lunda Norte (Luzamba, Cuango); numbers 3576–3792. **Herb.:** GRA (main), NBG, P, PRE. **Ref.:** Smith & Willis, 1999; Glen & Germishuizen, 2010; Dold & Cocks, 2012; T. Dold, pers. comm., October 2018.

[Donis, Camille Albert]
(1917–1988)
Note: Head of the Forest Division of the *Institut National pour l'Étude Agronomique du Congo Belge* in the D.R. Congo in the 1950s. Recorded by Figueiredo & al. (2008) based on a record in the Herb. PRE database of a specimen from 1948. There is no evidence that he collected in Angola and it is likely a database error.

Douglas Fox, Mildred Susan Harris [Mrs F.]
(fl. 1911–1913)
Bio.: Née Harris. She was the wife of Francis Douglas Fox (b. 1868), a civil engineer who was sent to Angola by the Benguela Railway Company to produce a report on the railway that was under construction from Lobito to the interior. The couple arrived in Lobito on 10 June 1911 and spent six weeks in the country, travelling from Lobito to Huambo. Apparently, Mildred did not manage to collect plants ("the ground was too hard to take anything up"), but after her return to England, a friend in Angola sent her some bulbs that she offered to Kew. One that flowered became the type specimen (K000365907) of *Moraea revoluta* C.H.Wright (Iridaceae). **Angola:** She apparently did not make collections herself. **Herb.:** K. **Ref.:** Fox (F.D.), 1912; Fox (M.), 1912; Wright, 1913.

Drummond, Robert ("Bob") Baily
(1924–2008)
Bio.: b. Petersfield, Hants, England, 27 February 1924; d. Harare, Zimbabwe, 3 June 2008. Botanist. After serving in the Royal Navy as a coder in World War II, he obtained a B.Sc. in botany and zoology at the Univ. of Leeds in

Fig. 40. Robert ("Bob") Drummond. Photographer unknown. South African National Biodiversity Institute. Reproduced with permission.

1948. He was employed by the Royal Botanic Gardens, Kew to work on the *Flora of Tropical East Africa* project (1949–1955) and in 1953 went to East Africa with J.H. Hemsley to collect specimens of species of which the original material had been destroyed in Berlin during World War II. In May 1955, he accepted an appointment as botanist at the National herbarium of Zimbabwe, Harare (Herb. SRGH), where he was Curator from 1966 onwards. He was a prolific collector in the *Flora Zambesiaca* region and amassed over 12,000 numbers. He also collected in South Africa (KwaZulu-Natal, North-West, Gauteng, Limpopo, and Mpumalanga provinces), Botswana, and Zambia. He is commemorated in several plant names. **Angola:** 1959; Moxico, across the border from Zambia; collected also with Cookson (q.v.). **Herb.:** BM, COI, EA, FHO, K, LISC (2 specimens), LMA, PRE, SRGH. **Ref.:** Gunn & Codd, 1981; Coates Palgrave & al., 2009; Darbyshire & al., 2015; Polhill & Polhill, 2015. Figure 40.

Duarte, Alexandre José
(fl. 1940s–1968)

Bio.: Agronomist. During the 1940s, he was director of the entomology department of the *Estação Agronómica Nacional* in Oeiras, Portugal. Later in the 1960s, he worked in Angola and published on insect pests in agriculture.

Angola: 1968; Cuando Cubango; collected with Leach (q.v.) and with (Brito) Teixeira (q.v.). **Herb.:** LISC (1 specimen). **Ref.:** Duarte, 1964, 1966.

Duparquet, Charles Victor Aubert (1830–1888)

Bio.: b. Laigle, Séez, France, 31 August 1830; d. Loango, Rep. of the Congo, 24 August 1888. Missionary with the Congregation of the Holy Spirit (Spiritans). After graduating with a B.A. degree at the Argentan College, France, he attended a seminary. He took orders with the Spiritans and was ordained as a priest on 2 June 1855. In the same year, he was sent to Dakar, Senegal, and afterwards to Gabon. In 1857, he was recalled to France but returned to Gabon in 1862 for a short stay. In 1865, the Congo Prefecture was entrusted to the Spiritans based on Duparquet's proposals. His first attempt to establish a mission in Angola was unsuccessful. He disembarked in Moçâmedes in 1866 and travelled inland. He had the intention of establishing a mission at Capangombe, about halfway between Moçâmedes and Lubango, but in 1867, he was expelled from the country. There were suspicions that he spied for France. In 1867, the Spiritans started a seminary in Santarém, Portugal, to advance their presence in Portuguese territories. The seminary was headed initially by Duparquet. Antunes (q.v.) was one of his students. Duparquet returned to France in 1869 and was sent to Bagamoyo (present-day Tanzania) and Zanzibar. He returned to France and in 1873 founded a mission in Landana (now in Cabinda, Angola), a region that at the time was not under Portuguese rule. Duparquet became vice-prefect of Cimbebásia (area between the Cunene and Orange rivers, i.e., Namibia, and Botswana) and in that position he disembarked in Cape Town in 1878 to proceed north overland. He reached Kimberley (Northern Cape Province, South Africa), but could not proceed further and had to return to Cape Town. In a second attempt, he went to Walvis Bay, Namibia, by sea and from there travelled by ox wagon to Omaruru where he established a mission. His travels continued northwards to Humbe, Cunene, Angola. In 1881, he was recalled to Lisbon and encumbered with the creation of a mission in Angola, the *Missão da Huíla* (see page 57). He arrived in Huíla on 7 December 1881, accompanied by the missionaries Antunes and C. Wunemburger. Shortly thereafter, he left to explore Humbe and Cuanhama. In 1883, he took another trip to the interior, advancing as far as Cassinga and exploring the Cuvelai and Colui rivers and making geographical observations. In 1886, he initiated a mission at Mafeking in South Africa and later returned to Europe with his herbarium of about 1500 specimens collected in various countries. He died in Loango in 1888, shortly after arriving in the Congo. His travel diary was published

Fig. 41. Charles Duparquet. Photographer unknown. Reproduced from Brásio (1940).

in 1953, and his extensive correspondence, consisting of several volumes, has been published since 2012. He is commemorated in c. 20 plant names including the genus *Duparquetia* Baill. (Fabaceae). **Angola:** 1867–1887; Cabinda (Landana), Huíla. **Herb.:** LISC (2 specimens), LY, P. **Ref.:** Armand, 1884; Estermann, 1941; Duparquet, 1953; Lanjouw & Stafleu, 1954; Bossard, 1993; Liberato, 1994; Romeiras, 1999; Figueiredo & al., 2008, 2019a; Vieira, 2012–2017; Anonymous, s.d.(c). Figure 41.

Duvigneaud, Paul Auguste (1913–1991)

Bio.: b. Marche-en-Famenne, Belgium, 13 August 1913; d. Ixelles, Belgium, 21 December 1991. Botanist. He graduated with a B.Sc. in chemistry (1935) and a B.Sc. in botany (1937) from the Univ. Libre de Bruxelles (ULB). From 1940 to 1950, he was an assistant at ULB and in 1949 also was associated with the *Institut agronomique de l'État* in Gembloux, Belgium. In 1950, he became professor of Botany at ULB, a position he held until 1978. From 1977 to 1983, he lectured at the Univ. Paris Diderot, in Paris, France. He undertook several expeditions to the D.R. Congo between 1935 and 1960, and collected in Zambia (1959–1960) and Kenya (1966). He made a few collections in Angola across the border from the D.R. Congo. His collecting numbers in the

Fig. 42. Paul Duvigneaud. Photographer unknown. Published by Leteinturier & Malaisse (2001). Reproduced with the permission of the Botanic Garden Meise, Belgium.

D.R. Congo alone amount to c. 5500. In 1956, he collected with Timperman (q.v.). He published extensively and authored several dozen species names, including some from the Angolan flora, and is commemorated in 18 species names. **Angola:** 1956; Moxico (Luau); a few collections. **Herb.:** BR, BRLU, K, P. **Ref.:** Lanjouw & Stafleu, 1954; Duvigneaud, 1956; Duvigneaud & Timperman, 1959; Bamps, 1973; Tanghe, 1992; Romeiras, 1999; Stafleu & Mennega, 2000: 224–225; Leteinturier & Malaisse, 2001; Figueiredo & al., 2008; Polhill & Polhill, 2015. Figure 42.

E

Ellenberger, Victor
(1879–1972)

Bio.: b. Masitise, Lesotho, 13 November 1879; d. Fontaine-Lavaganne, France, 7 October 1972. Pastor and anthropologist. Son of the Swiss missionary Daniel [David?] Frédéric Ellenberger (1835–1920). Victor's parents relocated to Masitise, Lesotho, after the Seqiti war of 1866 and he was born in a San cave. The cave became the Masitise Mission, now the Masitise Cave House Museum. Victor was consecrated as a pastor in 1902 and was stationed in

Zambia from 1904 to 1910, and later in Lesotho. He published extensively on language and anthropology of Lesotho and on the San. He is recorded as collecting in Angola in 1925, but no specimens have been located. **Angola:** No specimens located. **Herb.:** P. **Ref.:** Ellenberger, 1938, 1953; Bossard, 1993; Romeiras, 1999; Figueiredo & al., 2008; Rosenberg & Weisfelder, 2013.

Estermann, Carlos
(1895–1976)

Bio.: b. Illfurth, Alsace (then Germany), France, 26 October 1895; d. Lubango, Angola, 21 June 1976. Missionary and ethnologist. After studying at the missionary schools of Saverne, France, and Knechtsteden, Germany, he served with the German army during World War I. He was wounded, captured by the British army, and taken to England as a prisoner of war. In 1919, after the armistice, he attended the seminary at Neufgrange, France. He took vows on 29 September 1920, and two years later he finished his missionary studies at Chevilly, near Paris, and was ordained as a priest. He was sent to Angola with the missionaries of the Congregation of the Holy Spirit (Spiritans), arriving in the country in 1924. He was initially stationed at Mupa (Cunene), but was later appointed to Omupanda, near Ondjiva (Cunene). There he initiated and headed

Fig. 43. Carlos Estermann. Photographer unknown. Missionários do Espírito Santo. Reproduced with permission.

a new mission, the Omupanda mission, which was established in 1928. In 1932, he became the superior of the *Missão da Huíla* (see page 57), and superior for the district and general vicar of Huíla. He remained in the region for the rest of his life: as a vicar at Lubango until 1961 and thereafter, until his death, at the Munhino mission. Estermann became well-known for his numerous papers and monographs on ethnological subjects of the peoples of southwestern Angola. His *magnum opus* on the ethnography of the region (Estermann, 1956–1961) remains a fundamental reference in this field. In 1974, he received an honorary doctorate from the Univ. of Lisbon. According to Bossard (1993), there are some collections deposited at LUA. **Angola:** 1924–1976, recorded as collecting in 1940; Cunene (Cuanhama, Humbe), Huíla. **Herb.:** LISC (1 specimen); LUA. **Ref.:** Estermann, 1956–1961; Bossard, 1993; Figueiredo & al., 2008, 2020; Anonymous, s.d.(d). Figure 43.

Estes, Richard Depard
(1927–)
Bio.: Biologist. He specialised on African mammals, particularly on wildebeest. From 1969 to 1970, he developed a study of the hippotragine antelope in Angola funded by the National Geographic Society and *Direcção dos Serviços de Veterinária de Angola*. In this expedition he was accompanied by his wife R.E. von K. Estes (q.v.) and J.A. Silva (q.v.). Some plant collections were made by Estes and his team. **Angola:** 1969–1970; Malanje. **Herb.:** PH. **Ref.:** Silva, 1972.

Estes, Runhild Elizabeth von Knapitsch
(1939–)
Bio.: b. Tanganyika (now Tanzania), 1939. Née von Knapitsch. In 1964, she married the zoologist Richard Despard Estes (q.v.). She collaborated with her husband and J.A. Silva (q.v.) in a study of the hippotragine antelope in Angola from 1969 to 1970, funded by the National Geographic Society and *Direcção dos Serviços de Veterinária de Angola*. During the expedition she collected plants. **Angola:** 1969–1970; Malanje. **Herb.:** PRE. **Ref.:** Silva, 1972; Figueiredo & al., 2008.

Exell, Arthur Wallis
(1901–1993)
Bio.: b. Birmingham, England, 21 May 1901; d. Cheltenham, England, 15 January 1993. Botanist. After attending school in Warwickshire and Birmingham, England, he studied at Emmanuel College, Cambridge, graduating with an M.A. degree. In 1924, he joined the botany department of the

British Museum, where he would spend his whole career. He specialised in the Combretaceae, but eventually worked on many other groups of plants in the African flora. His first expedition to Africa took place in 1932, when he spent some months in the Gulf of Guinea in West Africa. He was the first (if not the only) botanist to collect on all four islands (São Tomé, Príncipe, Bioko, Annobon) in the Gulf. His extensive collections led to the compilation of a catalogue of the flora of São Tomé, Príncipe, and Annobon (Exell, 1944) and later a checklist of the flora of the four islands (Exell, 1973). He visited Coimbra, Portugal, in 1934, with his wife Mildred Exell (q.v.), to study the collections from São Tomé and Príncipe held at the herbarium of the Univ. of Coimbra (Herb. COI). While there, he met Carrisso (q.v.), who discussed with him the studies on the flora of Angola that had been on-going at Coimbra since the 1927 expedition to the country (*Missão Botânica I*, q.v.). Following a suggestion of Exell, a collaboration between the Univ. of Coimbra and the British Museum was later established to develop a catalogue of the flora of Angola. In Coimbra, Exell also met F.A. Mendonça (q.v.) and Gossweiler (q.v.), who would be close friends for the rest of their lives. Together with Carrisso, they undertook an expedition to Angola in 1937 (*Missão Botânica II*, q.v.). Exell spoke several languages and was able to deliver impromptu speeches in German, French and, more remarkably, Portuguese. His knowledge of Portuguese resulted in him being requisitioned by the British Foreign Office to serve in the Intelligence Office during World War II. He was Foreign Office Civilian at Bletchley Park from February 1940 to the spring of 1942. Then he was based at Elmers School, and finally at Berkeley Street until 1945. He was part of the Diplomatic Section as "Head of Portuguese and Brazilian". The service involved reporting and decoding enemy diplomatic communications in Portuguese. He served there until 1950. In that year, after having returned to the British Museum, he co-founded the *Association pour l'étude taxonomique de la flore d'Afrique tropicale* (AETFAT), an association of botanists that is still in operation. He initiated the project *Flora Zambesiaca* with Wild (q.v.) to produce a Flora for the area of south tropical Africa drained by the Zambesi River, the present-day countries of Botswana, Malawi, Mozambique, Zambia, Zimbabwe, and the Caprivi Strip region in northern Namibia. In 1962, the Univ. of Coimbra awarded him an honorary doctorate. In that year, he retired from the British Museum, as Deputy Keeper of Botany, a position he had held since 1950. Afterwards, he and his wife moved to the village of Blockley in the Cotswolds. Up until the early 1990s, when the by then widowed Exell had to leave his house and move to a care home, the village of Blockley became a mecca for numerous botanists who travelled there to visit him as his guests. He is commemorated

Fig. 44. Arthur Exell in the herbarium of the Univ. of Coimbra in 1934 with John Gossweiler and possibly Mildred Exell. Photographer unknown. From Centro de Botânica/Instituto de Investigação Científica Tropical (IICT), Universidade de Lisboa. Reproduced with permission.

Fig. 45. Arthur Exell in front of his house at Blockley, England in 1991. Photographer and private collection of Estrela Figueiredo.

in the names of two dozen plant species, and in the genus *Exellia* Boutique (Annonaceae). **Angola:** 1937; most provinces with exception of Uíge and Zaire. **Herb.:** B, BM (main), BR, COI, K, LISC (1873 specimens), LMA, M, MO, P, SRGH, US. **Ref.:** Gossweiler, 1939; Lanjouw & Stafleu, 1957; Exell & Hayes, 1967; Stafleu & Cowan, 1976: 808; Vegter, 1976 (sub "Mendonça"); Bossard, 1993; Launert, 1993; Romeiras, 1999: Stafleu & Mennega, 2000: 444–447; Figueiredo & al., 2008, 2018a; Polhill & Polhill, 2015; Bletchley Park, 2021a. Figures 44 and 45; see also Figures 82 and 133.

Exell, Mildred ("Milly") Alice Haydon (1905–1990)

Bio.: b. Wandsworth, London, England, 25 January 1905; d. North Cotswolds, Gloucestershire, England, August 1990. Née Haydon. A student and collaborator of Edmund Gilbert Baker (1864–1949), who was assistant keeper of the herbarium of the British Museum (Herb. BM) from 1887 to 1924, she got engaged to A.W. Exell (q.v.) in Berlin in 1926 and they were married in 1929. Together they visited the herbarium of the Univ. of Coimbra (Herb. COI) in 1934, a historical occasion that would eventually lead to the establishment of the *Conspectus florae Angolensis*. At Coimbra, while her husband studied collections from São Tomé and Príncipe, she studied collections of legumes made by Gomes e Sousa (q.v.) in Mozambique and later published a paper

Fig. 46. Mildred Exell in 1937. Photographer unknown. Private collection of Estrela Figueiredo.

on the subject. After the visit to Coimbra, she participated in an expedition to Angola (*Missão Botânica II*, q.v.) in 1937. During that expedition she made some collections. While in Lunda Sul, the group split, with A.W. Exell and F.A. Mendonça (q.v.) travelling to Muriege, and Mildred remaining at Saurimo. She then made collections that are "in Exell & Mendonça", including a specimen that would be the type of *Heeria mildredae* Meikle (Anacardiaceae), a name that commemorates her. While left alone for about two weeks in June 1937 with the collections made at Lubango (Huíla), she took charge of changing the blotting paper in the plant presses while the plants were being dried. World War II started a couple of years after this expedition and as both Arthur and Mildred had an excellent knowledge of Portuguese, they were valuable assets for British intelligence. She served as a Foreign Office Civilian at Bletchley Park (the intelligence headquarters) at the outstation of Berkeley Street in London. There she was part of the Diplomatic Section, where enemy diplomatic communications were reported and decoded. She worked on Brazilian translations and continued after the war, staying in Canada for six months, while her husband remained in England. When her husband retired in 1962, the couple moved to the Cotswolds, where they spent the rest of their lives. **Angola:** 1937; in Exell & Mendonça. **Herb.:** BM, COI. **Ref.:** Carrisso, 1937; Exell, 1937a, 1937b, 1939; Stafleu & Mennega, 2000: 447; Polhill & Polhill, 2015; Figueiredo & al., 2018a; Bletchley Park, 2021b. A. Exell, pers. comm. (letters dated 1990–1992 in the private archives of E. Figueiredo). Figure 46; see also Figures 31 and 44 (possibly).

Expedição portuguesa ao interior da África austral

May 1877–March 1880: The *Expedição portuguesa ao interior da África austral* that took place from May 1877 to March 1880 was developed under the auspices of the *Sociedade de Geografia de Lisboa*. Two navy officers, Capelo (q.v.) and Ivens (q.v.), and an army officer, Serpa Pinto (q.v.) were appointed to undertake the expedition. The main objective was the study of the Cuango River, and its relation to the Congo River and to the territories occupied by the Portuguese in West Africa, as well as the regions at the sources of the Zambezi and Cunene rivers, north to the Cuango and Cuanza rivers. After completing preparations for the expedition, Capelo and Serpa Pinto sailed from Lisbon to Angola on 5 July 1877, arriving at Luanda on 6 August 1877. There they failed in contracting porters, so Serpa Pinto travelled to Ambriz and from there to the Congo River for that purpose. He went up to Ponta da Lenha (name no longer in use, corresponding to a locality between Banana and Boma in the D.R. Congo) and there heard that Henry Morton

Stanley (1841–1904; see under Phillips), who had just journeyed downriver, was camped at Boma while on his way to Cabinda. Serpa Pinto travelled to Cabinda to meet Stanley and invited him to Luanda. On 20 August, Serpa Pinto sailed back to Luanda with Stanley and his entourage of 114 people. In the meantime, Ivens had also arrived in Luanda. The three explorers were stationed in Luanda with Stanley as their houseguest for 45 days while planning for their explorations and obtaining advice from the Welsh-born American explorer. The Portuguese expedition had been initially planned to travel east from Luanda by following the Cuango River to the Cassibi River and then to the Congo River, which would then be negotiated downriver to its mouth. As part of this itinerary had already been explored by Stanley, and because the Portuguese expedition could not find porters locally, they decided to rather start the exploration at Benguela. On 6 October 1877, Serpa Pinto was the first to leave to negotiate contracts with porters. In this he was assisted by the local trader and explorer António Francisco Ferreira da Silva Porto (1817–1890). Capelo and Ivens embarked on a warship to Sumbe and from there to Benguela where they arrived on 19 October. On 12 November, the three finally set off into the interior, travelling in a southerly direction towards Dombe and then eastwards to Quilengues. They left Quilengues on 1 January 1878 and proceeded to Caconda where they met Anchieta (q.v.), who was living there. After exploring the surroundings, they proceeded with the expedition, with Serpa Pinto the first to leave for Bié on 8 February; Capelo and Ivens left in mid-February. The expeditions of Serpa Pinto, and Capelo and Ivens took different routes that resulted in Capelo and Ivens being the first to arrive at Bié on 8 March 1878. They were received at Belmonte, at the house of Silva Porto. Some 30 years earlier, in 1845, Silva Porto had settled in Bié near the Cuíto River and named the settlement Belmonte. The town that developed in the area was known as Silva Porto until Angolan independence in 1975; now it is called Cuíto. Capelo and Ivens were later joined by Serpa Pinto, who arrived ill with fever. By then the incompatibility of the group was manifest and they decided to split, Capelo and Ivens continuing with the objectives of the original expedition, while Serpa Pinto aimed at crossing the continent (see under Serpa Pinto). The expedition of Capelo and Ivens did not follow a linear route as their purpose was to obtain detailed geographical information of the region. On 19 May, they left Bié in the direction of the location then known as Feira de Cassanje. When they reached the Cuango River, they parted ways following two different routes to Cassanje. They met there in mid-October 1878. On 18 February 1879, they departed from Cassanje towards Malanje, passing the Bale and Cuiji rivers. They camped near Malanje at Bango Hill where they were visited by Buchner (q.v.), and afterwards they went to the Fort Duque de Bragança (Calandula) and the waterfall with the same name. On 28 April, they travelled north along the Lucala River, reaching the Cugo River on 23 May. By then they were in the Yaka territory and reached the Cuango Riverm, which they crossed on 28 May 1879. However, a lack of provisions compelled them to abandon the objective of reaching its confluence with the Congo River. They turned back to the Lucala River and to Duque de Bragança where they were stationed at the end of July. In August, they travelled to Ambaca and then to Pungo Andongo, spending time at their base at Pedras Negras. On 15 September, they were near the Cuanza River and on 20 September they met Mechow (q.v.), who was on his way to Malanje. They finally arrived at Dondo and from there embarked for Luanda on a steamship. On 13 October 1879, Capelo and Ivens were back in Luanda, from where they sailed to Moçâmedes to

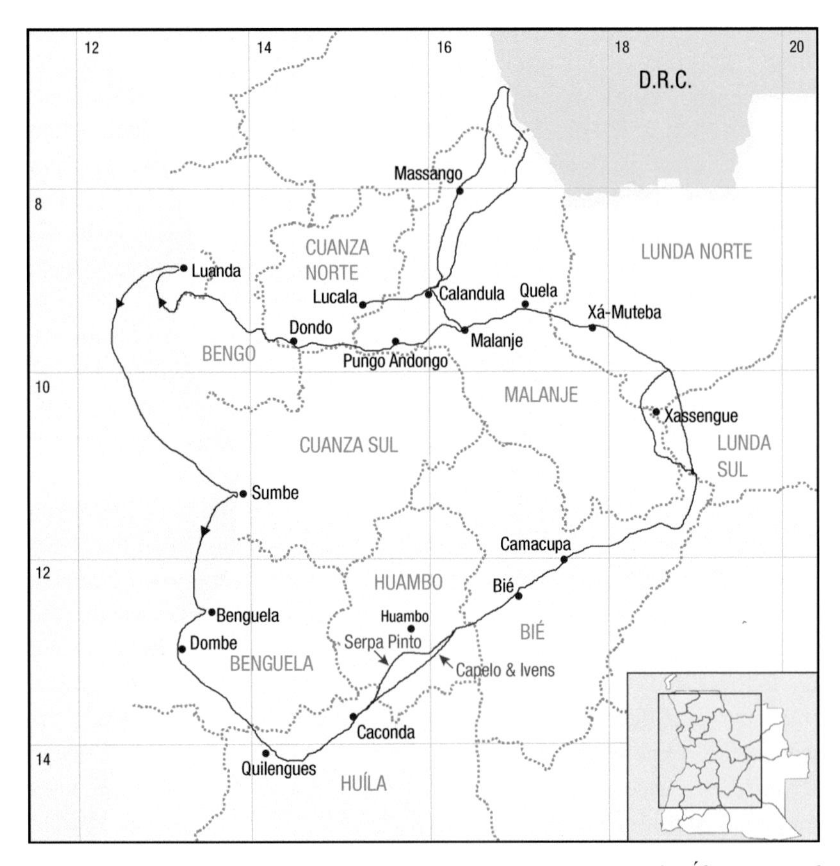

Fig. 47. The itinerary of the *Expedição portuguesa ao interior da África austral* 1877–1880 in Angola.

convalesce until 19 January 1880, at which time they returned to Luanda. A week later they left for Europe, arriving in Lisbon on 1 March 1880. Their expedition resulted in a large volume of precise geographical data that was used to fill in blanks on maps of Angola; they also collected data on geology, natural history, and anthropology. Nevertheless, their arrival in Lisbon was preceded by Serpa Pinto's earlier arrival and a conference that was attended by large crowds, establishing his status as a national hero. In 1881, Capelo and Ivens published an account of their expedition, with this work appearing after Serpa Pinto's more popular account.

January 1884–September 1885: On 6 January 1884, Capelo and Ivens embarked in Lisbon for their second expedition to Angola. Their objectives were to traverse the African continent from west to east; to find a trade route between Angola and Mozambique; to investigate the Congo and Zambezi River basins; and to gather geographical information while travelling through regions that were still lacking such data on maps. Shortly after arriving at Luanda, Capelo left for Sumbe to contract the services of porters. He returned on 13 February. On 12 March, with a contingent of 72 people, Capelo and Ivens left for Pinda (Tômbua, Namibe) on a steamship under the command of Capelo's brother, Guilherme Augusto de Brito Capelo (1839–1926). The initial intention was to start the journey on the Curoca River that has its mouth at Tômbua Bay. They soon realised that advancing overland along that route was not easy, and a group of 49 of their porters deserted the party. Ivens rode on horseback to Moçâmedes and managed to recover about half of them. Nevertheless, being abandoned by the men was an indication of the unfeasibility of the route. Capelo and Ivens then decided to travel to Moçâmedes by ship and from there in an easterly direction to Huíla. Before setting off, Ivens and Guilherme Capelo, the captain of the steamship on

Fig. 48. The itinerary of the *Expedição portuguesa ao interior da África austral* 1884–1885 in Angola.

which they travelled, proceeded up the Curoca River to investigate a possible connection of this river to the Cunene River. They reached Garganta do Diabo on 31 March 1884. Three weeks later, on 24 April, Capelo and Ivens finally left Moçâmedes. From 3 to 29 May, they undertook a detour through Huíla ascending Chela Mountain and visiting the recently created *Missão da Huíla* (see page 57). From there, the second expedition's accumulated botanical collections were dispatched to Lisbon. They then descended towards Humbe along the Caculovar River. They arrived in Humbe on 12 June and then travelled upriver along the Cunene to Quiteve. They crossed the Cunene on 25 June and travelled in an north-easterly direction towards Cuvelai. Many rivers were crossed: Cuebe, Cueio, Cuatir, Cubango, Cuíto, Lomba, and Cuando. They passed through the area between the Cuando and Mussesse rivers and then the Ninda region, and afterwards crossed the present-day border, continuing in territory that is now in the D.R. Congo. On 21 June 1885, they reached Quelimane, on the Mozambican coast. They were back in Lisbon on 20 September 1885. As they had done for the first expedition, the next year they published an account of their travels (Capelo & Ivens, 1886). Both expedition accounts contain an appendix with an overview of their plant collections authored by the Conde de Ficalho (1837–1903). **Ref.:** Capelo & Ivens, 1881, 1882, 1886; Serpa Pinto, 1881a, 1881b; Nowell, 1982; Santos, 1988. Figures 47, 48, and 49.

Fig. 49. *Expedição portuguesa ao interior da África austral* 1884–1885. The expedition at the Cape of Good Hope. Engraved by E[vert] v[an] Muyden and Th[eodore] Girardet from a photograph. Photographer unknown. Reproduced from Capelo & Ivens (1886).

F

Faro, João Cabral Pereira Lapa e – see **Lapa e Faro**

Faulkner, Helen
(1888–1979)
Bio.: b. Yorkshire, England, June 1888; d. Mwanbeni, Tanzania, 26 January 1979. Artist and collector. She studied art and ballet in Paris and married Major Hamlyn George Faulkner, an Australian veteran of World War I, in 1921. After a stay in Tasmania, her husband took a position managing sisal estates in Tanga, Tanzania. In 1940, they moved to Benguela Province in Angola where her husband worked for Wigglesworth & Co. It was at Alto Catumbela that she developed a deeper interest in the flora and started painting it. On the advice of Gossweiler (q.v.), she also started collecting, sending specimens (numbered with the prefix "A") to the National Herbarium in Pretoria (Herb. PRE), South Africa, over the five-year period that she was stationed in Angola. In 1942, the couple relocated to Mozambique where her husband held a position with the Namagoa Plantations at Mocuba (Zambézia). They spent the next eight years there, and during that time Helen collected more specimens, at first mostly for Herb. PRE, but after 1947 also for the Royal Botanic Gardens, Kew (Herb.

Fig. 50. Helen Faulkner. Photographer unknown. Reproduced from A.F. Gomes e Sousa (1949).

K) in England. The couple returned to Tanga in 1950, and after her husband's retirement in 1954, they settled at Mwanbeni. She died at home in 1979. She collected c. 5000 numbers in total. The main Kew series consists of 4700 numbers. Her watercolour sketches, executed from 1940 to 1967, are numbered 1–813, in 20 books that are kept at Kew. Sketches numbered 1–374 are of Angolan plants. She is sometimes listed as Helen G. Faulkner (e.g., Polhill & Polhill, 2015), and in specimen labels as H.G. Faulkner, but the initial "G." was derived from being "Mrs H.G. [Hamlyn George] Faulkner". She is commemorated in numerous names. **Angola:** 1940–1942. **Herb.:** B, BM, BR, COI, E, EA, FT, IFAN, K, LISC (7 specimens), LUA, P, PRE, SRGH, NY, U, US. **Ref.:** A.F. Gomes e Sousa, 1949; Lanjouw & Stafleu, 1957; Polhill, 1980; Bossard, 1993; Romeiras, 1999; Figueiredo & al., 2008; Glen & Germishuizen, 2010; Polhill & Polhill, 2015. Figure 50.

Fenaroli, Luigi
(1899–1980)
Bio.: b. Milan, Italy, 16 May 1899; d. Bergamo, Italy, 8 May 1980. Agronomist, geneticist, and botanist. He graduated from the Univ. of Milan in 1921 and worked with, or directed, several experimental stations in Italy; the *Stazione Sperimentale di Selvicoltura* in Florence (1933), the *Istituto*

Fig. 51. Luigi Fenaroli. Photographer Guido Serra. Reproduced under licence CC BY 3.0.

Aperimentale di Pioppicoltura in Casale Monferrato, Piedmont (1943), the *Stazione Sperimentale di Maiscultura* in Bergamo (1949), and the *Istituto Sperimentale di Assestamento Forestale e per l'Alpicoltura* in Trento (1968–1974). He also lectured at the Univ. of Milan and was a visiting professor at other universities. He published over 250 papers on various topics. In 1928 and 1930, early in his career therefore, Fenaroli collected in Angola while part of an expedition of the Italian Geographic Society with the purpose of studying the possibility of colonising the Benguela plateau (now the Bié plateau). This would appear to have been a recurring issue for Italy, as it was the third time that an expedition was sent to the plateau for that purpose. Previous expeditions were undertaken by Taruffi (q.v.) in 1912 and by Mazzocchi-Alemanni (q.v.) in 1923. **Angola:** 1928, 1930; Benguela (littoral), Bié (Chinguar), Huambo (Bailundo, Cachiungo), Luanda. **Herb.:** COI, FI, LISC (4 specimens). **Ref.:** Gossweiler, 1939; Lanjouw & Stafleu, 1957; Bossard, 1993; Liberato, 1994; Romeiras, 1999; Dorr & Nicolson, 2008: 128–129; Figueiredo & al., 2008. Figure 51.

[Fernando, F.C.]
Note: South African collector listed in Figueiredo & al. (2008) based on a record in the Herb. PRE database that has since been removed.

Fernando, José
(fl. 1971)
Angola: 1971; Huambo. **Herb.:** LISC (18 specimens). **Ref.:** Bossard, 1993; Figueiredo & al., 2008.

Ferreira, José Maria
(fl. 1960)
Angola: 1960; Huambo, Huíla; collected with (Brito) Teixeira (q.v.). **Herb.:** LISC, LUA. **Ref.:** Bossard, 1993; Romeiras, 1999; Figueiredo & al., 2008.

Ferreira, Maria Fernanda Duarte Pinto Basto da Costa – see **Pinto Basto da Costa Ferreira**

Figueira, João Maia – see **Maia Figueira**

Filipe
(fl. 1968)
Angola: 1968; Cunene, Huíla; collected with (R.) Santos (q.v.) and Schultz (q.v.). **Herb.:** COI, LISC, PRE. **Ref.:** Figueiredo & al., 2008.

[Fontes e Sousa]
Note: Listed in Figueiredo & al. (2008), but likely an error for Gomes e Sousa (q.v.).

Fontinha, Mário José dos Reis
(1918–1997)
Bio.: b. Évora, Portugal, 7 December 1918; d. Oeiras, Portugal, 15 December 1997. Technician. He went to Angola and worked at *Museu do Dundo, Companhia de Diamantes de Angola* in Lunda Norte, assisting its curator José Redinha (1905–1983). After Redinha left in 1959, Fontinha became involved in management as acting curator until at least 1969. Later, he returned to Portugal. He published on ethnology and collected for the Diamang Collections (q.v.). **Angola:** 1949–1953; Lunda Norte; collected also with Barros Machado (q.v.) and for Gossweiler's (q.v.) collection, appearing on the latter labels as "Fontinha in Gossweiler". **Herb.:** CANB, COI, DIA†?, LISC (5 specimens), US. **Ref.:** Fontinha & Videira, 1963; Fontinha, 1983; Bossard, 1993; Porto & al., 1999; Romeiras, 1999; Figueiredo & al., 2008; David, 2013. Figure 52.

Fig. 52. Mário Fontinha (third from left) in the expedition to Chingufo on 31 May 1969, with Matende Sanjinje (first from right, foreground). Photographer unknown.

Forst, Gisela
(fl. 1930s–1974)
Bio.: Artist. Likely Swiss-born, Gisela Forst arrived in Angola sometime before World War II and provided private tutoring at a farm near Ganda (Benguela), a region where there was a large German community at the time. Up to 1971, she lived part time at Calulo (Cuanza Sul) on a farm that belonged to Johannes Höpfner Mannhardt (1880–1973), a German farmer and settler in Angola. Forst was a graphic artist (painter) and exhibited at Lubango in 1966 during the festivities of *Nossa Senhora do Monte*. After 1971, she was based at Ganda. She taught at the *Deutsche Schule Benguela* and likely also at Chicuma (Benguela). She left Angola for Germany before the Angolan civil war broke out in 1975. According to Bossard (1993), she collected plants and sent c. 200 collections with watercolours of most of these plants to the herbarium of the *Instituto de Investigação Agronómica*, in Huambo, Angola (Herb. LUA). Sampaio Martins (q.v.), who worked at Herb. LUA at the time, recalls the existence of the plant collections. However, neither he nor Pinto Basto (q.v.), who also worked at Herb. LUA, recalls the existence of such watercolours. It is not known if the specimens collected by Forst and her paintings are extant. Other artwork produced by Forst exists in private collections. **Angola:**

Fig. 53. *Spathodea campanulata* P.Beauv. (Bignoniaceae) by Gisela Forst. Private collection of Ilse Tuebben. Reproduced with permission.

1966–1974 (Bossard, 1993); Benguela (Cubal, Ganda), Cuanza Sul (Bungo, Calulo, Libolo, Luati River). **Herb.:** LUA? **Ref.:** Bossard, 1993; Romeiras, 1999; Mannhardt, 2007; Figueiredo & al., 2008, 2020. M.F. Pinto Basto and E. Sampaio Martins, pers. comm., March 2020; Conny Lind, Ines Mannhardt, Amilcar Salumbo and Sabine Spiesser, pers. comm., April 2020. Figure 53.

Fox, Mildred Susan Harris [Mrs F.] Douglas – see **Douglas Fox**

Frade, Emílio Carita
(fl. 1959)
Bio.: Forester. He worked for the *Instituto de Investigação Científica de Angola* and co-authored a book with (Romero) Monteiro (q.v.) on timber of species from the forests of Dembos. **Angola:** 1959; Cabinda. **Herb.:** LISC (5 specimens), LUAI. **Ref.:** Monteiro & Frade, 1960; Frade, 1965; Bossard, 1993; Romeiras, 1999; Figueiredo & al., 2008.

Freitas, Augusto Santiago Barjona de
(1882–unknown)
Bio.: b. Lisbon, Portugal, 20 February 1882. Agronomist. He was the leader of the *Missão Agronómica de Angola*. In March 1908, he wrote to Júlio Henriques (1838–1928), the director of the Botanical Garden of Coimbra, informing Henriques of his upcoming trip to Angola and requested instructions on how to collect plant specimens. Freitas planned to depart on 22 April 1908 with an intention of visiting the regions of Bié and Bailundo. In that year, he published an index to Francisco Ficalho's *Plantas uteis da Africa Portugueza* and some papers on agricultural issues in the colonies. **Angola:** 1931; Bié. **Herb.:** LISC (2 specimens). **Ref.:** Freitas, 1908a, 1908b; Romeiras, 1999; Figueiredo & al., 2008.

[Frines, A.N.]
Note: A record from Herb. K database and JSTOR Global Plants (JSTOR, 2021) that is due to misreading of the label of a collection of Antunes (q.v.).

Fritzsche, Bertha Caroline Marie Bolle
(1863–unknown)
Bio.: b. Toddin near Hagenow, Mecklenburg-Schwerin, Germany, 4 April 1863. Née Bolle. On 15 April 1886, she married the medical doctor Hermann Richard Fritzsche in Ludwigslust, Germany. In 1887, they went to Angola, where they had two sons, born in 1889 and 1892. Hermann died at the age of 45, shortly after the birth of their second son. Bertha remained in Angola for

several years. In 1929, she was living in Portugal and had remarried. Her older son, Volkmar Fritzsche, married in 1921 and in the 1940s was a businessman in Lisbon working as a "confidential clerk" with a mining company that supplied tungsten to Germany. He was likely a *Sicherheitsdienst des Reichsführers-SS* agent, in other words, a spy. The younger son became a medical doctor and died on 25 June 1925. Bertha collected a few hundred plant specimens in Humpata from 1902 to 1905. She sent 302 collections to the herbarium of the Botanic Garden and Botanical Museum Berlin-Dahlem (Herb. B) in Germany, the first collections being sent through the mediation of Baum (q.v.). These collections are labelled with the heading "Reise nach Angola. Mossamedes-Humpata". The numbering is chronological, and the collection dates of the specimens examined range from 18 April 1903 to 3 November 1905. Many of her collections were likely destroyed at Berlin during World War II. She is commemorated in *Hibiscus fritzscheae* Exell & Mendonça (Malvaceae), a species endemic to Angola, and in *Cleome fritzscheae* Gilg & Gilg-Ben. (Capparaceae) and *Hybanthus fritzscheanus* Engl. (Violaceae). Several of her collections became types for new names, for example *Kalanchoe lindmanii* Raym.-Hamet (Crassulaceae), unfortunately named after the Swedish botanist Carl Lindman, who was in charge of the herbarium of the Univ. of Stockholm (Herb. S) and who made Fritzsche's collections available to Raymond-Hamet. **Angola:** 1902–1905; Huíla (Humpata, Jau, Munhino), Namibe (Bumbo). **Herb.:** B (main), G, GB, M, NY, S, US, W. **Ref.:** Hamet, 1913; Urban, 1916; Willgeroth, 1929; Gossweiler, 1939; Lanjouw & Stafleu, 1957; Bossard, 1993; Romeiras, 1999; Figueiredo & al., 2008, 2020; Glen & Germishuizen, 2010.

G

[Geraldes, A.]

Note: Listed by Bossard (1993), Romeiras (1999), and Figueiredo & al. (2008). Recorded as active in Bié with specimens at Herb. LISJC (integrated into Herb. LISC). The LISC database of Angolan collections does not include any specimens from this collector and we were unable to determine the identity of the person. It is possible that the name refers to the agronomist Carlos Eugénio de Melo Geraldes (1878–1962), who started his career in Angola and later became a professor at the *Instituto Superior de Agronomia* in Lisbon and was one of the founders of the *Jardim Colonial* (later the institution holding Herb. LISJC) in the same city. He published on agriculture and flora of the colonies. **Ref.:** Anonymous, s.d.(e).

Gerez, Armando Gonçalves
(fl. 1963–1964)
Bio.: He was commissioned by the *Instituto de Investigação Científica de Angola* to collect. **Angola:** 1963–1964; Cabinda, Uíge; collected with (Brito) Teixeira (q.v.). **Herb.:** LISC. **Ref.:** Bossard, 1993; Figueiredo & al., 2008.

Gerrard Reis, Jacqueline
(fl. 1980s–1991)
Bio.: British agronomist. She settled in Angola in the 1980s. She collected with Alcochete (q.v.), (E.) Matos (q.v.), and Newman (q.v.) in 1991. **Angola:** 1991; Cunene, Huíla, Namibe. **Herb.:** K, LISC. **Ref.:** Gito, 2016; Goyder & Gonçalves, 2019.

Ghesquière, Jean Hector Paul Auguste
(1892–1982)
Bio.: b. Gembloux, Belgium, 1892; d. Menton, France, September 1982. Agronomist and entomologist. He left Gembloux in 1912, having graduated in agronomy. Between the two World Wars, he actively collected and studied fauna and flora of tropical regions, particularly in the D.R. Congo, where he was stationed for many years and directed the *Division de Phytopathologie et d'Entomologie du Congo Belge*. During the occupation of Belgium by Axis powers in World War II, he conducted studies at the Museum of Tervuren near Brussels. He was then involved in the Resistance, communicating to the Allied forces information on the army officials who passed through the museum. In 1944, he was arrested between Louvain and Tirlemont (Tienen) while dressed in a Belgian army uniform and was sent to a camp east of Berlin. He was released by the Russian army in 1945. He published on several insect families and was particularly concerned with insect pests, being one of the founders of the *Organisation Internationale de Lutte Biologique*. In the 1950s, he settled at Menton, France, where he spent the last 30 years of his long life. He established a research station on his property at Menton, but the station was never officially recognised. **Angola:** October 1934; Benguela (Lobito); a few numbers. **Herb.:** A, BR, K, MO, P. **Ref.:** Lanjouw & Stafleu, 1957; Bamps, 1973; Decelle, 1982; Bossard, 1993; Romeiras, 1999; Pauly, 2001; Polhill & Polhill, 2015.

Gibson, Gordon Davies
(1915–2007)
Bio.: b. Vancouver, British Columbia, Canada, 22 June 1915; d. Escondido, California, U.S.A., 18 September 2007. Anthropologist. He graduated from

the Univ. of Chicago with master and doctoral degrees in anthropology. During World War II, he taught in the Army Special Training Program at Kalamazoo College in Michigan and later worked for the Naval Ordnance Test Station at Inyokern in California. After the War, he lectured at the Univ. of Utah and in 1958 he joined the Smithsonian Institution, becoming a curator at the National Museum of Natural History from 1958 to 1983. He specialised in the Herero and Himba people of southwestern Africa and undertook four expeditions to the region between 1952 and 1973, doing field work in Botswana, Namibia, and Angola. After retiring he moved to California and remained active, particularly in botany. **Angola:** 1972–1973; Cunene, Namibe. **Herb.:** US. **Ref.:** Frank, 2007.

Gillett, Jan Bevington
(1911–1995)

Bio.: b. Oxford, England, 28 May 1911; d. Kew, Richmond, England, 17 March 1995. Botanist. Godson of, and named for, General Jan C. Smuts (1870–1950), celebrated South African politician. Gillett attended King's College in Cambridge. After an interruption in 1932 to serve as a botanist in the Somaliland-Ethiopian Commission, he finished his degree, graduating with an M.A. in 1934. He worked as a schoolteacher for a while and got involved in politics, joining the Communist Party. He was conscripted in

Fig. 54. Jan Gillett. Photographer unknown. South African National Biodiversity Institute. Reproduced with permission.

1941 to serve during World War II. He held the rank of Captain in the Royal Armoured Corps of the British army. After the war ended, he was a botanist in the agriculture department, Iraq (1946–1949), and thereafter employed by the Royal Botanic Gardens, Kew (Herb. K) to work on the *Flora of Tropical East Africa* (1949–1964). In 1959, he applied for the post of botanist in charge of the East African Herbarium (Herb. EA), Nairobi, Kenya, and was rejected because of his communist background. Later, after the independence of Kenya, he was appointed to that position and was in charge of Herb. EA from 1964 to 1971. He carried on working at that Herbarium until 1984, when he returned to England. He visited Herb. K regularly until his death. He is commemorated in numerous plant names. **Angola:** Cabinda? (fide Bossard, 1993). **Herb.:** BOL, BR, EA, FI, K, PRE, S, SRGH, WAG. **Ref.:** Lanjouw & Stafleu, 1957; Gunn & Codd, 1981; Bossard, 1993; Polhill, 1995; Beukes, 1996; Polhill & Polhill, 2015; Pieterse, 2019. Figure 54.

Gladman
(fl. 1690)
Bio.: One of the four pre-Linnaean collectors recorded in Angola [the others being Brown(e) (q.v.), Kirckwood (q.v.), and Mason (q.v.)], with collections in James Petiver's herbarium at the herbarium of the Natural History Museum (Herb. BM), London, England. **Angola:** 1690. **Herb.:** BM (Sloane Herb.). **Ref.:** Exell (A.W.) 1939, 1962; Lanjouw & Stafleu, 1957; Bossard, 1993; Liberato, 1994; Romeiras, 1999; Figueiredo & al., 2008.

Gomes, António Lopes
(1939–)
Bio.: b. 1939. Forester. He worked with the *Instituto de Investigação Agronómica de Angola* and published on forestry from 1972 to 1973. After returning to Portugal, he was attached to the Univ. of Trás-os-Montes e Alto Douro. **Angola:** 1970; Cuanza Norte, Huambo; collected with Murta (q.v.) and (Manuel) Silva (q.v.). **Herb.:** COI, LISC (22 specimens). **Ref.:** Figueiredo & al., 2008.

Gomes e Sousa, António de Figueiredo
(1896–1973)
Bio.: b. Lisbon, Portugal, 5 February 1896; d. Johannesburg, South Africa, 23 October 1973. Agronomist. He studied at the *Instituto Superior de Agronomia*, in Lisbon, graduating as a specialist engineer in tropical agriculture. In 1922, he travelled to Angola, where he remained until 1927 while working first as an agronomist with the *Serviços de Agricultura* and later as a botanist

in a project aimed at the eradication of African trypanosomiasis (sleeping sickness). In 1930, he headed the *Missão de Reconhecimento Botânico* with Gossweiler (q.v.) among his staff. At the start of his career, Gomes e Sousa was based at the *Estação Experimental do Cazengo* in Cuanza Norte, but he soon was subjected to a disciplinary hearing that resulted from a candid report on the labour supply to that *Estação* that he had submitted to the governor of the district. Consequently, he was transferred to the province of Malanje in Angola. During his stay in Angola, he developed field work activities in several districts (Benguela, Cabinda, Cuanza Norte, Cuanza Sul, Malanje, Moxico, Namibe, Uíge, and Zaire), conducting agricultural, forestry or botanical surveys. His work in Moxico resulted in a portfolio that consisted of 85 drawings derived from plants that he collected, as well as chorological sketches and photographs he took; these were presented to the Univ. of Coimbra, Portugal. He was in Cabinda before taking a position to conduct a botanical survey and collect herbarium specimens for one year (October 1927–October 1928) in the then Portuguese province of Guinea (Guinea-Bissau). From his work in Guinea, he produced 44 drawings and a phytogeographical sketch that, together with an extensive report, were published in 1930. After

Fig. 55. António Gomes e Sousa. Photographer unknown. Reproduced from Anonymous (1950).

a stay at the Univ. of Coimbra, he returned to Angola in 1929 and worked in two more districts (Namibe and Huíla). In 1930, for political reasons, he asked to be transferred to Mozambique, where he became the first ecologist in the territory and the first to record aspects of the physiognomy of the vegetation. His career progress was not smooth, and he took early retirement. However, a few years later, from 1953 onwards, he was commissioned to do a dendrological study of the province of Mozambique, which resulted in the book *Dendrologia de Moçambique* published in 1966–1967. Afterwards he continued publishing articles on forests and nature conservation until his death in 1973. He collected 4908 numbers. **Angola:** 1922–1930; Benguela, Huíla, Namibe. **Herb.:** A, B, BM, BR, COI, EA, K, LISC (9 specimens), LISU, LMA, MO, P, PRE, SRGH. **Ref.:** Gomes e Sousa, 1930; Gossweiler, 1939; Anonymous, 1950; Lanjouw & Stafleu, 1957; Mendonça, 1962b; Gomes e Sousa, 1966–1967; Bossard, 1993; Fernandes, 1993; Liberato, 1994; Romeiras, 1999; Figueiredo & al., 2008, 2017; Glen & Germishuizen, 2010. Figure 55.

Gossweiler, John
(1873–1952)
Bio.: b. Regensdorf, near Zürich, Switzerland, 26 December 1873; d. Lisbon, Portugal, 19 February 1952. Horticulturalist and botanist. He studied horticulture in Zürich and at the Institute of Horticulture in Stuttgart, and later in Dresden, Germany. In 1896, he went to London, spending about two years at the Royal College of Science and another two at the Royal Botanic Gardens, Kew. In 1899, at the age of 25, the Portuguese government appointed him to develop a botanical garden in Angola. He then moved to that country, where he spent the rest of his professional career (with several periods of leave in Portugal). He started collecting plant specimens immediately. His first number is from Moçâmedes, dating from 1900. His early collections are mostly from the Luanda area (1900–1902), and afterwards from Cazengo in Cuanza Norte (1902) and Malanje (1903). In 1905, he was commissioned to study rubber-producing plants and moved to what was then Forte Princesa Amélia at Vila da Ponte (later Vila Artur de Paiva, now Cuvango) in the Ganguelas region, where he collected many new species. In 1907, the Cazengo colonial garden, later to be known as *Estação Experimental do Cazengo*, was established at the Granja de S. Luis. At that time, Gossweiler exhaustively explored the Cazengo flora with excursions to Ambriz (Bengo) in 1907 and Amboim (Cuanza Sul) in 1908. From 1915 to 1918, he made several visits to the Mayombe forests in Cabinda and for months on end camped with his wife in a remote area near the source of the Zanza River. He interrupted his civil service from 1919 to 1926 when he worked for *Fomento Geral de Angola*. He then collected extensively

in Cuanza Norte, Cuanza Sul, Malanje, along the Congo and Bengo rivers, and at Quissama. In 1927, he directed the *Estação Experimental do Algodão* at Catete. Two years later the *Missão de Reconhecimento Botânico*, headed by Gomes e Sousa (q.v.), was established with Gossweiler appointed as agrarian technician. In this capacity, he explored and collected in the Dembos forests. In 1930, he was in charge of studying the cultivation of cotton and coffee in several areas, in Malanje, Benguela, and Cuanza Sul (Seles and Amboim). In 1932, he explored the area of Gabela and Capir (Cuanza Sul). From 1933 to 1934, he spent some time in Portugal to study his collections. In Coimbra, he met A.W. Exell (q.v.), who at that time was also visiting the herbarium of the Univ. of Coimbra (Herb. COI). In 1937, they both participated in an expedition to Angola (*Missão Botânica II*, q.v.). As head of the *Gabinete de Botânica* of the *Serviços de Agricultura*, Gossweiler continued with collecting activities and floristic and vegetation studies, as well as agronomical investigations. From 1942 to 1946, he was in Portugal, labelling his set of collections at Herb. COI, at the herbarium of the *Museu Nacional de História Natural e da Ciência* (Herb. LISU), and at the herbarium of the then *Jardim Colonial* (Herb. LISJC) in Lisbon. From 1946 to 1948, Gossweiler made his last major explorations, mostly at Dundo (Lunda Norte). Altogether he amassed a large collection of herbarium specimens – more than 14,600 numbers with numerous duplicates – from all the provinces of Angola. Sampaio Martins (1994a) considered the main set of the collection to be the one at Herb. LISJC (now LISC). However, according to Gossweiler (1939) the main set ("princeps") is at the herbarium of the Natural History Museum, London (Herb. BM). The holotypes of names published by Ronald Good (1896–1992) and Exell

Fig. 56. John Gossweiler. Photographer unknown. From Centro de Botânica/Instituto de Investigação Científica Tropical (IICT), Universidade de Lisboa. Reproduced with permission.

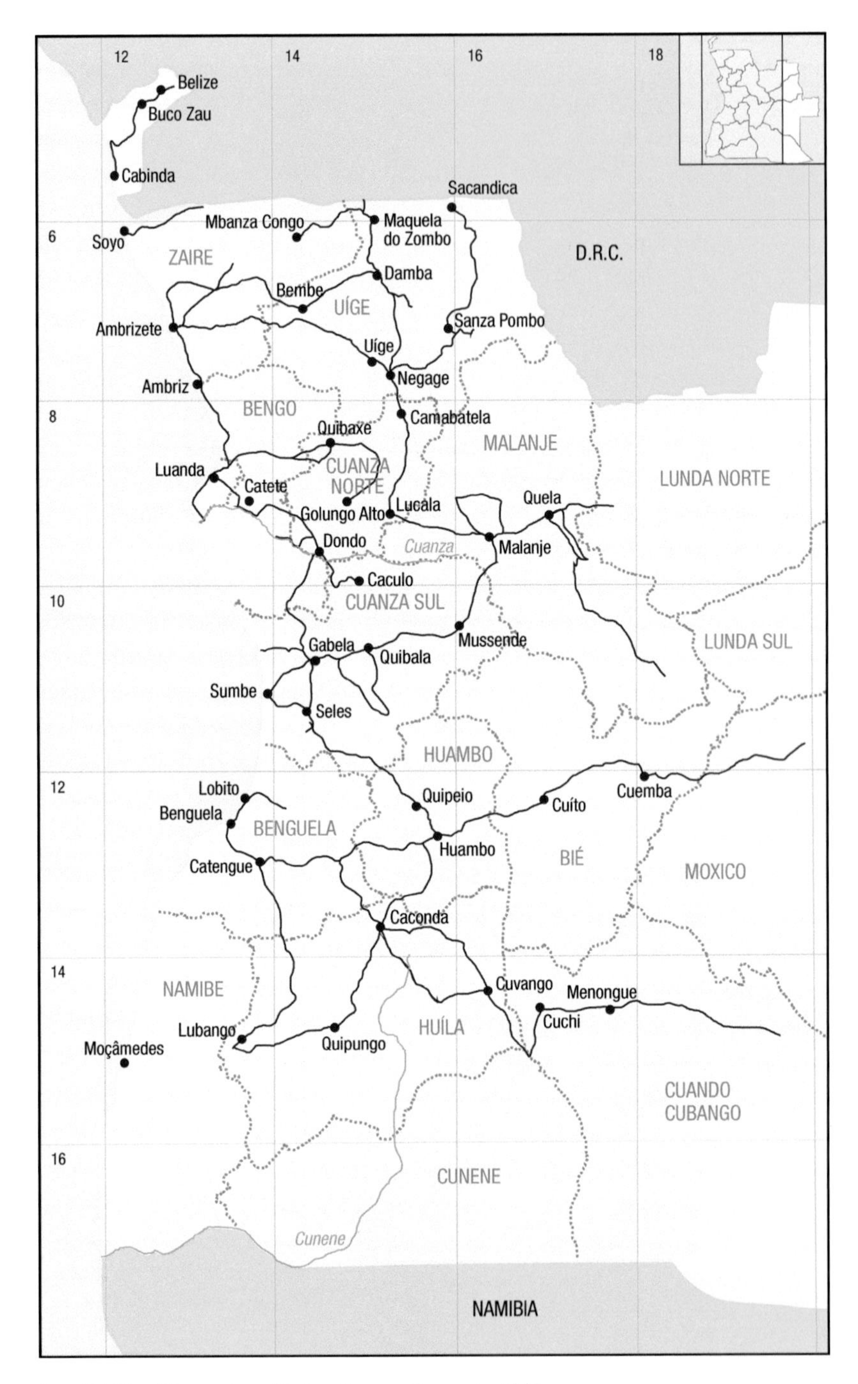

Fig. 57. The itineraries of John Gossweiler up to 1939.

between 1926 and 1933 for some of the taxa collected by Gossweiler are also at Herb. BM. Many duplicates of his collections were distributed to numerous herbaria. His field notes have been digitised at the Univ. of Lisbon. Gossweiler made a huge impact on plant diversity and vegetation studies in Angola, and several of his publications became standard references. *Carta fitogeográfica de Angola* (1939, with Mendonça), "Flora exótica de Angola" (1948–1950), and "Nomes indígenas das plantas de Angola" (1953) remain cornerstones of the study of the Angolan flora and vegetation. He is commemorated in over 130 species names, and in the genera *Gossweilera* S.Moore (Asteraceae) and *Gossweilerodendron* Harms (Fabaceae). **Angola:** 1900–1950. **Herb.:** B, BM, BR, C, COI, E, K, LISC (c. 12,160 specimens), LISU (2695 collections), LUA, LUBA, M, MO, P (c. 600 numbers), PC, US, WAG, WRSL. **Ref.:** Gossweiler, 1939, 1948, 1949, 1950, 1953; Gossweiler & Mendonça, 1939; A.F. Gomes e Sousa, 1949; Exell, 1952; Fernandes, 1954; Lanjouw & Stafleu, 1957; Cavaco, 1959; Mendonça, 1952, 1962a; Leeuwenberg, 1965; Bossard, 1993; Liberato, 1994; Sampaio Martins, 1994a; Romeiras, 1999; Figueiredo & Smith, 2008: 2, 3; Glen & Germishuizen, 2010. Figures 56 and 57; see also Figures 31 and 44.

[Gossweiler, Mrs J.]
Note: Wife of John Gossweiler (q.v.). She was recorded in JSTOR Global Plants (JSTOR, 2021) as collector but this is likely a mistake.

Gouveia, António Carlos Pissarra
(fl. 1963–1972)
Bio.: Agronomist. He collected c. 1000 numbers in Cunene from 1963 to 1965, and later in Malanje. In 1966, he published an account of the vegetation of Cunene. **Angola:** 1963–1972; Cunene (1963–1965), Malanje (1972). **Herb.:** COI, LISC (52 specimens), LUAI, LUBA. **Ref.:** Gouveia, 1966; Bossard, 1993; Liberato, 1994; Romeiras, 1999; Figueiredo & al., 2008.

Gregory, John Walter
(1864–1932)
Bio.: b. London, England, 27 January 1864; d. Urubamba River, Peru, 2 June 1932. Geologist. He studied at the London Mechanics' Institute (now Birkbeck College, Univ. of London), graduating with a B.Sc. in 1891 and a D.Sc. in 1893. In 1887, he was appointed as an assistant in the geological department of the then British Museum (Natural History) in London. He undertook many expeditions to North and Central America, Africa, and Asia. In 1912, he was the leader of a mission to Angola, with the purpose of assessing the potential of the area for Jewish colonisation. While in Angola, he collected a few

Fig. 58. John Gregory. Photographer unknown. Univ. of Melbourne Archives, photo number UMA/I/1217. Public domain.

hundred plant specimens. From 1900 to 1904, he held a position at the Univ. of Melbourne, Australia, but returned to Europe in 1904 to accept the chair in geology at the Univ. of Glasgow, Scotland, a post he held until he retired in 1929. He published 20 books and over 300 papers, being well-known, in particular, for his study of the Great Rift Valley, as he named it. He died in Peru when the canoe he was travelling in overturned. He is commemorated in several plant names. **Angola:** 1912; Benguela, Huambo (Bailundo). **Herb.:** BM, K. **Ref.:** Gossweiler, 1939; Lanjouw & Stafleu, 1957; Lovering, 1983; Bossard, 1993; Liberato, 1994; Romeiras, 1999; Figueiredo & al., 2008; Polhill & Polhill, 2015. Figure 58.

Grey, Frederick William (1805–1878)

Bio.: b. Howick, Northumberland, England, 23 August 1805; d. Sunningdale, Berkshire, England, 2 May 1878. Naval commander. He was one of the 16 children of the British Prime Minister Charles Grey, 2nd Earl Grey (1764–1845). Frederick Grey joined the Royal Navy at an early age, in 1819, and six years later, in 1825, he had his first post serving as lieutenant on H.M.S. *Sybille* in the Mediterranean. He gained his first command with the sloop H.M.S. *Heron* in South America in 1827. Later he served in the East Indies and China Station (1835–1843) being active in the First Opium War. Afterwards he was posted to Finland and saw duty in the Black Sea (1854–1855) during the Crimean War. He was promoted to the rank of rear admiral and, in 1857, was appointed as commander-in-chief for the Cape of Good Hope and the West Coast of Africa, a Royal Navy Station in South Africa. At that time and until 1860, he commanded H.M.S. *Boscawen*. Returning to England, he was appointed First Lord of the Admiralty and Commissioner of the Admiralty (1861–1866) in the Second Palmerston Ministry of the liberal government. He eventually resigned when the Liberal government left power. While stationed in southern Africa from 1857 to 1860, Grey collected in South Africa and Angola. He is commemorated in *Erica greyi* Guthrie & Bolus (Ericaceae) and in *Phylica greyi* Pillans (Rhamnaceae) of which he collected the type specimens. He is not to be confused with the contemporaneous Sir George Grey (1812–1898), who was born in Lisbon and was governor of the Cape Colony from 1854 to 1861. Sir George was recorded by Jackson (1901) as a collector in the Cape from 1850 to 1863, but he was incorrectly referred to as "Admiral Hon. George Grey". Note though that Sir George only governed the then Cape Colony from 1854 onwards. In turn, Sir George is not to be confused with F. Grey's brother, the Admiral George Grey (1809–1891). **Angola:** 1857–1860. **Herb.:** K, P. **Ref.:** Jackson, 1901; Lanjouw & Stafleu, 1957; Gunn & Codd, 1981; Bossard, 1993; Romeiras, 1999.

Gründler, H.
(fl. c. 1904)
Note: Specimens were sent to the Royal Botanic Gardens, Kew, England, by "Mme. H. Grundler". There is a reference in the literature to a missionary in Angola in 1909 named Hilda Gründler, but there is also a reference to a German named H. Gründler who owned a copper mine at Zenza do Itombo (Cuanza Norte) in 1909; Mme H. Gründler was likely the wife of the mine owner. **Angola:** "communicated" 1904; Luanda. **Herb.:** K. **Ref.:** Couceiro, 1948: 329; Satre, 2005; Figueiredo & al., 2008.

Grutterink, Barend Johannes
(1886–unknown)
Bio.: b. Wageningen, The Netherlands, 1886. Forester. He was with the Dutch East Indian Forest Service from 1913 to 1934, when he retired and returned to the Netherlands. He is recorded in the literature as collecting in Angola in 1910, with specimens deposited at the herbarium of the *Muséum national d'Histoire naturelle* in Paris, France (Herb. P). **Angola:** 1910. **Herb.:** P. **Ref.:** Van Steenis-Kruseman, 1950; Lanjouw & Stafleu, 1957; Bossard, 1993; Romeiras, 1999; Figueiredo & al., 2008.

Gruvel, Jean Abel
(1870–1941)
Bio.: b. Le Fleix, Dordogne, France, 14 February 1870; d. Dinard, Ille-et-Vilaine, France, 18 August 1941. Zoologist and marine biologist. With a doctorate in natural sciences, he started his career as a lecturer at the Univ. of Bordeaux, where he founded an entomological research laboratory. He later joined the *Muséum national d'Histoire naturelle*, Paris, and was the first to hold the chair *Pêches et productions coloniales d'origine animale* at the *Muséum*. He was eventually succeeded by his former assistant Monod (q.v.). Gruvel had a special interest in groups of economic importance and produced monographs on Cirripedia (barnacles) and Palinuridae (spiny lobsters). He was active in different subjects and published popular books on fish products, such as oils and roe, and contributed to the creation of parks and reserves in French colonies. From 1932, Gruvel was the driving force for the development of the marine station of the *Muséum*, including during the period when the station was transferred to Dinard in 1935. At present, the station is a research and education centre on coastal systems. From 1905 to 1936, Gruvel undertook c. 30 expeditions to Africa in connection with fisheries and the establishment of aquaculture. From 1909 to 1910, he undertook an expedition to the west coast of Africa, with Angola being one of the places visited. He collected a few terrestrial plant specimens in Namibe. **Angola:** May–June 1910; Namibe (Bero River, Moçâmedes). **Herb.:** P, PC. **Ref.:** Gruvel, 1911; Lanjouw & Stafleu, 1957; Romeiras, 1999; Figueiredo & al., 2008.

Guerra, Guilherme
(fl. 1940s–1960)
Bio.: Agronomist. He was director of the *Serviços Agrícolas de Angola* in the late 1940s and at least up to 1960. In 1952, he was based at the *Posto do Cunje* (Bié) and sent Reynolds (q.v.) material of an aloe that was later described as *Aloe guerrae* Reynolds (Asphodelaceae). The holotype is a Reynolds collection, *Reynolds 9218* (Herb. PRE), with isotypes at Herb. K and Herb. LUA. **Angola:** 1952; Bié. **Herb.:** It is not known if any specimens collected by Guerra were preserved. **Ref.:** Reynolds, 1960, 1966.

Güssfeldt [also as "Güßfeldt"], Paul
(1840–1920)
Bio.: b. Berlin, Germany, 14 October 1840; d. Berlin, Germany, 17 January 1920. Explorer and mountaineer. He studied natural sciences and mathematics in Heidelberg, Berlin, Gießen, and Bonn, and obtained his habilitation at Bonn in 1868. He was active in the Franco-Prussian War from 1870 to 1871

as a volunteer. Afterwards he was appointed by the *Deutsche Gesellschaft zur Erforschung Aequatorial-Afrikas* to lead an expedition to the Kingdom of Loango (Rep. of the Congo and Cabinda, Angola) with the purpose of extending the exploration of the African interior. His team included Pechuël-Loesche (q.v.), Mechow (q.v.), and Soyaux (q.v.). Güssfeldt was in charge of geographical and topographical observations and logistics. He embarked on 30 May 1873. The ship was wrecked on 14 June 1873 near Freetown (Sierra Leone), which resulted in the loss of his equipment that he had acquired at great cost, and this delayed his arrival to Banana (D.R. Congo) until 25 July. He explored the Loango coast, and the Kouilou (Rep. of the Congo), Nyanga (Rep. of the Congo, Gabon), and Chiloango (Cabinda) rivers. In March 1874, he travelled to Luanda. To contract porters, he travelled to Quicombo and Sumbe in Cuanza Sul. However, the physical condition of these porters would be one of the reasons for the failure of the expedition. Out of the 100 porters brought from Sumbe, only 20 were fit to travel, the others were ill, absconded or died. The lack of rain and subsequent drought and famine, and the onset of a measles epidemic created further difficulties. When a date to start the expedition was finally set, most of the remaining porters escaped together with their guards. Güssfeldt returned to Germany, arriving in Berlin on 24 August 1875. After his report to the sponsors and following his recommendation, the expedition was terminated, and the other members recalled. In the view of his

Fig. 59. Paul Güssfeldt. Artist unknown. Reproduced from Lange (1874). From Wikimedia Commons. Public domain.

opinionated companion, Pechuël-Loesche, Güssfeldt was an hypochondriac and prone to bouts of melancholy, and noted as lacking warmth and with a tendency to brood, characteristics that are disadvantageous in the leader of an expedition. After this venture, Güssfeldt travelled to Egypt and Arabia with Georg Schweinfurth (1836–1925), and in 1882 to the Andes and Bolivia. He became known as a pioneer of mountaineering in the Chilean and Argentinian Andes. He climbed the Maipo volcano in the Andes and was the first European to climb Aconcagua, tracing the route up to an elevation of 6500 m above sea level. He also explored mountains in Europe such as Mont Blanc, and some mountain features bear his name. From 1883 to 1885, he was secretary-general of the *Gesellschaft für Erdkunde zu Berlin*. In 1892, he was appointed as professor and lectured in Berlin. He retired shortly before the onset of World War I. According to available records, he made collections during the Loango expedition that were deposited at the herbarium of the Botanic Garden and Botanical Museum Berlin-Dahlem (Herb. B). However, these specimens have not been located. He is commemorated in *Dianthus guessfeldtianus* Muschl. (Caryophyllaceae), *Oxalis guessfeldtii* R.Knuth (Oxalidaceae), and *Icacina guessfeldtii* Asch. (Icacinaceae). **Angola:** 25 July 1873–July 1875; Cabinda. **Herb.:** B. **Ref.:** Güssfeldt & al., 1888; Gossweiler, 1939; Lanjouw & Stafleu, 1957; Ronge, 1966; Bossard, 1993; Liberato, 1994; Romeiras, 1999; Figueiredo & al., 2008, 2020; Heintze, 2010, 2011. Figure 59.

H

Hallé, Nicolas
(1927–2017)
Bio.: b. Haute-Normandie, France, 8 February 1927; d. France, September 2017. Botanist, entomologist, engineer, and artist. Brother of the French botanist Francis Hallé (b. 1938). In 1947, Nicolas was attached to the *Laboratoire d'Entomologie agricole coloniale* of the *Muséum national d'Histoire naturelle* (MNHN), in Paris. The following year he graduated as an engineer at the *École Supérieure d'agriculture et de viticultures* at Angers, France. From 1951 to 1953, he was an illustrator for archaeology, prehistory, and zoology at the *Institut Français d'Afrique Noire* in Dakar, Senegal, which was followed by a position as illustrator for agronomy, botany, and herpetology at the *Institut d'Adiopodoumé* (ORSTOM) in Abidjan, Côte d'Ivoire, from 1954 to 1956. He was *Chargé de Recherches* at the *Centre National de la Recherche Scientifique* from 1957 to 1958 and obtained a doctorate from the Univ. of

Paris Sorbonne in 1958. Afterwards, he was a researcher at the MNHN from 1959 to 1992, where he was in charge of the phanerogam and pteridophyte herbaria, and in 1987 sub-director of the *Laboratoire de Phanérogamie*. In 1990, he became a professor at the MNHN and retired two years later. From 1994 to 2004, he was archivist at the *Société Nationale des Sciences Naturelles et Mathématiques* of Cherbourg, France. He travelled extensively, especially in Africa, Polynesia, and New Caledonia. It is estimated that he produced about 3500 drawings, illustrations, and paintings. **Angola:** 1973; Cuanza Sul (Cuvo River, Conda, Seles); P (c. 15 specimens). **Herb.:** BR, G, K, LISC (3 specimens); MO, P, UCJ, WAG. **Ref.:** Dorr, 1997; Anonymous, 2015.

Hardy, David ("Dave") Spencer (1931–1998)

Bio.: b. Pretoria, South Africa, 24 September 1931; d. Pretoria, South Africa, 31 May 1998. Cultivator and collector, mainly of succulents. Initially a technician at the Veterinary Research Institute, Onderstepoort, Pretoria (1951–1958), he then transferred to the Botanical Research Institute, Pretoria, and remained there until his retirement in 1991. He freely shared material he collected and was responsible for the introduction of several species into general cultivation, for example *Aloe ×inopinata* Gideon F.Sm. & al. (Asphodelaceae) and *Kalanchoe*

Fig. 60. Dave Hardy. Photographer unknown. South African National Biodiversity Institute. Reproduced with permission.

sexangularis N.E.Br. (Crassulaceae). He is commemorated in the names of several succulents, including *Aloe hardyi* Glen. **Angola:** 1972. **Herb.:** K, MSUN, P, PRE (main), WIND. **Ref.:** Bornman & Hardy, 1971; Chaudhri & al., 1972: 298 (sub "Ihlenfeldt, H. & al."); Gunn & Codd, 1981; Hardy & Fabian, 1992; Dorr, 1997; Glen, 1998; Figueiredo & al., 2008; Smith & al., 2016, 2019a. Figure 60.

Hauser, Walter
(fl. 1935)
Note: In the 1930s, Walter Hauser sent or took living material of plants he collected in Angola to the *Botanischer Garten und Botanisches Museum Berlin* in Germany. Only one such collection is known to have been made by him and recorded in Berlin. This collection was made in 1935 near the town of Ganda, Benguela highlands. One year later, the plant flowered in Berlin and was described as *Kalanchoe hauseri* Werderm. (Crassulaceae). **Angola:** 1935; Benguela (Ganda). **Herb.:** B, one specimen recorded, likely destroyed. **Ref.:** Werdermann, 1937: 1–2; Gunn & Codd, 1981: 374; Figueiredo & al., 2020; Smith & Figueiredo, 2021.

[Henam, F.]
Note: Listed in Figueiredo & al. (2008) based on a record in the Herb. PRE database that has since been removed.

Henriques, Carlos Alves
(fl. 1960–1974)
Bio.: Technical assistant at the *Instituto de Investigação Científica de Angola*. He collected in several provinces. **Angola:** 1960–1972; Bengo, Benguela, Bié, Cuanza Norte, Cuanza Sul, Cunene, Huambo, Huíla, Luanda, Malanje, Moxico, Namibe, Uíge; over 1500 numbers alone; collected also with Barbosa (q.v.; 1960, Bengo, Cuanza Norte); Barbosa & (F.) Moreno (1966, Huíla); Brites (q.v.; 1967, Huíla, Namibe); Menezes (q.v.; 1962, Cunene); (F.) Moreno (q.v.; 1962, Namibe); (R.) Santos (q.v.; 1961 and 1963, Huíla, Namibe); and (A.F.) Sousa (q.v.; 1962, Huíla). **Herb.:** BM, COI, K, LISC (1216 specimens), LISU, LUAI, LUBA, P, PRE. **Ref.:** Bossard, 1993; Liberato, 1994; Romeiras, 1999; Figueiredo & al., 2008.

Henriques, José Cristóvão
(1917–1976)
Bio.: b. Aldeia da Mata, Alentejo, 8 December 1917; d. 4 April 1976. Forester. In 1927, at the age of 10, he went to Lisbon, where he attended school. In 1935, he entered the *Instituto Superior de Agronomia*, graduating in 1940 with a

degree in forestry. In 1946, he took a position as forester in Angola. During his career he also worked in Mozambique and Timor. After returning to Portugal, he worked for the *Ministério do Ultramar* (Overseas Ministry) at the *Direção-Geral de Economia* from 1961 to 1975. After his early death at the age of 58 years, a *Festschrift* was published in his honour in 1981. **Angola:** 1951–1962; Bengo (Quiçama), Luanda, Uíge (Sanza Pombo); collected also with Menezes (q.v.). **Herb.:** COI, LISC (23 specimens), LISU, LUA, LUAI, PRE. **Ref.:** Henriques, 1968; Anonymous, 1981; Bossard, 1993; Romeiras, 1999; Figueiredo & al., 2008.

Hess, Hans Ernst
(1920–2009)
Bio.: b. Rubigen, Berne, Switzerland, 10 April 1920; d. Zürich, Switzerland, 15 March 2009. Botanist. He studied agriculture at the Univ. of Zürich, Switzerland, and was awarded a doctorate in plant pathology. He was the curator of the herbarium of the Botanical Garden of Zürich (Herb. ZT) from 1965 to 1967, and a professor of "Spezielle Botanik" at the Swiss Federal Institute of Technology in Zürich, retiring in 1987. He was one of the authors of the *Flora der Schweiz und angrenzender Gebiete*. In 1950, sponsored by the CIBA Company (Basel, Switzerland), he undertook a six-month expedition

Fig. 61. Hans Hess with his wife Esther Hess-Wyss in Angola 1951/1952. Photographer unknown. ETH-Bibliothek, University Archives, Akz.-2006-37. Reproduced with permission.

to Africa, working especially in Cameroon, the D.R. Congo, and Angola, studying the genus *Strophanthus* DC. (Apocynaceae). A second expedition to Angola took place from 24 November 1951 to 21 July 1952. During this expedition he was accompanied by his wife, Esther Hess-Wyss (q.v.). His research resulted in several papers on the flora of Angola, particularly on the aquatic and swamp plants. His Angolan collection consists of c. 3000 numbers and is deposited at Herb. ZT, which was combined with Herb. Z in 1991. These herbaria work together as Herb. Z+ZT. **Angola:** 1950; 1951–1952; Benguela, Cuando Cubango, Cunene, Huambo, Huíla, Luanda. **Herb.:** K, M, NY, P, WAG, Z+ZT. **Ref.:** Hess, 1952, 1953, 1955; Figueiredo & al., 2008; Landolt, 2010; May & Dossenbach, 2018; Alessia Guggisberg, pers. comm., November 2018. Figure 61; see also Figure 36.

Hess-Wyss, Esther
(1923–2010)
Bio.: b. Zürich, Switzerland, 23 September 1923; d. Zürich, 16 May 2010. Wife of Hans Hess (q.v.). She obtained a qualification in pharmacy and afterwards worked as a pharmacist, always part-time, until the age of 72. She accompanied her husband on his second expedition to Angola from 24 November 1951 to 21 July 1952 and collected with him. **Angola:** 1951–1952. **Herb.:** K, M, NY, P, WAG, Z+ZT. **Ref.:** Alessia Guggisberg, pers. comm., October 2021. Figure 61.

Hoepfner [also as "Höpfner"], Karl [or "Carl"]
(1857–1900)
Bio.: b. Friedrichslohra, near Nordhausen, Germany, 8 February 1857; d. Denver, Colorado, U.S.A., 14 December 1900. Geologist and electrochemical engineer. He studied medicine, physics, chemistry, mineralogy, and geology at the Univ. of Berlin, graduating in 1881. From 1882 to 1883, he was in Namibe (Angola) and Hereroland (Namibia) engaged in prospecting activities. He travelled from Moçâmedes to Humpata, then in a southerly direction crossing the Cunene River into Namibia, to the region then known as Damaraland, and from there to Walvis Bay, which was a British colony at the time. From 1884 to 1885, Hoepfner returned to Africa to undertake an expedition in Namibia. In 1884, he registered a patent for extracting metals from ore. He continued investigations and pioneered several techniques for manufacturing various products and the extraction of metals, mainly by electrolysis. In 1899, he established the short-lived Hoepfner Refining Company in Canada. The following year, he died of typhoid fever while investigating the extraction of silver in Denver, Colorado. Most of his collections were likely destroyed in Berlin during World War II. At the herbarium of the Botanic Garden and

Botanical Museum Berlin-Dahlem (Herb. B) there were 142 numbers collected by Hoepfner from 1882 to 1883 in southwestern Africa. Some of these were collected in Angola, for example one syntype of *Sida hoepfneri* Gürke (Malvaceae), *Hoepfner 20*, which dates from April/May 1882 and was collected in Moçâmedes. Hoepfner is also commemorated in other names, such as *Justicia hoepfneri* Lindau (Acanthaceae), which was published as "*höpfneri*". **Angola:** 1882–1883; Namibe (Moçâmedes). **Herb.:** B (main), CORD, HT, PRE, US. **Ref.:** Armand, 1884; Belck, 1901; Urban, 1916; Gossweiler, 1939; Lanjouw & Stafleu, 1957; Fischer, 1972; Gunn & Codd, 1981; Bossard, 1993; Romeiras, 1999; Figueiredo & al., 2008, 2020.

[Horning]
Note: Recorded in JSTOR Global Plants (JSTOR, 2021), apparently based on a specimen at Herb. K that was databased as Horning, but that it is in fact "Hornig in Waibel" and from Namibia.

Humbert, Jean-Henri
(1887–1967)
Bio.: b. Paris, France, 24 January 1887; d. Bazemont, Yvelines, France, 20 October 1967. Taxonomist, ecologist, phytogeographer, and collector extraordinaire. His earliest herbarium collections date from 1900, when he was thirteen. In 1910, he graduated with a B.Sc. in Paris, and in 1913 was appointed as a lecturer at the Univ. of Clermont-Ferrand (now Univ. of Clermont Auvergne). He was mobilised during World War I and even then he made collections (e.g., from the battle areas of Verdun and Argonne). After the war, he continued at the Univ. of Clermont-Ferrand, preparing his Ph.D. dissertation; the degree was conferred in Paris in 1923. He held the chair of botany (1920) and taught botany at the Institute of Chemistry and Industrial Technology from 1920 to 1922. In 1922, he moved to Algiers (Algeria) to be head of the Faculty of Science of the Univ. of Algiers, a post he held until 1931 when he returned to France and was appointed professor, holding the chair of botany at the *Muséum national d'Histoire naturelle* in Paris (1931–1958). Humbert initiated several Floras, including the *Flore de Madagascar et des Comores* in 1936, which he directed until 1967. He complemented his herbarium studies with many collecting and mountaineering expeditions, including trips to less explored areas and the higher mountains, in several continents. Between 1912 and 1960, he undertook ten long-term expeditions in Madagascar, as well as in Africa. He collected in several African countries, with a focus on mountains, such as in the D.R. Congo (Mounts Nyiragongo, Kahusi, and Karisimbi) and Uganda (Ruwenzori, Mount Mugule) from 1928 to 1929, and in (present-day)

Tanzania (Kilimanjaro and the Rift Valley) from 1933 to 1934. During the 1930s, he explored most of the southern African countries, including, from 1933 to 1934, mountainous areas such as Inyangani Mountain (Zimbabwe), and Mont-aux-Sources (Drakensberg, South Africa). Unlike many of his contemporaries, he is known to have undertaken these expeditions with respect for local people and their beliefs, providing good conditions for his porters, visiting the local chiefs and taking care not to infringe on prohibitions. Overall, his explorations resulted in over 50,000 numbers: c. 30,000 from the intertropical regions and c. 20,000 from France and the Mediterranean area. His collections resulted in the discovery of hundreds of new taxa and include thousands of type specimens. He published extensively and was involved in the creation of protected areas. Humbert is commemorated in over 250 species names and six genus names. **Angola:** 1937; Benguela, Huíla, Moxico, Namibe; over 600 numbers with duplicates distributed to several herbaria such as Herb. BR, K, LISC, US, and WAG. **Herb.:** A, AL, B, BM, BR, C, E, EA, EGR, G, GH, K, L, LIL, LISC (1 specimen); MO, MPU, NA, NCY, NY, P (main, c. 616 numbers), PRE, RAB, REN, S, TAN, TEF, U, US, WAG. **Ref.:** Gagnepain, 1944; Lanjouw & Stafleu, 1957; Heim, 1968; Keraudren & Aymonin, 1968;

Fig. 62. Henri Humbert. Photographer unknown. Photograph supplied by L.J. Dorr. Reproduced with permission.

Wurmser, 1971; Stafleu & Cowan, 1979: 363; Gunn & Codd, 1981; Bossard, 1993; Dorr, 1997; Romeiras, 1999; Figueiredo & al., 2008; Polhill & Polhill, 2015; Smith & Figueiredo, 2019; Smith & al., 2019b. Figure 62.

Hundt, Otto
(1878–unknown)
Bio.: b. Köthen (Anhalt), Germany, 21 April 1878. Merchant and farmer. In the 1920s, he lived in Berlin-Charlottenburg, Germany and made a living as a merchant. On 14 January 1928, he embarked from Hamburg, Germany for Lobito, Angola, and settled in Benguela as a farmer. In the 1930s, he lived on a farm called Xangorolo (also spelled "Xongorola" or "Chongorollo"), a coffee plantation where cattle also were herded. He sent specimens to the Botanic Garden and Botanical Museum Berlin-Dahlem (Herb. B) in Germany. His collecting numbers amount to, at least, 1019 and include several type specimens, many of which were destroyed during the World War II bombing of Berlin. He is commemorated in several species names including *Hibiscus ottoi* Exell, *H. hundtii* Exell & Mendonça (both Malvaceae), *Alectra hundtii* Melch. (Orobanchaceae), and *Vigna hundtii* Rossberg (Fabaceae). **Angola:** 1930–1935; between Benguela (Ganda), and Huíla (Caconda). **Herb.:** AAU, B (main), BM, COI, G, LISC (14 specimens), MO, P, PRE. **Ref.:** Jessen, 1936; Gossweiler, 1939; Lanjouw & Stafleu, 1957; Bossard, 1993; Liberato, 1994; Romeiras, 1999; Figueiredo & al., 2008, 2020.

Huntley, Brian John
(1944–)
Bio.: b. Durban, KwaZulu-Natal, South Africa, 20 February 1944. Botanist. He graduated with a B.Sc. from the Univ. of Natal (now Univ. of KwaZulu-Natal) in 1964, and a B.Sc. (Hons) and M.Sc. from the Univ. of Pretoria, in 1967 and 1968, respectively. He was appointed plant ecologist on the first scientific expedition to the sub-Antarctic Prince Edward Islands (December 1964), spending 15 months exploring and describing these islands. Later, he was ecologist in game reserves of northern South Africa (1967–1971) and for Angola's five major national parks, establishing the country's protected area expansion strategy (1971–1975). From 1975 to 1989, he coordinated South Africa's Council for Scientific and Industrial Research's (CSIR) Cooperative Scientific Programmes. He was CEO of the National Botanical Institute (South African National Biodiversity Institute after 2004) from 1990 to 2007, when he retired. From 1996 to 2004, he established and led the SABONET project. After retiring, he became advisor to the South African government's Department of Environmental Affairs and Tourism (2007–2009) and then a self-employed

Fig. 63. Brian Huntley. Photographer unknown. Private collection of B. Huntley. Reproduced with permission.

independent ecological consultant (from 2009). He is commemorated in *Aloe huntleyana* van Jaarsv. & Swanepoel (Asphodelaceae), *Androcymbium huntleyi* Pedrola & al. (Colchicaceae), and *Brianhuntleya intrusa* (Kensit) Chess. & al. (Aizoaceae). **Angola:** 1971–1975; Bengo (Quiçama), Huambo (Morro Moco), Huíla, Luanda, Namibe (Bicuar and Iona National Parks); c. 1500 numbers; most specimens were lost in 1975, but several hundred were sent to PRE; collected also with Roberts (q.v.) and (J.D.) Ward (q.v.). **Herb.:** CANB, E, K, PRE, US. **Ref.:** Gunn & Codd, 1981; Figueiredo & al., 2008; Huntley, 2017; Huntley & al., 2019; B. Huntley, pers. comm., October 2018. Figure 63.

I

Ivens, Roberto
(1850–1898)

Bio.: b. S. Miguel, Azores, Portugal, 12 June 1850; d. Dafundo, Oeiras, Portugal, 28 January 1898. Naval officer. A descendant of an English merchant who lived in the Azores, Ivens moved to continental Portugal as a child and entered the Navy in 1867. After graduating from the *Escola Naval* in 1870, he served in Goa (India) and in Angola, where he was engaged in patrolling the maritime area between Moçâmedes and the island of São Tomé in the Gulf of Guinea. During that time, he disembarked in Cabinda and Congo for punitive incursions. He returned to Lisbon in 1874 and the following year was commissioned as a second lieutenant. Afterwards, he saw service on a warship travelling to São Tomé and Brazil, and later to the U.S.A., where he was seconded to support the Centennial International Exhibition of 1876. After returning to Lisbon, he was sent to Angola to undertake a survey of Baía dos Tigres. Afterwards, he travelled on a steamship up the Congo River to Noqui and produced a map of the area between Boma and Noqui, as well as drawings that were later published. Based on the experience gained during these explorations, he was promoted to the rank of first lieutenant in May 1877 and appointed, with Capelo (q.v.) and Serpa Pinto (q.v.), to undertake the first Portuguese expedition into the interior of southern Africa; the *Expedição portuguesa ao interior da África austral* (q.v.), organised by the *Sociedade de Geografia de Lisboa*. In March 1880, almost three years later, Ivens and Capelo were back in Lisbon. On 12 March 1884, again with Capelo, Ivens left Lisbon on the second Portuguese expedition into the southern African

Fig. 64. Roberto Ivens in 1885. Photographer Alfred Fillon. Public domain.

interior, returning to Portugal on 20 September 1885, some 18 months later. Each expedition was the subject of a book (Capelo & Ivens, 1881, 1882; Capelo & Ivens, 1886). Ivens was promoted to the rank of *Capitão de Fragata* on 7 December 1895. He was appointed to the *Comissão de Cartografia* and was aide-de-camp to the King D. Carlos I. He was an accomplished author and artist, and the engravings of his drawings were published in the books he co-authored with Capelo. Both works became classic literature references on the exploration of Africa. Although his fieldbooks of the first expedition have not been located, those of the second expedition, which include the original drawings, are deposited at *Sociedade de Geografia de Lisboa*, where other materials pertaining to the expeditions are also kept. **Angola:** 1877–1885; Bengo, Benguela, Bié, Cuando Cubango, Cuanza Norte, Cuanza Sul, Cunene, Huambo, Huíla, Luanda, Lunda Norte, Lunda Sul, Moxico, Namibe, Uíge. **Herb.:** COI, LISU (main). **Ref.:** Capelo & Ivens, 1881, 1882, 1886; Noronha, 1936; Gossweiler, 1939; Lanjouw & Stafleu (1954, sub "Cardoso"); Exell, 1960; Chaudhri & al., 1972; Nowell, 1982; Santos, 1988; Bossard, 1993; Silva, 1996, 2005; Figueiredo & al., 2008; Taquelim, 2008; Lima, 2012. Figure 64.

J

Jessen, Otto
(1891–1951)
Bio.: b. Kronprinzenkoog, Schleswig-Holstein, Germany, 18 February 1891; d. Munich, Germany, 9 June 1951. Geographer. Shortly after he obtained his doctorate at the Univ. of Munich in 1914, he served as an army officer in World War I. Being severely injured at the Somme, he then served as a war geologist in Alsace-Lorraine. After the war, he was with the Univ. of Tübingen, where he obtained his habilitation in 1921. He became a professor in 1924. He moved to the Univ. of Cologne, and later, in 1933, to the Univ. of Rostock. He also served during World War II. After the latter war ended in 1946, he was with the Univ. of Würzburg and in 1949 he returned to the Univ. of Munich to accept a chair. He died in Munich a couple of years later. From 4 June 1931 to 16 December 1931, he travelled in Angola accompanied by his wife. With a driver and two African helpers, they explored the area between Luanda and Tômbua. He took eleven different routes from the coast inland, recording observations on geomorphology, phytogeography, and human geography. During this expedition he made plant, mineral, and ethnographic collections and took 1250 photographs. He left Lobito on 6 January 1932 and

Fig. 65. Otto Jessen in the early 1930s. Photographer unknown.

published the results of his study in 1936. According to Gossweiler (1939), he collected a few hundred plant specimens that were kept at the Botanic Garden and Botanical Museum Berlin-Dahlem (Herb. B). It is not known how many of these remained intact following the World War II bombing of Berlin. **Angola:** 1931; Cuanza Sul, Huambo, Huíla, Namibe. **Herb.:** B (main), COI. **Ref.:** Jessen, 1936; Jaeger, 1937; Gossweiler, 1939; Chaudhri & al., 1972 (as "Jessen, D.O."); Louis, 1974; Bossard, 1993; Romeiras, 1999; Figueiredo & al., 2008, 2020; Heintze, 2010. Figures 65 and 66.

John, David Michael
(1942–)

Bio.: b. England, 3 April 1942. Algologist. He graduated from Durham Univ., England, and, after continuing his studies, was awarded a Ph.D. in 1968. In that year he applied for a lectureship in botany at the Univ. of Ghana, which would allow him to work closely with Lawson (q.v.), a leading authority on the seaweeds and seashore ecology of tropical West Africa. He left Ghana in 1980 to join the Natural History Museum, London, where he would remain for the rest of his career, first as head of the Freshwater Algae Division (1980–1989), then as head of the Division of Algae (1989–1995), and then as head of the

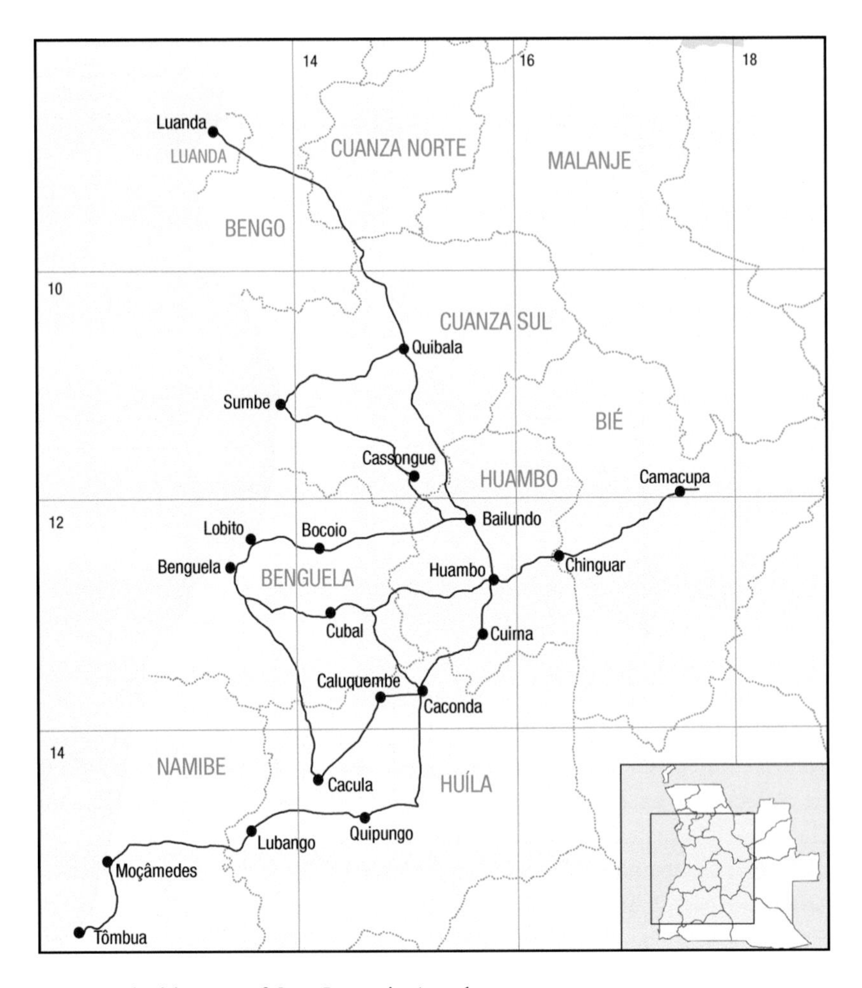

Fig. 66. The itinerary of Otto Jessen in Angola.

Environmental Quality Research Theme (1995–1998). Afterwards, he was a phycological researcher (1998–2002) and after retiring he continued an involvement with the Museum as scientific associate (2002–2019), remaining active in research and presenting training courses in algal identification. In 1988, he was awarded a D.Sc. by Durham Univ. for his contribution to research on freshwater and marine algae in West Africa and the British Isles. He published extensively on algae, including several catalogues. While in Ghana, John participated in several expeditions to African countries, ranging from Western Sahara in the north to Angola in the south. In 1974, with Lawson (q.v.) and Price (q.v.), he undertook a six-week expedition to Angola to survey and collect seaweeds along the coast and in the shallow subtidal zone where he could scuba-dive. They received local support from the *Instituto de Investigação Científica de Angola* (IICA) and were provided with two vehicles and staff to undertake a trip to the coast of southern Angola. They were accompanied by R.M. Santos (q.v.), C.A. Henriques (q.v.), and the zoologist M.S. Vasconcelos, from the IICA, and the scuba-diver F. de Figueiredo from the Univ. of Luanda. During the expedition they could not collect from many sites north of Luanda and south of Tômbua because of the then on-going hostilities in the Angolan war for independence. Duplicates of the seaweed collections were deposited at the Natural History Museum in London, the botany department at the Univ. of Ghana, and the botany department at the Univ. of Luanda. **Angola:** January–February 1974; with Lawson and Price. **Herb.:** BM, US. **Ref.:** Lawson & al., 1975; Lawson & John, 1987; John & al., 2003; John, 2017; D.M. John, pers. comm. October 2021.

Johnston, Henry ("Harry") Hamilton (1858–1927)

Bio.: b. London, England, 12 June 1858; d. Worksop, England, 31 July 1927. Explorer, artist, colonial administrator, and linguist. He studied languages at King's College, London, and art at the Royal Academy Schools. When he turned 18 years old in 1876, he started travelling, at first in Europe, then in Tunisia (1879). In 1881, his friend, the zoologist William Alexander Forbes (1855–1883), suggested that he join an expedition that the Earl of Mayo (Dermot Robert Wyndham-Bourke; 1851–1927) was organising to southwestern Africa. Mayo was enthralled by descriptions of Angola he had heard in London from the *Dorslandtrekkers* (see page 160). Mayo was initially proposing to cross Angola from Moçâmedes (Namibe), in the south, to the then almost unknown southern Congo River basin or less ambitiously "at any rate as far as the upper course of the Cunene River" (in reality they only travelled on the Cunene River for a short distance north of Humbe, likely up to Cafu). Johnston decided to join Mayo in this venture, with the purpose of studying the languages and making natural history collections. They left Liverpool in April 1882 on the steamer *Benguela*. After calling at Sierra Leone, Nigeria, and Cameroon, concerned with Johnston's health Mayo allowed him to be transferred from the "sordid" *Benguela* to the gunboat *Rambler* of the West Africa Squadron (see page 102), which they had encountered off the coast of Nigeria. The journey proceeded with more comfort, the *Rambler* calling at several places, including Banana (D.R. Congo) and Luanda (Angola), and finally arriving at Moçâmedes (Angola). After a short stay at the house of Lapa e Faro (q.v.), they proceeded to the interior, riding mules or walking. They reached the

Chela Mountains in three to four days. For a while Johnston got separated from his companions and spent a night at the house of a young Portuguese man in exile for manslaughter. The next day, he rejoined Mayo and the day after that the party reached Humpata (Huíla). From there they travelled down the embankments of the Caculovar River for several months. The Caculovar joins the Cunene River at Humbe, which was then a trading post belonging to a trader/slave dealer. It was also there that Duparquet (q.v.) was managing the Mission of the Sacred Heart of Mary. At Humbe, Johnston left Mayo and returned to Luanda to pursue his own project of journeying in the Congo River and meeting Henry Morton Stanley (1841–1904). While in Luanda, he explored the country with short journeys inland along the northern coast and took a trip up the Cuanza River on a steamer of a line owned by Robert Scott Newton, a Scotsman who was the British vice-consul at Luanda. Johnston left Luanda in October 1882 to continue to the Congo. He took a steamer to Ambriz and from there travelled overland in a hammock to Quicembo, then boarded the steamer again at Musserra, continued to Ambrizete, and disembarked at Banana (D.R. Congo) in November 1882. There he met Pechüel-Loesche (q.v.). In December 1882, Johnston departed on a steamer to Quissanga, on the south bank of the Congo River (i.e., now Angola), where he was hosted by a Portuguese trading agent. From there he continued to Ponta da Lenha (name no longer in use, corresponding to a locality between Banana and Boma, in the D.R. Congo) and Boma, both localities on the northern side of the river in territory of the D.R. Congo, and further into the interior. The expedition lasted c. eight months, and he succeeded in meeting Stanley at Kinshasa (D.R. Congo). This first trip to Africa was important in shaping Johnston's career. Afterwards he undertook an expedition to Kilimanjaro (1884–1885) and in 1886 was appointed as the British vice-consul in the region of the Niger delta and Cameroon. He was later involved in the diplomatic conflict for the control of the Shire Highlands (part of present-day Malawi), meeting Serpa Pinto (q.v.) in 1889, and afterwards becoming commissioner for Malawi from 1891 to 1896 and serving in Uganda from 1899 to 1901. In 1902, he was awarded a doctorate by the Univ. of Cambridge. He became well-known for his involvement in the scramble for Africa as advisor to British Prime Minister Lord Salisbury and supported the policy of extending British influence "from the Cape to Cairo". He was also a prolific author, publishing books on several subjects, including novels, and numerous articles. Johnston is commemorated in numerous names such as *Hermannia johnstonii* Exell & Mendonça (Malvaceae) and *Urginea johnstonii* Baker (Asparagaceae), two endemic Angolan species that are only known from the types collected by him. Between 1883 and 1888, Johnston collected and sent to the Royal Botanic Gardens, Kew (Herb. K), 1436 plants from Sierra Leone, the D.R. Congo, Angola, the Nile River, and Tanzania. His collections from Angola that have been observed online are labelled with the date September 1883. However, this appears to be the date on which they were received at Herb. K, not when they were collected, as at the end of 1882 Johnston had left Angola and was travelling on the Congo River. Some collections are labelled "coll. & com." (i.e., collected and communicated) by Johnston, while others are only stated as "com." by Johnston. The latter collections, from localities such as Humbe (Cunene), Gambos (Huíla), and Palanca (Huíla, near Humpata), have original labels handwritten in Portuguese with dates from May 1882 to May 1883. At least one of the examined labels is signed by Newton (q.v.; not Vice-Consul R.S. Newton mentioned before, who appears not to be a direct relation). These collections are thus correctly cited as "leg. Newton in Johnston". Harry Johnston is not to be confused with his contemporary, the Scottish botanist Henry Halcro Johnston (1856–1939). **Angola:** 1882; Cuanza Norte (Cuanza River), Cunene (Cunene River), Namibe (Curoca River as "R. Croque", Chela), Huíla (Humpata). **Herb.:** B, BM, CGE, COI, GRA, K, PRE, W. **Ref.:** Johnston, 1895, 1923; Jackson, 1901; Gossweiler, 1939; Chaudhri & al., 1972; Bossard, 1993; Liberato, 1994; Romeiras, 1999; Figueiredo & al., 2008; Polhill & Polhill, 2015. Figure 67.

Fig. 67. Harry Johnston. Photographer unknown. From Wikimedia Commons. Public domain.

Dorslandtrekkers

Meaning "Thirstland Trekkers". Starting in May 1874, several *trekke* or *treks* left the then Zuid-Afrikaansche Republiek, later the Transvaal Province of the Union of South Africa, in a northwesterly direction through Botswana and the Kalahari, a dry area, hence the name applied to the trekkers. After crossing Namibia, some families reached the Cunene River at the southern border of Angola. While there, in 1880, they met the missionary Duparquet (q.v.), who was travelling by ox-wagon on his way to Humbe (Cunene, Angola). Duparquet suggested that they settle at Huíla, in Angola. After negotiations with Portuguese authorities the trekkers were allowed into Angola to establish a community in Humpata (Huíla). At the time, the Huíla region had about 20,000 inhabitants, of which 100 were white (75 European and 25 African). As there were no roads from the Cunene to Humpata, the trekkers had to make hundreds of kilometres of trails to pass with their ox-wagons. Led by Jacobus Botha, this group of trekkers consisted of 57 families (270 persons), 61 wagons, 840 draft oxen, 2160 cattle, 120 horses, and 3000 sheep and goats, as well as many native servants and mixed-race descendants. They finally reached Humpata and settled there in January 1881, six years after leaving the Transvaal. **Ref.:** Duparquet, 1953; Stassen, 2010; Azevedo, 2014.

Jordan, Heinrich Ernst Karl
(1861–1959)

Bio.: b. Almstedt, near Hildesheim, Hanover, Germany, 7 December 1861; d. Hemel Hempstead, Hertfordshire, England, 12 January 1959. Entomologist. He graduated in botany and zoology from the Univ. of Göttingen. In 1893, after military service and five years as a grammar schoolteacher, he was invited to join the staff of Walter Rothschild's (1868–1937) Zoological Museum at Tring (Hertfordshire, England) as entomologist and immigrated to England. He was hired to curate Coleoptera but, in 1894, Rothschild switched his focus to Lepidoptera and Jordan became a specialist in the latter group. He also worked on Siphonaptera (fleas), publishing with Miriam Rothschild (1908–2005), who was Walter's niece. After Rothschild's death, this Museum became part of the Natural History Museum, London. In 1911, Jordan became a naturalised British citizen. He published over 400 papers and described over 3000 new species and was still scientifically active well into his nineties. He is commemorated in a renowned award for entomologists, the Karl Jordan Medal of The Lepidopterists' Society. In 1934, Jordan undertook an expedition to Namibia and Angola. In Namibia, he was based at a farm near Windhoek. After a stint in Namibia, he sailed from Walvis Bay to Lobito, Angola. From there he set off with the Angola-resident amateur ornithologist Rudolf Braun, two chauffeurs, and a cook ("so-called", in his words). They left Lobito in March 1934 in a car and a van. They undertook two separate trips through the provinces of Benguela, Huambo, and Cuanza Sul. On Whit-Sunday (20 May 1934) they were back at Bocoio and, shortly thereafter, at Lobito. The expedition is described in an account published in 1936, which includes maps and itineraries, many observations of animals and habitats and some mention of plant species. His little-known botanical collections are at the herbarium of the Natural History Museum (Herb. BM). As far as we know, there are no collections databased and no information available on the total number of Jordan specimens, nor of the precise dates of these collections. **Angola:** 1934; Benguela (from March), Bié, Cuanza Sul, Huambo. **Herb.:** BM (main), MO. **Ref.:** Jordan, 1936; Hering, 1959; Riley, 1960; Chaudhri & al., 1972; Johnson, 2003, 2012; Figueiredo & al., 2008, 2020. Figures 68 and 69.

Fig. 68. Heinrich Jordan. Photographer unknown. George Grantham Bain Collection (Library of Congress). Public domain.

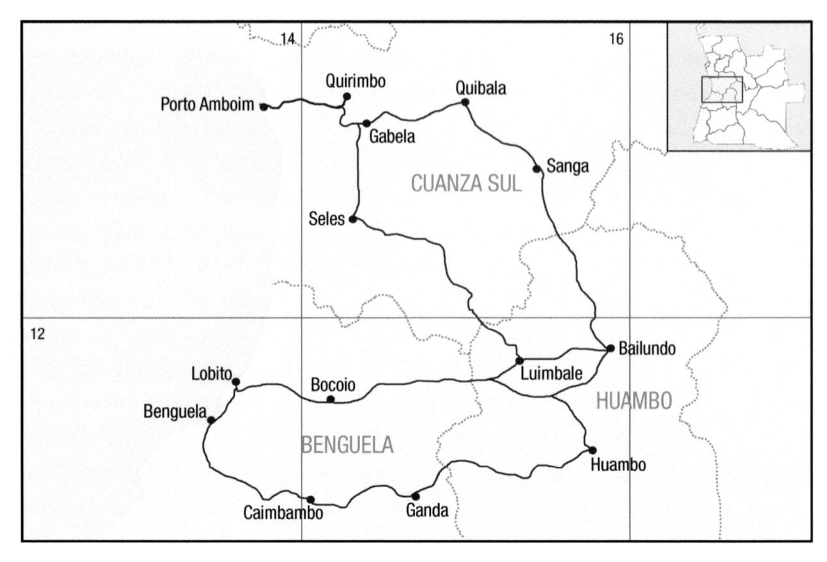

Fig. 69. The itinerary of Heinrich Jordan in Angola.

Jorge, Carlos
(fl. 1957–1960)
Note: Likely Carlos Martins Jorge, who in 1960 was a member of the staff at *Companhia de Diamantes de Angola* in Lunda, in charge of constructions. **Angola:** 1957; Lunda Norte. **Herb.:** DIA†?, LISC (2 specimens). **Ref.:** Romeiras, 1999; Figueiredo & al., 2008; G. Valente, pers. comm., January 2019.

K

[Kampashi, R.]
Note: Recorded in JSTOR Global Plants (JSTOR, 2021). This refers to the Kampashi River.

Kers, Lars Erik
(1931–2017)
Bio.: b. Leksand, Sweden, November 1931; d. Sweden, 2017. Botanist. He studied geography and geology at the Univ. of Gothenburg, and later studied botany and morphology with Folke Fagerlind (1907–1996) at the Univ. of

Stockholm. Kers travelled to Africa to undertake field work for his doctoral thesis on the genus *Cleome* L. (Capparaceae). In 1968, he travelled to Angola, visiting the southwest of the country. His collections are databased at the herbarium of the Swedish Museum of Natural History (Herb. S). **Angola:** 1968; Cunene, Huíla, Namibe; collection numbers 3090–3664. **Herb.:** BM, COI, LISC (118 specimens), MO, P, PRE, S (main), SBT, US, WIND. **Ref.:** Chaudhri & al., 1972; Jönsson, 2017.

Kestilä Liljeblad, Alma Helena
(1877–1965)
Bio.: b. Turku, Åbo, Finland, 6 September 1877; d. Helsinki, Finland, 20 March 1965. Née Kestilä. Missionary teacher. She graduated as a teacher from a training college at Sortavala, Finland. After practising nursing in hospitals in Turku and Helsinki, she went to Amboland (Namibia) in 1902 as a missionary teacher. In 1904, she married Karl Emil Liljeblad (1876–1937), who had been in Amboland as a missionary of the Finnish Ovamboland Mission since 1900, and thereafter they worked jointly. They left for Finland in 1907, returning to Namibia in 1912 for a stay of a further seven years. By then, her husband had qualified to be a religious minister and teacher. In 1913, he founded the teacher seminary in Oniipa and was its head teacher until he left missionary work in 1919, and they returned to Finland. She collected c. 100 specimens and recorded vernacular names. She is commemorated in *Scilla kestilana* Schinz (Asparagaceae). Specimens labelled "Ukuanyama" refer to the Ovambo people in north Namibia and Angola (Cuanhama, in Cunene). **Angola:** 1903; Cunene. **Herb.:** H, PRE, Z+ZT. **Ref.:** Chaudhri & al., 1972; Roivainen, 1974; Gunn & Codd, 1981; Bossard, 1993; Romeiras, 1999; Figueiredo & al., 2008.

Kirckwood, John
(fl. 1696)
Bio.: Naval surgeon. One of the four pre-Linnaean collectors in Angola, the others being Brown(e) (q.v.), Gladman (q.v.), and Mason (q.v.). Kirckwood collected a few plants for James Petiver's herbarium, which is now at the Natural History Museum, in London (Herb. BM). The specimens date from 1696 and originated from Cabinda. **Angola:** 1696; Cabinda. **Herb.:** BM (Sloane Herb.). **Ref.:** Exell (A.W.), 1939, 1962; Chaudhri & al., 1972; Bossard, 1993; Liberato, 1994; Romeiras, 1999; Figueiredo & al., 2008; Glen & Germishuizen, 2010.

Kotze, T.J.
(fl. 1962)
Bio.: Agronomist in the service of the South African government. In what was then South West Africa (now Namibia), he inter alia worked on pastures, grazing, and cattle feed intake, as well as on karakul sheep farming. He paid visits to Angola that were later reported on in *Farming in South Africa* (Kotze, 1965). Herb. PRE holds 11 specimens that were collected by T.J. Kotze at the Ruacaná Falls and Epupa Falls (Quedas do Monte Negro) in southern Angola. Eight specimens are of representatives of the Poaceae and three are of representatives of the Fabaceae. The collecting labels were written in Afrikaans. Costa & al. (2004), in their checklist on the grasses of Angola, do not mention Kotze. Kotze is also not mentioned in the works on Namibian grasses by Müller (1983, 1984) nor Klaassen & Craven (2003). In the last work, a voucher specimen is cited for each grass species recorded, but none refers to Kotze. **Angola:** April–May 1962; Cunene (Epupa, Ruacaná). **Herb.:** PRE. **Ref.:** Kotze, 1965; Figueiredo & al., 2008; B. van Wyk, pers. comm., 2021.

L

Lains e Silva, Hélder José
(1921–1984)
Bio.: b. Alcobaça, Portugal, 19 January 1921; d. Lisbon, Portugal, 16 March 1984. Agronomist. After graduating from the *Instituto Superior de Agronomia* in Lisbon, in 1947, he joined the *Companhia da Zambézia* in Mozambique, where he worked until 1951 at which time he returned to Portugal. In 1953, he again went to Africa, accepting a four-year appointment with the *Junta de Exportação do Café* in Luanda, Angola. He returned to Portugal in 1957, and at the time became involved with and directed several projects for overseas agronomical studies, such as the *Brigadas de Estudos Agronómicos do Ultramar* (1958–1960) and the *Missão de Estudos Agronómicos do Ultramar* (1960–1965). Afterwards, he held several management positions. In 1976, he joined the *Junta de Investigações Científicas do Ultramar* (later *Instituto de Investigação Científica Tropical*) where he remained until his death. He produced several publications, being particularly well-known for those on the cultivation of coffee in East Timor and São Tomé and Príncipe. **Angola:** 1970; Cubal (Benguela). **Herb.:** LISC (31 specimens). **Ref.:** Lains e Silva, 1956a, 1956b, 1958; Gonçalves, 1985; Romeiras, 1999; Figueiredo & al., 2008; Castelo, 2011.

Lam, Herman Johannes
(1892–1977)

Bio.: b. Veendam, Netherlands, 3 January 1892; d. Leiden, Netherlands, 15 February 1977. Botanist. He attended school in Rotterdam from 1904 to 1911, and then enrolled at the Univ. of Utrecht, graduating with a B.Sc. just before World War I hostilities began. From 1914 to 1918, he was in military service part of the years, being allowed to work on his thesis during the winter months. He obtained his Ph.D. in 1919 and in that same year he was appointed as botanist at the herbarium of the Botanic Gardens at Bogor, in Java, Indonesia. In 1932, he became professor of botany at the academic high school at Jakarta and chief of the Treub Laboratory at Bogor. Soon thereafter, he returned to the Netherlands to assume the position of director of the Rijksherbarium and professor of plant taxonomy and geography at the Univ. of Leiden in 1933. He revived Malesian botany, started the journal *Blumea*, and raised interest in New Guinean botany. With World War II and the occupation of the Netherlands in 1940, the Univ. of Leiden was closed. Lam resigned his functions in 1942, resuming them after the war, in 1945. Earlier in his career, he focused on the Sapotaceae and Burseraceae; later he was interested in phytogeography and phylogeny but was mostly occupied with administration and teaching. He also served as Secretary to the Univ. Senate and as Rector

Fig. 70. Herman Lam. Photographer unknown. Reproduced from Van Steenis (1976) with the consent of the Flora Malesiana Foundation and Naturalis Biodiversity Center.

Magnificus. He is commemorated in dozens of plant names and at least one animal name, and was the subject of several biographical accounts, including a book (Jacobs, 1984). In 1938, Lam joined a Dutch collecting expedition to Namibia, South Africa, the Mascarenes, and Madagascar. He was assisted by Meeuse (q.v.). They collected at Lobito (Benguela, Angola) in August 1938 while on their outward journey. **Angola:** 1938; Benguela; collected with Meeuse (q.v.). **Herb.:** BK, BM, BO (main), BR, G, K, L (main), MO, NY, P, PRE, S, U, WAG. **Ref.:** Chaudhri & al., 1972; Van Steenis, 1976; Vegter, 1976 (sub "Meeuse"); Stafleu & Cowan, 1979: 729; Gunn & Codd, 1981; Jacobs, 1984; Visser, 2013; Dorr, 1997. Figure 70.

[Lang]
Note: Listed by Romeiras (1999) and Figueiredo & al. (2008) as collector in 1842 with collections at Herb. BM and Herb. L. We could not locate a source for this information. There are no records in the Dutch Herbaria database at Naturalis nor the Herb. BM database.

Lapa e Faro, João Cabral Pereira
(1818–1886)
Bio.: b. Mangualde, Portugal, 1 June 1818; d. Moçâmedes, Namibe, Angola, 28 September 1886. Surgeon. After studying in Porto and Lisbon, he joined the Portuguese navy. In January 1857, he moved to Moçâmedes in Namibe, Angola, to be in charge of the newly created hospital. He was among the earliest settlers and the first surgeon in Moçâmedes, a town that had been established in 1849. In 1858, in the same publication where Welwitsch (q.v.) published some of his botanical work, Lapa e Faro published a paper on the climate of Namibe, where he listed plants cultivated there. In 1868, in his absence, the clinic was run by Anchieta (q.v.). Lapa e Faro's house was built in that same year; this iconic building in Moçâmedes is still extant. It was likely in this house that Johnston (q.v.) stayed in 1882 when he visited Angola and was hosted by Lapa e Faro in his "large and clean" house with a family of many pleasant women whose "manners were irreproachable". The visit made a strong impression on Johnston, who, in his autobiography published forty years later, was still reticent to provide the full name of Lapa e Faro (referring to him as "L.— e F.—") because of the nature of the story he was disclosing. As described by Johnston, this "elderly, upright, rather well-seeming Portuguese, of considerable learning in medicine" had an enormous household that included his five daughters of mixed race, who in turn had become the wives of their father and had had children by him, the old man being at the time of Johnston's visit involved with his granddaughter. **Angola:** No specimens located. **Herb.:** COI.

Ref.: Lapa e Faro, 1858; Anonymous, 1862; Johnston, 1923; Andrade, 1985; Romeiras, 1999; Figueiredo & al., 2008; Azevedo, 2014.

Lawson, George William
(1926–2014)
Bio.: b. Sunderland, England, 18 April 1926; d. London, England, 18 August 2014. Ecologist and algologist. The son of a miner, he grew up in Sunderland and left school at 16. During World War II, he saw service with the Royal Navy, working in a shipyard and later in a coal mine office. During that time, he attended evening classes and qualified to enrol for a university degree. After military service, he attended King's College, London, and in 1951 graduated with a B.Sc. He then accepted a lectureship at the Univ. College of the Gold Coast (now Univ. of Ghana) and conducted research for which he was awarded a Ph.D. by the Univ. of London. In 1962, he became a professor of botany and also head of the department at the Univ. of Ghana. During his time in West Africa, he surveyed the whole of the coast from Western Sahara in the north to Namibia in the south. He left Ghana in 1971 and was appointed as head of the botany department at the Univ. of Nairobi, Kenya (1971–1972), and as a visiting professor at the Univ. of Tanzania and Bangor Univ., Wales, then as professor (1976–1979) and head of department (1979–1982) at the Univ. of Lagos, Nigeria, and as professor (1983–1987) at Bayero Univ., Nigeria. In 1987, after having spent 35 years in Africa, he retired and moved to London, where he joined the Natural History Museum. He was a leading authority on the biogeography of western African seaweeds and was awarded a D.Sc. by the Univ. of London for his contributions to African botany. Over the next 20 years, at the Museum, he was instrumental in the publication of a series of catalogues of West African algae. In addition to numerous publications on algae, Lawson also authored papers on savannas and forests, and a popular work entitled *Plant life in West Africa* that was aimed at senior secondary school pupils and university students. The first edition was published in 1966, while a second edition appeared in 1985 (Lawson, 1985). Given Lawson's interest in coastal research, chapter six of the book dealt with "Life on the seashore". He is commemorated in the names of algae. From January to February 1974, Lawson, John (q.v.), and Price (q.v.) undertook an exploration along the coast of Angola. They surveyed 22 locations off the coast from Ponta Spilimberta (Bengo) in the north to Tômbua (Namibe) in the south. The survey resulted in a doubling of the number of marine algal species known for Angola. **Angola:** January–February 1974; collected with John (q.v.) and Price (q.v.). **Herb.:** BM, GC, US. **Ref.:** Chaudhri & al., 1972; Lawson & al., 1975; Lawson, 1985; Lawson & John, 1987; Dravers, 2014; John, 2014.

Leach, Leslie ("Larry") Charles (1909–1996)

Bio.: b. Southend-on-Sea, Essex, England, 18 November 1909; d. Polokwane, Limpopo, South Africa, 18 July 1996. Businessman and amateur botanist. He started his working career in the British army. In January 1938, he moved to Zimbabwe where he began a business. During World War II he served in the Royal Air Force at the military base in Harare (then Salisbury). In 1950, he started developing a garden of succulents in a suburb of Harare. After his first wife died in 1956, he sold his business and concentrated on the taxonomy of selected succulent groups. He had an interest especially in the Euphorbiaceae, Apocynaceae, and Asphodelaceae (*Aloe* in particular). He was an honorary botanist at the National Herbarium (Herb. SRGH) in Harare, Zimbabwe (1972–1981). Afterwards, he was based at the Karoo Desert National Botanical Garden, Worcester, South Africa (1982–1989), and then until his death attached to the Univ. of the North (now Univ. of Limpopo), South Africa. The herbarium of the Univ. of Limpopo, Polokwane (Herb. UNIN), was named the Larry Leach Herbarium. He collected extensively throughout the *Flora Zambesiaca* region, over 14,000 numbers. He is commemorated in several plant names. On his expeditions he collected many new species of succulents in *Euphorbia*, *Aloe*, *Hoodia*, and *Huernia*, most of which he described himself. **Angola:** July–September 1967, September–November 1970, 1973; c. 500 numbers; collected with Cannell (q.v.) and with Duarte (q.v.).

Fig. 71. Larry Leach (left) in 1988 with Gideon F. Smith. Photographer Pieter D. Theron. Original in the private collection of Gideon F. Smith.

Herb.: BM, BOL, BRLU, COI, E, EA, FHO, G, K, LISC (144 specimens), LMA, LUA, M, MO, NBG, P, PRE, SAM, SRGH, UNIN, US. **Ref.:** Verdoorn & Codd, 1966; Chaudhri & al., 1972; Gunn & Codd, 1981; Kimberley, 1988; Bossard, 1993; Archer, 1997; Smith & Williamson, 1997; Figueiredo & al., 2008; Polhill & Polhill, 2015. Figure 71.

Lebre, António Tavares
(1882–1966)
Bio.: b. Aveiro, Portugal, 21 March 1882; d. Verdemilho, Aveiro, Portugal, 11 February 1966. Veterinarian and military officer. From 1914 to the 1930s, he worked in Angola in several positions including as director of the *Serviços de Pecuária e Indústria Animal*. Sometime during World War I, he also was deployed to Mozambique. In 1925, while in Cuanhama (Cunene), he met the veterinarian Carlos Baptista Carneiro and they took the initiative of creating the *Missão Veterinária da Huíla*. In the 1930s, Lebre published a few papers on Angolan subjects; livestock husbandry and local people. After returning to Portugal, he was an active member of the society in Verdemilho where he lived on his estate, the 18th century *Quinta de Nossa Senhora das Dores*, and where he is commemorated in a street name. He was first listed as a collector by Bossard (1993, sub "Dr A. Lebre"), based on information from the herbarium of the *Instituto de Investigação Agronómica* in Huambo (Herb. LUA). **Angola:** 1915–1916; Benguela, Luanda, Malanje (Pungo Andongo). **Herb.:** LUA. **Ref.:** Bossard, 1993; Romeiras, 1999; Martins (D.P.), 2005; Mendes, 2005; Figueiredo & al., 2008.

[Letouzey, René Gustave]
(1918–1989)
Bio.: b. Bonneval, Eure-et-Loire, France, 21 October 1918; d. Versailles, France, 4 August 1989. Botanist. He studied in Paris, France, at the *Institut National Agronomique*, graduating as an agronomist in 1939 just as World War II started. He was mobilised and could only resume his professional career in 1944 when he joined the *École des eaux et forêts*. After training, and short stays at Senegal and Indochina (Cambodia, Laos, and Vietnam), he settled in Cameroon in 1945. In 1948, he became the head of the *Section de recherches forestières* at Yaounde, Cameroon, and remained in that post until 1961. During that time, he obtained a degree in natural sciences. In 1955, he became *Conservateur des eaux et forêts* and two years later, in 1957, *Maître de recherches* with the *Institut national de la recherche agronomique*. In 1961, he joined the *Centre national de la recherche scientifique* in France and was detached to the *Laboratoire de Phanérogamie* of the *Muséum national*

d'Histoire naturelle in Paris, where he worked for almost three decades. There he focused on systematics, but his interests were wider, extending to phytogeography, ecology, and ethnobotany. His research on the phytogeography of Cameroon became the subject of the thesis with which he obtained a doctorate in 1966. The work was eventually published as *Carte phytogéographique du Cameroun, 1 : 500 000* in 1985. He collected extensively, over 15,000 specimens, mostly in African countries, and especially in the francophone ones. He initiated the National Herbarium of Cameroon (Herb. YA) and in 1963, he started the *Flore du Cameroun*. Letouzey is recorded in JSTOR Global Plants (JSTOR, 2021) as a collector in Angola, however we could not locate any specimens. **Herb.:** B, BR, COI, G, HBG, K, P, U, WAG, WRSL, YA. **Ref.:** Chaudhri & al., 1972; Normand, 1988; Villiers, 1989; Boulvert, 2011a.

Lewis, Walter Hepworth
(1930–2020)
Bio.: b. Carleton Place, Ontario, Canada, 26 June 1930; d. St. Louis, Missouri, U.S.A., 17 November 2020. Botanist. In 1957, he was awarded a Ph.D. by the Univ. of Virginia, U.S.A., and afterwards he lectured for four years at the Stephen F. Austin State College (now Univ.) in Nacogdoches, Texas. He received grants to travel and develop research in Europe and Africa, and for the ensuing years spent time working at the herbarium of the Royal Botanic Gardens, Kew, at the Univ. of Leeds, and at the Swedish Academy of Sciences in Stockholm. At the time, he also undertook a six-month expedition to Africa to collect material for cytological studies, visiting Kenya, Tanzania, Zambia, and South Africa. While in Zambia, he crossed the border to Angola and made some collections there. When he returned to the U.S.A., he accepted an appointment with the Missouri Botanical Garden and became a professor at Washington Univ. in St. Louis. From 1964 to 1972, he directed the herbarium of the Missouri Botanical Garden (Herb. MO). **Angola:** 1962; Moxico (Caianda); a few specimens. **Herb.:** BM, BRIT, C, COL, DAO, DUKE, F, GH (main), K, MO (main), NMW, NY, P, TEX, UC, US (main), USMK. **Ref.:** Chaudhri & al., 1972; Gereau & al., 2021.

Liljeblad, Alma Helena Kestilä – see **Kestilä Liljeblad**

Loeb, Ella-Marie Karr
(1916–1989)
Bio.: b. Ketchikan, Alaska, U.S.A., 14 July 1916; d. Berkeley, California, U.S.A., 26 June 1989. Née Karr. Artist and cartographer. She collected for ethnobotanical studies during an African expedition from 1947 to 1948, undertaken

by her husband, the anthropologist Edwin Meyer Loeb (1894–1966), to study the tribes of Namibia. They were accompanied by Rodin (q.v.) as botanist of the expedition. With Carl Koch, the couple published on the ethnobotany of the Kuanyama (Cuanhama in Angola). **Angola:** 1948. **Herb.:** C, E, PRE, UC. **Ref.:** Loeb & al., 1956; Toffelmier, 1967 (re. "Edwin Meyer Loeb"); Chaudhri & al., 1972; Gunn & Codd, 1981; Rodin, 1985; Figueiredo & al., 2008.

Lopes, R.
(fl. 1971)
Angola: 1971; Bié, Cuanza Sul, Huambo, Malanje; collected with (F.) Moreno (q.v.), Murta (q.v.) and (Manuel) Silva (q.v.). **Herb.:** COI, LISC (5 specimens). **Ref.:** Romeiras, 1999; Figueiredo & al., 2008.

Loureiro, D.
(fl. 1963)
Angola: 1963; Huíla. **Herb.:** LISC (1 fragment), LUAI.

Lúcio, Lino
(fl. 1963–1964)
Angola: 1963–1964; Huambo. **Herb.:** LISC (10 specimens). **Ref.:** Figueiredo & al., 2008.

Lugard, Edward James
(1865–1957)
Bio.: b. Worcester, England, 23 March 1865; d. Abinger Common, Surrey, England, 3 January 1957. Army officer. A lieutenant in the British army, he served in the Anglo-Boer War (1899–1902) as a special services officer and in World War I in the Naval Intelligence Division (1915–1919). In 1896, he participated in an expedition of the British West Charterland Company led by his brother Frederick John Dealtry Lugard (1858–1945) to Ngamiland (north Botswana), where both collected plants. From 1897 to 1898, he revisited Ngamiland, this time with his wife Charlotte Eleanor ("Nell") Lugard (1859–1939), née Howard, an artist who made drawings and painted watercolours of some of the plants they encountered. At least 245 drawings were made, and they collected material of 374 species. The map with the itinerary of the expeditions shows that they crossed the border to Angola at Mucusso (Cuando Cubango). E.J. Lugard also collected plants in Nigeria (1905–1907), while assistant to the high commissioner of northern Nigeria (his brother), and later in Kenya (1930–1931). He is commemorated in several plant names, and his wife is commemorated in *Ceropegia lugardiae* N.E.Br. (Apocynaceae), a

species that she discovered. **Angola:** c. 1897–1898; Cuando Cubango. **Herb.:** B, BM, GRA, K, MO, Z+ZT. **Ref.:** Lugard, 1909, 1941; Chaudhri & al., 1972; Gunn & Codd, 1981; Polhill & Polhill, 2015.

Luja, Pierre-Edouard (1875–1953)

Bio.: b. Luxembourg, 11 February 1875; d. 14 September 1953. Horticulturalist. He undertook his first expedition to Africa in 1899, with Émile Duchesne (fl. 1899–1902), the *Mission Luja-Duchesne* to the then *État indépendent du Congo* (now the D.R. Congo), to collect living material of ornamental plants to be exhibited at the World Exhibition in Paris in 1900. He travelled from Lisbon to the mouth of the Congo River (border between Angola and the Congo). At Kinshasa (D.R. Congo), Luja and his companion Duchesne separated, with Luja exploring the Lower Congo and Kasai. In 1901, Luja undertook an expedition to Mozambique commissioned by the *Companhia da Zambézia*, with the task of introducing plants of economic interest originating from other tropical regions and testing their agronomic suitability and improvement. Soon after, he travelled to Sankourou (D.R. Congo) where he was director of an agronomical station at Kondue from 1903 to 1914. He was also in Brazil (1924) and Kivu, D.R. Congo (1928). He is recorded as having collected in Angola, but no specimens were

Fig. 72. Pierre-Edouard Luja. Photographer unknown. From Luja (1951). Digitised by Christian Ries. Reproduced under licence CC-GFDL.

located; it is possible that he collected there during the *Mission Luja-Duchesne*. He is commemorated in many names. **Angola:** No information recorded. **Herb.:** BR, K, P. **Ref.:** Gilbert, 1951; Luja, 1951; Heuertz, 1953, 1955; Chaudhri & al., 1972; Figueiredo & al., 2008; Glen & Germishuizen, 2010. Figure 72.

Luna de Carvalho, Eduardo Augusto (1921–2006)

Bio.: b. Lisbon, Portugal, 11 February 1921; d. Algueirão, Portugal, 5 March 2006. Technician, artist, and entomologist. He had an interest in natural history since his childhood. He studied at *Escola Patrício Prazeres* (then a professional school) and took some drawing courses at *Escola António Arroyo* (an art school) in Lisbon. He started working at an early age, in several different occupations, as an accountant for mining companies at Queirã (Viseu), in several companies in Lisbon, and as a ceramic artist at the factory *Fábrica de Louças de Sacavém*. Throughout his career, he retained an interest in studying and drawing fauna, particularly insects. Without formal training, he was initially mentored by the entomologist Antero Frederico de Seabra (1854–1952) and later by Barros Machado (q.v.). In 1947, he went to Mozambique employed by the *Junta das Missões Geográficas e de Investigações Coloniais*. Later, in 1954, he moved to Angola where he worked as a technician at the

Fig. 73. Eduardo Luna de Carvalho. Photographer unknown. Private collection of David Luna de Carvalho. Reproduced with permission.

Fig. 74. Eduardo Luna de Carvalho (right) with António Barros Machado (left). Photographer and private collection of David Luna de Carvalho. Reproduced with permission.

Laboratório de Investigação Biológica of the *Museu do Dundo, Companhia de Diamantes de Angola*, in Lunda, Angola, assisting Barros Machado. After the independence of Angola, he returned to Portugal to join the *Instituto de Investigação Científica Tropical* (IICT). He became a specialist on the Paussinae (ant nest beetles) and Strepsiptera (parasitic insects), among other groups. He published numerous papers on entomology and obtained a degree at the Univ. of Marseille, France, in 1990. He retired from IICT in 1991, still as a technician. The following year, his degree was considered to be the equivalent of a Master's degree. In 2000, at the age of 79, he was recontracted as a researcher by the IICT and his services were retained for a few more years. He collected for the Diamang Collections (q.v.). **Angola:** 1953–1961; Lunda Norte, Lunda Sul, Moxico; collected with (D.) Machado (q.v.). **Herb.:** DIA†?, LISC (21 specimens), LUAI, P. **Ref.:** Bossard, 1993; Romeiras, 1999; Figueiredo & al., 2008; D.L. Carvalho, pers. comm, January 2021. Figures 73 and 74.

[Lupula, R.]
Note: Recorded in JSTOR Global Plants (JSTOR, 2021). This refers to the Lupula River.

Lux, Anton Erwin
(1847–1908)
Bio.: b. Venice, Italy, 23 December 1847; d. Stockerau, Austria, 31 May 1908. Army officer and geographer. After studying at the military academy of *Weißkirchen*, Mähren (now Hranice, Czech Republic), he joined the Austrian army in 1868 as an artillery officer. He published a few geographical studies and was contracted as topographer during the 1874–1876 expedition to Angola of the *Deutsche Afrika-Gesellschaft*, from 1875 to 1876. This expedition was led by Alexander von Homeyer (1834–1903) and included the botanist Soyaux

(q.v.) and the hunter-explorer Pogge (q.v.). On 2 May 1875, Lux arrived in Luanda to join the others who already had been in the country for about two months. In Luanda, Lux met Soyaux, who had travelled from Pungo Andongo back to the coast to obtain funds, and they left together to join the others. On 12 May, they travelled up the Cuanza River to Dondo, then continued on foot to Pungo Andongo where Homeyer and Pogge were based. Soyaux and Homeyer had taken ill with malaria and had to return to Luanda on 27 August. Lux and Pogge continued to Malanje. On 14 June, the expedition departed from Malanje with 114 men, and by 26 August it was beyond the Cuango River and reached Mona Quimbundo. Lux fell ill and had to turn back on 14 September 1875. Pogge continued alone with the expedition. On his way back to Malanje, Lux took a different itinerary through the country of the Shinje and Mbangala people. On 2 November, he arrived in Luanda and returned to Europe with Mechow (q.v.). In 1880, he published a book on his travels. Afterwards he became a geography teacher at the *Militär-Unterrealschule* (military secondary schools) at Güns (now in Hungary) and Eisenstadt and was later stationed in Vienna and Peterwardein (now in Serbia) and, in 1900 in Przemýsl (now in Poland). In 1903, he retired from the military service and later reached the rank of major-general. Lux has been cited in the literature (e.g., Bossard, 1993)

Fig. 75. Anton Lux. By Anton Gratl. From Wikimedia Commons. Public domain.

as having collected with Pogge during the 1875–1876 expedition. Out of the 547 numbers reported as having been collected during that expedition, the c. 90 that can be examined online do not reference Lux, and none was collected on a date earlier than when Pogge and Lux arrived in Mona Quimbundo, the place from where Lux turned back to return to Malanje. It is unlikely that Lux collected any specimens by himself. **Angola:** Lux accompanied Pogge, but Lux likely did not collect. **Herb.:** B? **Ref.:** Lux, 1880; Bossard, 1993; Romeiras, 1999; Figueiredo & al., 2008, 2020; Heintze, 2010. Figure 75.

Lynes, Hubert
(1874–1942)
Bio.: b. England, 27 November 1874; d. Holyhead, Isle of Anglesey, Wales, 10 November 1942. Naval officer, amateur naturalist, and ornithologist. He joined the Navy in 1887, where he had a distinguished career, particularly during World War I, retiring in 1919 with several decorations and being promoted to rear admiral. After retiring, he pursued his interests in ornithology undertaking 12 expeditions to Africa, some lasting one year or longer. He collected in Angola on several occasions between 1919 and 1934. After publishing his *magnum opus*, a revision of the genus of insectivorous birds *Cisticola* (Cisticolidae), he returned to Africa four times, until 1938, when having contracted shingles, he had to return to England. He served during World War II until his death in 1942. He was an extraordinary field ornithologist and collected plant specimens that are deposited in several herbaria. His collection notebooks are at the herbarium of the Botanic Garden Meise, Belgium (Herb. BR; Bamps, 1973). He is commemorated in several names. **Angola:** 1919–1934; Benguela (Lobito), Lunda Norte (Dundo), Lunda Sul (Dala, Saurimo), Moxico (Luena); 144 collections at BR, numbers 263–403 (fide Bamps, 1973). **Herb.:** BM, BR, COI, K, LISC, MO. **Ref.:** Gossweiler, 1939; Palmer, 1943; Witherby, 1943; Chaudhri & al., 1972; Bamps, 1973; Gunn & Codd, 1981; Bossard, 1993; Romeiras, 1999; Figueiredo & al., 2008; Polhill & Polhill, 2015.

M

M.E.F.A. – Missão de Estudos Florestais a Angola
(1957–1960)
A project organised by the *Jardim e Museu Agrícola do Ultramar* in Lisbon, Portugal, which was directed by D'Orey (q.v.). It consisted of four expeditions. The first, in 1957, as well as the second, in 1958, were led by J.A.T.

Carvalho (q.v.). The third expedition took place in 1959 and was led by Semedo (q.v.). The fourth and final expedition, which took place in 1960, was led initially by Semedo and later by D'Orey. A total of c. 950 collection numbers of herbarium specimens were secured, along with several duplicates, in the forests of Cabinda, Angola. Timber samples for anatomical studies and other analyses were also collected. D'Orey retired from the directorship of the *Jardim* in 1975 and was succeeded by Semedo. When Semedo died in 2003, almost 50 years after the expeditions, studies on the samples were still being carried out as part of an on-going project at the (by then called) *Jardim-Museu Agrícola Tropical*, which he still directed. **Angola:** Cabinda. **Herb.:** COI, K, LISC (1981 specimens), LUAI. **Ref.:** Anonymous, 1963; Liberato, 1994; Instituto de Investigação Científica Tropical, 2003; Figueiredo & al., 2008.

Machado, António Barros – see **Barros Machado**

Machado, Dora Lustig
(1907–1986)
Bio.: b. Berlin, Germany, 1907; d. 1986. Née Lustig. She married A. Barros Machado (q.v.) in 1944 and accompanied him to Lunda, Angola, where he directed the *Laboratório de Investigações Biológicas* of the *Museu do*

Fig. 76. Dora Machado with António Barros Machado. Photographer unknown.

Dundo, Companhia de Diamantes de Angola from 1947 to 1973. While they were based there, she made a few plant collections for the Diamang Collections (q.v.), some with Luna de Carvalho (q.v.). **Angola:** 1954–1955; Lunda, Moxico. **Herb.:** DIA†?, LISC (8 specimens). **Ref.:** Anonymous, 1996–; Romeiras, 1999; Figueiredo & al., 2008; Ceríaco & al., 2020. Figure 76.

Machado, Filomeno Guerreiro
(fl. 1962–1978)
Bio.: Technician. He was an auxiliary technician (*auxiliar técnico de 1ª classe*) in the 1960s and worked in the fisheries industry. He collected some plants connected to aquaculture. It is likely that he later moved to Portugal, as in 1978 he published an article on ichthyology in that country. **Angola:** 1962–1963; Bié, Huambo (Sacaála), Moxico. **Herb.:** LUA. **Ref.:** Machado, 1971, 1978; Bossard, 1993; Figueiredo & al., 2008.

Maia Figueira, João
(c. 1933–unknown)
Bio.: b. Portugal, c. 1933; d. Portugal. Technician. He went to Angola and joined the *Instituto de Investigação Agronómica de Angola* at Chianga, Huambo, where he was an auxiliary collector in 1964. After independence of the country, he returned to Portugal where he died. He appears in labels as "Maia" or "J.M. Figueira". **Angola:** 1959–1974; Benguela, Bié, Cuanza Norte, Cunene, Huambo, Huíla, Malanje, Zaire; collected also with Bamps (q.v.) & Sampaio Martins (q.v.); Cardoso de Matos (q.v.) & Raimundo (q.v.); and (Brito) Teixeira (q.v.). **Herb.:** COI, LISC (7 specimens), LUA, PRE. **Ref.:** Bossard, 1993; Figueiredo & al., 2008; M.F. Pinto Basto, pers. comm., April 2019.

Mamede, Norberto
(fl. 1890–1905)
Note: Likely Norberto Pais Mamede, a pharmacist in Luanda from 1890 to 1905. **Angola:** 1905; Golungo Alto (Cuanza Norte). **Herb.:** COI. **Ref.:** Anonymous, 1905; Figueiredo & al., 2008.

[Manning, Stephen D.]
(fl. 1986–1998)
Note: Botanist. According to JSTOR Global Plants (JSTOR, 2021) he collected in Angola, but no collections have been located. **Herb.:** MO (main), P, WAG. **Ref.:** Figueiredo & al., 2008; Goyder & Gonçalves, 2019.

Marcelino, Fernando Augusto Branco (1931–1992)

Bio.: b. Luso, Portugal, 1931; d. Huambo, Angola, 20 October 1992. Agronomist. He attended school in Huambo, Angola, and later, in 1957, graduated as an agronomist from the *Instituto Superior de Agronomia* in Lisbon. In 1959, he settled in Huambo after having first worked for the *Junta dos Cereais.* He then joined the *Instituto de Investigação Agronómica de Angola* (IIAA), which he eventually directed. In 1975, he had to leave Huambo as a result of escalating hostilities in the Angolan civil war and then helped to organise a camp for refugees. He returned to Huambo in 1976 as the only graduate technician on a staff of several hundred who remained in the institute after the diaspora. He was instrumental in safeguarding the herbarium of the IIAA (Herb. LUA) when its collections were damaged by bombing during the war. He developed research on the breeding and improvement of maize through hybridisation and was a lecturer at the *Faculdade de Ciências Agrárias* in Huambo, which he helped to establish. In 1992, the car in which Marcelino, his wife, sister, and a family friend travelled came under machine gun fire in an ambush; only the family friend survived. Although there are none recorded, it is likely that he made some collections for Herb. LUA. **Angola:** 1959–1992. **Herb.:** LUA. **Ref.:** Daskalos, 2000; Figueiredo & al., 2008.

Marques, Agostinho Sisenando (1847–1923)

Bio.: b. 1847; d. 1923. Pharmacologist. From about 1870 to 1882, he was a government pharmacist with the health services of São Tomé and Príncipe. During that time, he directed the hospital pharmacy and was also in charge of its meteorological station, undertaking and recording all the weather observations. He prepared a collection of natural products to be sent to the Paris World's Fair in 1878 and for an exhibition in Luanda in 1880. He also collected insects for the Univ. of Coimbra. With these interests and experience, he was invited in January 1884 by Henrique Augusto Dias de Carvalho (1843–1909), whom he had previously met in São Tomé, to join an expedition to the Kingdom of Lunda, in Angola. Marques was appointed sub-chief of the expedition, in charge of collecting plants and natural history specimens and minerals and making meteorological observations. On 6 May 1884, Carvalho, Marques, and the army captain and photographer Manuel Sertório de Almeida Aguiar left Lisbon for Angola. They disembarked in Luanda and proceeded to Dondo, Cazengo, Ambaca, Pungo Andongo, and finally Malanje, which they reached on 6 July. In Malanje, they met Wissmann (q.v.) and his group and eventually saw them off on 27 July as the Germans ventured into the interior. Finally, on 11 October, Carvalho's expedition left Malanje towards Lunda. They crossed the Cuango River and by December 1885 were near the Chicapa River, and later the Luachimo River in eastern Lunda. On 8 November 1886, they had reached the Luembe River, east of present-day Saurimo (Lunda Sul), a locality that was known as Henrique de Carvalho for some time. A lack of provisions did not allow the whole expedition team to proceed, resulting in Carvalho deciding to split the group. He sent Marques and Aguiar back to Malanje with a group of soldiers from Ambaca and porters contracted in Luanda. Carvalho continued towards the Musumb (the capital of the kingdom) with a small group of men. On 24 January 1887, Marques and his group arrived at Malanje after a hard journey that lasted 78 days. Carvalho only returned to Malanje on 26 October 1887, after having successfully reached the Musumb in January 1887. Shortly after their return to Lisbon, Carvalho started publishing the eight volumes of the work *Expedição portuguesa ao Muatiânvua 1884–1888*. One of the volumes was authored by Marques; it includes information on the plants encountered with their local names and uses. In addition to these volumes, there is an album of photographs taken by Carvalho and Aguiar that consists of 287 photographs with text, including some of vegetation. Only four copies of this album entitled "Album da expedicao ao Muatianvua" are known to exist, of which one is available online (https://purl.pt/23726). The main set of Marques's collections is deposited at the herbarium of the *Museu Nacional de História Natural e da Ciência* (Herb.

Fig. 77. Agostinho Sisenando Marques. Engraved by Rafael. Public domain.

LISU), with duplicates elsewhere. The collections consist of c. 350 numbers and were studied by Júlio Henriques (1838–1928) at the Univ. of Coimbra, who published a catalogue of the 221 species collected. A catalogue of the collections with information on habitat, habit, uses, and vernacular names is deposited at Herb. LISU: however, the parts referring to numbers 1–53, 67–112, and 124–150 are missing. Marques's name has been spelled "Sizenando" and "Sesinando"; however, in the signature reproduced in his book he used the spelling "Sisenando". He is commemorated in the genus *Marquesia* Gilg (Dipterocarpaceae) and several species names such as *Dissotis sizenandoi* Cogn. (Melastomataceae) and *Plectranthus marquesii* Gürke (Lamiaceae), both endemic to Angola. **Angola:** 1884–1888; Cuanza Norte, Lunda Norte, Malanje. **Herb.:** B, BM, BR, COI, G (fragment ex COI), K (fragment ex B), LISU (main, 345 numbers). **Ref.:** Carvalho & Aguiar, 1887; Marques, 1889; Henriques, 1899; Gossweiler, 1939; Romariz, 1952; Vegter, 1976; Bossard, 1993; Liberato, 1994; Romeiras, 1999; Figueiredo & al., 2008. Figure 77.

Martins, A.F.
(fl. 1973)
Angola: 1973; Malanje. **Herb.:** LISC (1 specimen).

Martins, Emílio Víctor
(fl. 1933)
Bio.: Veterinary doctor who was the director of the *Estação Zootécnica* of Pereira d'Eça (now Ondjiva). **Angola:** 1933; Cunene. **Herb.:** COI, PRE (incl. TRV). **Ref.:** Gossweiler, 1939; Vegter, 1976; Ferrão, 1993; Bossard, 1993; Romeiras, 1999; Figueiredo & al., 2008.

Martins, Eurico Sampaio – see **Sampaio Martins**

Martins, João Vicente
(1917–unknown)
Bio.: b. Moimenta, Vinhais, Portugal, 12 October 1917. Mineral prospector and anthropologist. In 1929, at a young age, he started working in Spain. Later he went to Angola and worked for the mining industry at *Companhia dos Diamantes* in Lunda, as a prospector, then as a curator in the *Museu do Dundo*. While in Angola, he finished secondary education at the Malanje High School and acquired an interest in local languages and ethnology. He returned to Portugal after Angola gained independence in 1975. After graduating in overseas administration at the Technical Univ. of Lisbon, he continued studying there, later graduating at the same university with a B.Sc. in

anthropology (1976), M.Sc. (1986), and finally Ph.D. (1997) with a thesis on the Tchokwe in northeast Angola, when he was 80 years old. He published papers and books on ethnology and collected for the Diamang Collections (q.v.). **Angola:** 1959–1970; Lunda Norte, Lunda Sul. **Herb.:** DIA†?, LISC (133 specimens), LUA, P. **Ref.:** Bossard, 1993; Martins (J.V.), 1993, 2001, 2005; Fonte, 1998; Romeiras, 1999; Figueiredo & al., 2008.

Martins, Teresa Gonçalves
(1958–)
Bio.: b. Malanje, Angola, 24 July 1958. Botanist. She attended school in Malanje but moved to Luanda in 1975 at the onset of the civil war. In 1979, she enrolled at the *Instituto Normal de Educação Garcia Neto*, graduating in 1983 as a teacher. She taught during the next years and then obtained a B.Sc. from the Univ. Agostinho Neto in Luanda in 1995. In that year, she joined the staff of the herbarium of the former *Centro Nacional de Investigação Científica* (Herb. LUAI) in Luanda. She was one of the trainees of the SABONET project and attended biodiversity and environmental courses in South Africa and Zambia. She co-authored the checklist of Poaceae of Angola published through SABONET and a paper on Angolan herbaria. She also trained at the *Centro de Botânica* of the *Instituto de Investigação Científica Tropical*,

Fig. 78. Teresa Martins (seated behind microscope) with Eurico Sampaio Martins. Photographer unknown. South African National Biodiversity Institute. Reproduced with permission.

in Lisbon. She is presently the curator of LUAI. **Angola:** No information recorded. **Herb.:** LUAI. **Ref.:** Anonymous, 2001; Sampaio Martins & Martins, 2002; Costa & al., 2004. Figure 78.

Mason
(fl. 1696)
Bio.: British naval surgeon. He made the first known terrestrial plant collections recorded for Angola. His collections, which were initially held in James Petiver's herbarium, are at the herbarium of the Natural History Museum, London (Herb. BM). The collections consist of c. 36 specimens that Petiver dated as being from 1696, and that were probably collected in the vicinity of present-day Luanda, at the mouth of the Congo River or at the port of Dondo (Cuanza Norte) on the Cuanza River where slave traffic took place. He is not to be confused with Francis Masson (1741–1805), a Scottish botanist and gardener, who collected for the Royal Botanic Gardens, Kew, including at the Cape. **Angola:** 1696. **Herb.:** BM (Sloane Herb.), P. **Ref.:** Exell (A.W.), 1939, 1962; Vegter, 1976; Bossard, 1993; Liberato, 1994; Romeiras, 1999; Figueiredo & al., 2008.

Mason, Leonard Maurice
(1912–1994)
Bio.: b. Fincham, King's Lynn, England, 26 May 1912; d. Fincham, King's Lynn, England, 1994. Farmer and amateur horticulturalist. He collected orchids and grew them and other exotic plants in his glasshouses at Fincham. He developed a garden with ten glasshouses, including several tropical ones. The family-run farms still operate, growing rapeseed, wheat, and sugar beet. He travelled extensively and collected material from various continents. In Africa, he collected in Madagascar, Ethiopia, and Angola. His orchid collection was donated to the Royal Horticultural Society, London, with some specimens going to the Royal Botanic Gardens, Kew. He is commemorated in *Begonia masoniana* Irmsch. ex Ziesenh. (Begoniaceae), which he had imported into England in 1952 and named "Iron Cross" not knowing if it was a species or hybrid. The species was later discovered in the wild, in China. **Angola:** No information recorded. **Herb.:** E, K. **Ref.:** Dorr, 1997; Gu & al., 2007; Figueiredo & al., 2008; Buchan, 2013.

Matos, Agostinho
(fl. 1940)
Angola: 1940; Huambo (Calombo). **Herb.:** LUA. **Ref.:** Bossard, 1993; Figueiredo & al., 2008.

Matos, António Oliveira de
(fl. 1940s)
Note: Listed by Bossard (1993) as one of those individuals who either collected specimens or provided information on plants from Angola, and as being active in Luanda and Malanje in the 1940s. **Angola:** No collections have been recorded so far. **Ref.:** Bossard, 1993; Figueiredo & al., 2008.

Matos, Elizabeth ("Liz") Merle
(1938–)
Bio.: b. London, England, 1938. Biologist. She graduated with a B.Sc. from Birkbeck College in London, in 1964, and later with an M.Sc. from the Univ. of Birmingham. She worked as a lecturer at the Univ. of Havana, Cuba, until 1970. In 1972, she joined the Univ. of Zambia, Lusaka, as a lecturer and in 1975 moved to Angola to be a lecturer at Univ. Agostinho Neto in Luanda. From 1993 to 1997, she was head of the genetics section in the biology department of the university. She founded and directed the Angolan National Plant Genetic Resources Centre. She held positions in several committees, retiring in 2008. She is known for the part she played in preserving the herbarium of the *Instituto de Investigação Agronómica*, in Huambo (Herb. LUA), during

Fig. 79. Elizabeth Matos. Photographer unknown. South African National Biodiversity Institute. Reproduced with permission.

the civil war, arranging for the collections to be flown on a military aircraft from Huambo to Luanda for safekeeping at the herbarium of the former *Centro Nacional de Investigação Científica* (Herb. LUAI). In the 1990s, Matos co-authored a review of the botanical diversity and its conservation in Angola. **Angola:** 1990–1995; collected with Alcochete (q.v.), Gerrard Reis (q.v.) and Newman (q.v.). **Herb.:** LUAI. **Ref.:** Huntley & Matos, 1994; Anonymous, 1997; Goyder & Gonçalves, 2019. Figure 79.

Matos, Gilberto Cardoso de – see **Cardoso de Matos**

Mazzocchi-Alemanni, Nallo
(1889–1967)
Bio.: b. Todi, Perugia, Italy, 4 May 1889; d. Todi, Perugia, Italy, 6 May 1967. Agronomist, rural economist, and planner. In 1908, he registered for studies at the *Regio Istituto Superiore Agrario* in Perugia, Italy. From August 1911 to June 1912, he travelled in German and British East African territories, Italian territories, and in Egypt and Somalia, to conduct agricultural experiments. In 1913, he visited Tripolitania (now Libya) and Tunis (Tunisia). He was appointed director of the *Istituto agricolo coloniale italiano* in Florence in 1920 and in 1923 travelled to Angola on a mission to investigate the suitability of collaboration between Portugal and Italy, and to promote Italian immigration to the Benguela plateau (now the Bié plateau). In his report to the Italian government, he suggested the creation of Portuguese-Italian societies to establish agricultural colonies and commercial enterprises. In 1930, he was in charge of agriculture at the Italian National Institute of Statistics. From 1939 to 1943, during World War II, he was based in Palermo. He then returned to Rome, but on learning of the invasion of Sicily by Allied forces in July 1943, and Benito Mussolini's fall in the same month, he distanced himself from the government and retired to his estate in Todi until the end of the war. After the war, he resumed work on land and agricultural reform in Italy. **Angola:** 1923. **Herb.:** FI, FT, K. **Ref.:** Chiovenda, 1924, 1925; Mazzocchi-Alemanni, 1924; Vegter, 1976; Bossard, 1993; Romeiras, 1999; Settesoldi & al., 2005; Figueiredo & al., 2008; Misiani, 2017.

Mechow, Friedrich Wilhelm Alexander von
(1831–1904)
Bio.: b. Lauban, Silesia (then a province of Prussia), Poland, 9 December 1831; d. Jugenheim/Bergstraße, Germany, 14 March 1904. Topographer and explorer. He had an active military career serving in the Prussian Campaign of 1866 and later, in 1870, in the Franco-Prussian war, when he was wounded

in action. He retired at the rank of major and settled in Berlin. In 1874, he was contracted by the *Deutsche Gesellschaft zur Erforschung Aequatorial-Afrikas* to join the expedition of Güssfeldt (q.v.) to the Kingdom of Loango (Rep. of the Congo and Cabinda, Angola). From January to November 1875, Mechow was in Cabinda, but he left the expedition apparently as a result of conflict with the rest of the team. Afterwards, Mechow was selected to lead an expedition to Cuango. His team included Teusz (q.v.) and also a naval carpenter, Bugslag, as they were carrying a steamboat that could be disassembled into several parts for ease of transportation. Mechow left Germany on 19 September 1878, arriving in Angola two months later and proceeded to the interior. On 23 December, he was in Dondo where he met Buchner (q.v.). On 26 January 1879, the two met again in Pungo Andongo where Mechow was stationed for several months. Mechow only arrived in Malanje on 25 June. About a year later, on 12 June 1880 the expedition started off with 115 porters. After many difficulties, including with transporting the steamboat, they arrived at the Cuango River south of its confluence with the Cambo River on 19 July 1880. On 25 August, they started exploring the Cuango and on 7 September they crossed to its right bank (D.R. Congo) and travelled to the Musumb (capital) near the Ganga River. Mechow was received by the Mwat Yamv (in Portuguese *Muatiamvo, Muatiânvua*, i.e., the king), called Muene Puto Kassongo, and proposed to the king that some members of the expedition could be left at the Musumb as a guarantee of his return. With this having been agreed, Mechow returned to the camp and, with Teusz, travelled

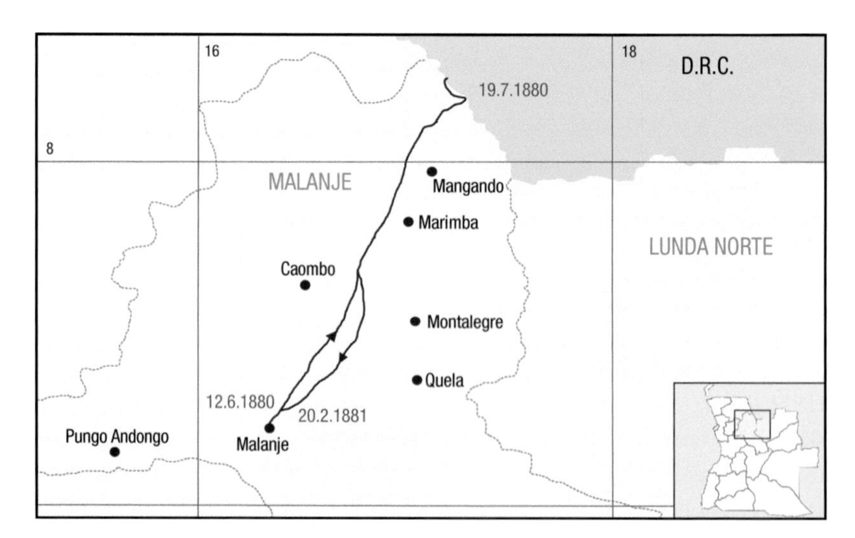

Fig. 80. The itinerary of Friedrich von Mechow in Angola.

again to the Musumb, where Teusz remained. Finally, on 20 September 1880, the expedition proceeded. Mechow, Bugslag, and 19 porters departed and travelled by boat and one canoe downstream on the Cuango River. They went as far north as c. 05°05′S and then turned back taking a route overland along the banks of the Cuango and the Cambo rivers. After being reunited with Teusz, they returned to Malanje on foot along the banks of the Cambo and Cuango rivers on a more easterly route than the one taken on the outbound journey. At the end of February 1881, Mechow was back at Malanje where he met Pogge (q.v.), Wissmann (q.v.), and Buchner. He returned to Luanda in April and from there embarked on the return trip to Germany, travelling with Lux (q.v.). He arrived in Germany in August 1881. He published a detailed map of the region in 25 parts, and his meteorological observations were the most complete at the time. Afterwards, he lived at Marksburg by the Rhine and later at Jugenheim where he died. In 1861, he and his brother received a nobility title, taking then the nobiliary particle "von". He is commemorated in numerous plant names of which the types were collected during his expedition, likely by Teusz. It appears that Mechow did not collect in Cabinda in 1875. Most material recorded as having been collected by Mechow from 1878 to 1881 may have been collected by Teusz, who was the natural history collector of the expedition. The collecting numbers of the expedition go up to at least number 579. The numbering only partially follows a chronological order. Many specimens were likely destroyed at B, but duplicates exist in several herbaria. **Angola:** 6 November 1878–April 1881; Cuanza Norte, Luanda, Lunda Norte, Malanje (Pungo Andongo and Malanje up to around collecting number 500; Tembo-Aluma [Mangango] Calala Canginga [near Sunginge]); the collections likely made by Teusz. **Herb.:** B (main), BR, C, G, GH, GOET, JE, K, L, M, P, PH, S, US, W, WU. **Ref.:** Kiepert, 1882; Mechow, 1882, 1884; Urban, 1916; Mendonça, 1962a; Vegter, 1976; Bossard, 1993; Liberato, 1994; Romeiras, 1999; Figueiredo & al., 2008, 2020; Heintze, 2010, 2018;. Figure 80.

Meeuse, Adriaan Dirk Jacob
(1914–2010)

Bio.: b. Sukabumi, Java, Indonesia, 18 October 1914; d. Netherlands, 15 September 2010. Botanist. Attended school in Bogor and Jakarta in Indonesia and later in The Hague, after his family moved to the Netherlands in 1931. He graduated from the Univ. of Leiden with a B.Sc. and M.Sc. in 1938. That same year he was invited by his professor, Lam (q.v.), to join a collecting trip to Namibia, South Africa, the Mascarenes, and Madagascar. On his return, he started working on his dissertation, but this was interrupted when he was mobilised during World War II. He obtained his Ph.D. in biology at the Univ.

of Leiden in 1941. In 1942, he was appointed scientific officer at the Fiber Research Institute in Delft, where he worked until 1952 when he moved to South Africa to take a position at the Botanical Research Institute and National Herbarium, Pretoria. He contributed to several African Floras, monographing several families. Returning to the Netherlands in 1960, he was appointed as professor of botany at the Univ. of Amsterdam and director of the botanical garden, positions he held until 1984. Thereafter, he was professor emeritus. He produced c. 200 publications including several books on the evolution of angiosperms. He was an eclectic collector and is commemorated in the names of animals, plants, and fossils. His botanical collection amounts to c. 6000 numbers. The material collected with Lam in Angola in 1938 is kept at the *Nationaal Herbarium Nederland* (Herb. L). **Angola:** 1938; Benguela; with Lam (q.v.); at Herb. L, duplicates have been located at Herb. BR and WAG. **Herb.:** BM, BO, BR, COI, DAO, EA, G, K, L (main), M, MO, P, PRE (main), S, SRGH, WAG. **Ref.:** Vegter, 1976; Gunn & Codd, 1981; Dorr, 1997; Meeuse, 2005; Figueiredo & al., 2008.

Mendes, Eduardo José dos Santos Moreira (1924–2011)

Bio.: b. Lisbon, Portugal, 26 November 1924; d. Lisbon, Portugal, 24 September 2011. Botanist. He was educated in Lisbon and in 1946 graduated from the Univ. of Lisbon in biological sciences. He remained there as a lecturer until 1956, developing research on cryptogams. In 1954, he was contracted as an assistant for the project *Missão Botânica de Angola e Moçambique* and moved into a different area of study, the African vascular plant flora. After participating in a six-month expedition to Angola led by Mendonça (q.v.), he joined the *Junta de Investigações do Ultramar* (later named the *Instituto de Investigação Científica Tropical*) in Lisbon in 1956. He was appointed as a researcher at the *Centro de Botânica*, a unit that he eventually directed from 1974 to 1986. He worked mostly on the floras of Angola and Mozambique. His revisions of the Angolan flora were published as papers in a series under the title *Additiones et adnotationes florae Angolensis*. He was editor or co-editor of several Floras, including *Conspectus florae Angolensis*. Mendes participated in two expeditions to Angola, the *Campanhas de Angola* (q.v.) in 1955/1956 and 1959/1960. His field books of these two expeditions are deposited at the Univ. of Lisbon and digitised at JSTOR Global Plants (JSTOR, 2021). He is commemorated in several species names, such as *Aloe mendesii* Reynolds (Asphodelaceae) and *Euphorbia eduardoi* L.C.Leach (Euphorbiaceae). **Angola:** 1955/1956 and 1959/1960; Bengo, Benguela, Bié, Cuando Cubango, Cuanza Sul, Cunene, Huambo, Huíla, Luanda, Namibe.

Fig. 81. Eduardo Mendes in 2010. Photographer and private collection of Gideon F. Smith.

Herb.: BM, BR, C, COI, DAO, L, LISC (main; 3406 specimens), LISU, LUA, LUAI, LUBA, M, MO, PRE, SRGH, US, WAG, WIND, WU. **Ref.:** Mendonça, 1962a; Vegter, 1976; Gunn & Codd, 1981; Bossard, 1993; Liberato, 1994; Sampaio Martins, 1994b; Romeiras, 1999; Figueiredo & al., 2008; Sousa & al., 2010; Smith & Figueiredo, 2011; Smith & al., 2012. Figure 81.

Mendonça, António Leonel
(fl. 1962)
Angola: 1962, Huambo. **Herb.:** LISC (2 specimens); LUA.

Mendonça, Estevão Barroso – **see Barroso Mendonça**

Mendonça, Francisco de Ascensão
(1899–1982)
Bio.: b. Conceição, Faro, Portugal, 30 May 1899; d. Belém, Lisbon, Portugal, 28 September 1982. Botanist. He studied in Lisbon, but only completed his degree much later at the Univ. of Coimbra in 1926. He was drafted into the army during World War I, but fortunately missed the train that took the other ill-fated troops to the front in France. For a short time, he taught in a secondary school. In 1921, he accepted a position as chief gardener of the *Instituto Botânico* at the Univ. of Coimbra. In 1927, he took part in the first expedition of the Univ. of Coimbra to Angola (*Missão Botânica I*, q.v.) with

Carrisso (q.v.). They collected 657 numbers. He held the position of gardener at the *Instituto Botânico* until 1929, when he became a naturalist in the same institution. It was in that position, in 1934, that he met A.W. Exell (q.v.), who was then visiting Coimbra to study material from São Tomé and Príncipe. As a result of discussions between Exell and Carrisso, the project *Conspectus florae Angolensis* was initiated. Shortly afterwards, Mendonça travelled to London to work with Exell on some revisions aimed for inclusion in the *Conspectus*. The first part of volume 1 of the *Conspectus florae Angolensis* was published in 1937; however, the second part only appeared in 1951 as a result of interruptions caused by World War II. Together with Exell, Carrisso, and Gossweiler (q.v.), Mendonça took part in Carrisso's second botanical expedition (*Missão Botânica II*, q.v.). During that expedition, he collected a total of 3217 numbers with Exell. With the experience thus gained on the African flora, Mendonça became the head of the *Missão Botânica de Moçambique* from 1942 to 1948. After World War II, Mendonça and Exell resumed their collaboration, and with Wild (q.v.) they led the intergovernmental negotiations for the establishment of the *Flora Zambesiaca* project. In 1952, Mendonça was appointed director of the new *Centro de Botânica* (Herb. LISC) in Lisbon. From February to April 1955, he travelled with Exell and Wild to Zambia, Zimbabwe, Malawi, and Mozambique, collecting and preparing for the *Flora Zambesiaca* project. In the same year, from 7 September 1955 to 6 November 1955, he led an expedition to Angola (*Campanhas de Angola*, q.v.) with Mendes (q.v.), for its initial months; during that expedition he collected 160 numbers. He retired in 1963. He was much admired for his

Fig. 82. Francisco Mendonça (right) with Arthur Exell. Photographer unknown. From Centro de Botânica/Instituto de Investigação Científica Tropical (IICT), Universidade de Lisboa. Reproduced with permission.

observational capacities and botanical instincts and considered a good field naturalist and superb collector. He is commemorated in 23 plant names. **Angola:** 1927, 1937, 1955. **Herb.:** B, BM, BR, COI (main), K, LISC (main; 163 specimens), LISU, LMA, LUA, M, MO, NY, P, SRGH. **Ref.:** Exell, 1960; Mendonça, 1962a, 1962b; Gomes e Sousa, 1971b; Quintanilha, 1975; Vegter, 1976; Stafleu & Cowan, 1981: 413; Exell, 1984; Bossard, 1993; Liberato, 1994; Romeiras, 1999; Figueiredo & al., 2008, 2018a; Polhill & Polhill, 2015;. Figure 82; see also Figures 31 and 133.

[Mendonça, S.A.]
Note: Listed by Figueiredo & al. (2008). Likely an error from a database and referring to F.A. Mendonça (q.v.).

Menezes, Óscar Jacob Azancot de
(1924–1994)
Bio.: b. São Tomé and Príncipe, 9 August 1924; d. Lisbon, Portugal, 11 November 1994. Agronomist. He attended school in Chibia and Lubango, in the Huíla Province of Angola, and high school in Lisbon. In 1943, he enrolled at the *Instituto Superior de Agronomia* in Lisbon and graduated as an agronomist in 1948. From 1950 to 1957, he worked at the then *Jardim e Museu Botânico do Ultramar* in Lisbon. He then returned to Angola to assume a managerial position in a commercial company in Lobito (Benguela). In 1960, he left that job to work for a while at the *Instituto de Investigação Científica de Angola* (IICA) and at the *Junta de Povoamento Agrário do Vale do Bengo*. Shortly afterwards, in 1962, he joined the IICA and two years later became the head of the botany section (*Secção de Fitogeografia e Botânica Sistemática*) of the *Centro de Estudos* of the IICA in Lubango. Later he became head of the *Centro de Estudos*. In 1974, he was appointed to the provisional government of Angola as a *Subsecretário de Estado de Fomento Agrário* (Undersecretary of State for Agricultural Development), serving until 1975. After the independence of Angola in 1975, he was in charge of the *Centro Nacional de Investigação Científica* (successor of the IICA), and from 1977 onwards he was also professor at the Univ. Agostinho Neto, lecturing in Huambo and in Luanda. From 1981 to 1985, he held a directorship position at the Ministry of Agriculture in Angola. He collected over 5000 numbers, authored several papers, and described a few new species of Cyperaceae. He is commemorated in *Terminalia menezesii* Mendes & Exell (Combretaceae). He also appears as "Meneses", for example as the author of plant names and in some publications. **Angola:** 1950–1990; Bengo, Benguela, Bié, Cabinda, Cuando Cubango, Cuanza Norte, Cuanza Sul, Cunene, Huambo, Huíla,

Malanje, Namibe; collected also with Barroso Mendonça (q.v.), Brites (q.v.), (J.C.) Henriques (q.v.), (R.) Santos (q.v.) and (J.A.) Sousa (q.v.). **Herb.:** BM, COI, K, LISC (4133 specimens), LISI?, LISU, LUA, LUAI, LUBA, P, PRE. **Ref.:** Meneses, 1956; Bossard, 1993; Liberato, 1994; Romeiras, 1999; Figueiredo & al., 2008, 2018a; Glen & Germishuizen, 2010; Anonymous, 2018a.

Mercier
(fl. 1949–1950s)
Bio.: Missionary. In the literature and on the collecting labels attached to herbarium specimens referred to as "Pe Mercier". He was a Catholic missionary who made some plant collections at *Missão de Capico* in Cuando Cubango. In the 1950s, he also collected ethnographic objects in the Capico region; he offered these to the *Musée de l'homme*, in Paris. He also collected for Hess (q.v.). **Angola:** 1949–1954; Bié, Cuando Cubango (Missão de Capico). **Herb.:** LISC (1 specimen), LUA, Z+ZT. **Ref.:** Bossard, 1993; Romeiras, 1999; Figueiredo & al., 2008.

Milne-Redhead, Edgar Wolston Bertram Handsley
(1906–1996)
Bio.: b. Frome, Somerset, England, 24 May 1906; d. Crouched Friars, Colchester, England, 29 June 1996. Botanist. In 1925, he attended Gonville and Caius College, Cambridge, reading natural sciences but instead of taking the tripos (examinations to qualify for a degree) he accepted an unpaid position at the Royal Botanic Gardens, Kew. In 1929, he was employed as a temporary subassistant in the herbarium (Herb. K). The next year, he was appointed to work on a project on aerial surveys of Zambia and spent four-and-a-half months collecting in the Mwinilunga District of that country. In 1935, he became a member of the Kew staff and soon head of the Tropical African Section, a position he held until 1959. In 1937, he returned to Zambia, again for c. four-and-a-half months, and also visited Angola. A total of c. 2150 collections was made during that expedition. He had been commissioned as a second lieutenant in the Territorial Army in 1929 and served for ten years with the 30th (Surrey) Searchlight Battalion, Royal Engineers. When World War II started, he was called up to serve and in 1940 became a gunner, rising to the rank of captain and temporary major. The war did not stop him from pursuing his interest in natural history, as even while stationed in Nigeria, Sierra Leone, and Ghana he collected plants and recorded some observations. He returned to England in early 1942 and became a sector searchlight control officer working on night interception of German bombers near Houghton Regis. There, likewise, he explored the flora and fauna of the area during

daylight hours. He returned to Kew after the war. Until his retirement in 1971, he was the main editor of the *Flora of West Tropical Africa*. He was also a founding member of the *Association pour l'Étude Taxonomique de la Flore d'Afrique Tropicale* (AETFAT) in 1950 with A.W. Exell (q.v.) and the Belgian botanist Jean Joseph Gustave Léonard (1920–2013). In 1959, Milne-Redhead became deputy keeper of the herbarium and editor of *Kew Bulletin*. He was deeply involved in various aspects of conservation and continued in this field after his retirement. He then also became involved in projects on the British flora. He is commemorated in 25 species names. During the 1937–1938 expedition, Milne-Redhead was in Lobito (Benguela, Angola) on 28 September 1937 where he made some collections. A few days later, on 3 October, he was in Zambia where he collected until 7 January 1938. That same day he crossed the border to Angola and again collected there until 20 January 1938. **Angola:** September 1937 (Benguela), January 1938 (Moxico). **Herb.:** B, BM, BR, CGE, DBN, K (main), LISC (27 specimens), MO, NMW, OXF, P, PRE, SRGH, US. **Ref.:** Exell, 1960; Vegter, 1976; Stafleu & Cowan, 1981: 505; Bossard, 1993; Verdcourt, 1998; Romeiras, 1999; Figueiredo & al., 2008; Glen & Germishuizen, 2010; Polhill & Polhill, 2015.

Missão Botânica I
(1 June–November 1927)
The *Missão Botânica da Universidade de Coimbra a Angola* that took place in 1927 was a 5-month expedition undertaken by Carrisso (q.v.), as leader, and F.A. Mendonça (q.v.). They embarked from Lisbon on 1 June 1927 and arrived in Luanda on 16 June. While at Luanda, they undertook some excursions in the surrounding area to Catete, Bengo Valley, and mangroves to the south of the town of Luanda. Then they travelled east by railway to Ndalatando (Cuanza Norte). They visited the agricultural station of Cazengo, and Dondo and Golungo Alto. Then, travelling by car, they proceeded to Pungo Andongo, Lucala, and Malanje (Malanje). The journey continued through Quela, then in a southeasterly direction to Cambundi and east to Xassengue (Lunda Norte), and Saurimo where they were based for a while. Up to that time, 500 specimens had been collected and 150 photos taken. From Saurimo, they undertook excursions to Dundo, Nordeste, Dala, Meconda, and Luma-Cassai. While in the Lunda region, they collected 2000 specimens and took 400 photographs. On 24 September, they left the region and returned to the coast via Luena (Moxico), Cuíto (Bié), Huambo (Huambo), Lubango (Huíla), and finally Moçâmedes (Namibe) from where they undertook excursions into the Namib desert up to Ponta Negra. They left Moçâmedes on a steamboat, headed to Luanda, stopping at Benguela and Lobito (Benguela). Back in Luanda, they

undertook a final excursion to the Zaire Province, visiting Ambriz, Soyo, and Noqui, and, in the D.R. Congo, to Boma, Matadi, and Kisantu. On the way back they visited Cabinda. They departed from Luanda on 28 November 1927. The main set of the collections made during *Missão Botânica I* is held at the herbarium of the Univ. of Coimbra (Herb. COI) and consists of 657 numbers. The photographs, some of which have been published, are deposited in the archives of the Univ. of Coimbra. Transcripts of the collections recorded during *Missão Botânica I* are at the Univ. of Lisbon and digitised at JSTOR Global Plants (JSTOR, 2021). **Herb.:** COI. **Ref.:** Carrisso, 1930, 1932; Paiva, 2005; Bernaschina & Ramires, 2007. Figure 83.

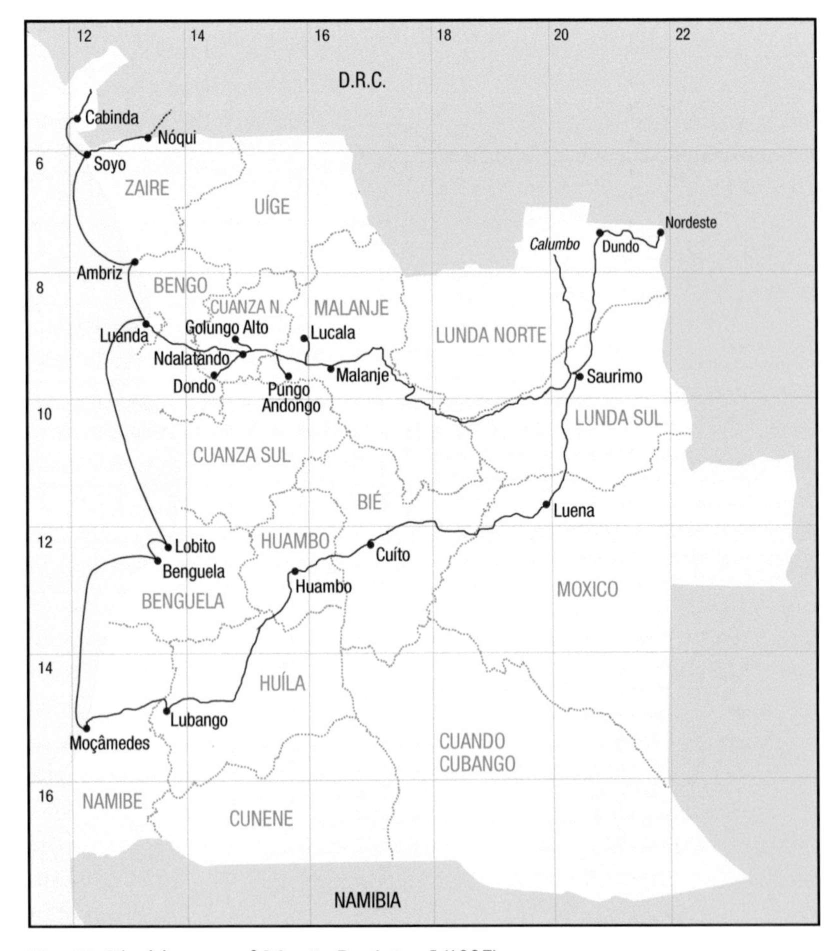

Fig. 83. The itinerary of *Missão Botânica I* (1927).

Missão Botânica II
(February–17 July 1937)
During the ten years after his initial two-man expedition to Angola in 1927 (*Missão Botânica I*, q.v.), Carrisso (q.v.) planned and secured funding for an ambitious new, six-month venture scheduled to take place in 1937. This new expedition included as main team: Carrisso as leader; F.A. Mendonça (q.v.), F. Sousa (q.v.), and an assistant, Manuel Jara de Carvalho, from the Univ. of Coimbra, Portugal; A.W. Exell (q.v.) from the British Museum, England; Gossweiler (q.v.) from Angola; and two accompanying wives, Ana Maria Carrisso (1892–1980) and M. Exell (q.v.). Additionally, numerous African men were contracted in Angola. The expedition started in Luanda in March 1937 and travelled to Malanje (Malanje) by railway and then by road, in one car, one van, and three lorries. At Saurimo (Lunda Sul) the group split. Carrisso undertook official visits to Dundo (Lunda Norte) with his group including Sousa, Carvalho, and A. Carrisso. From Saurimo, the expedition returned to the coast through the provinces of Huambo and Huíla. On their way from Huambo to Cuima, a car accident resulted in a further split of the team. Carrisso and his group returned to Huambo, whereas Mendonça, Gossweiler, A.W. Exell, and M. Exell continued to Lubango (Huíla). There, Mendonça and his group met Moreno Júnior (q.v.), who was then in command of the local barracks. They continued to Moçâmedes (Namibe) and travelled as far south as Tômbua. They left Moçâmedes following a little-used road in a southeasterly direction, to the Tampa River, Cainde, and then to Lubango. From there, they travelled south to the border with Namibia, returning to Lubango on 14 June. Carrisso and his group travelled in Huíla, then followed to Moçâmedes, and travelled in the provinces of Namibe and Cunene. On 14 June, Carrisso died in the Namibe desert. When Mendonça's group received the tragic news, they travelled to Moçâmedes to attend the funeral. Afterwards, Mendonça became the leader of the expedition. Exell and Gossweiler returned to Lubango to rejoin M. Exell and from there they travelled to Luanda via Huambo, Quibala (Cuanza Sul), and Dondo (Cuanza Norte). In Luanda, after the Portuguese team left for Lisbon on 8 July, A.W. Exell, Gossweiler, and M. Exell undertook a trip to Cuanza Sul, passing Gabela, Amboim, and Porto Amboim. They also undertook excursions to Cazengo, Golungo Alto, and Muxaula in Cuanza Norte. Overall, the expedition was very successful and resulted in a collection of 5030 numbers and about 25,000 specimens. Rather confusingly, the collections were made under three different series: (1) A.W. Exell & Mendonça; (2) Carrisso & Sousa; and (3) Gossweiler, regardless of the collectors who were present when the gatherings were made. This division was in effect since the first collecting day, 14 March 1937. The Carrisso & Sousa collection consists

of 354 numbers collected up to 14 June. The A.W. Exell & Mendonça collection includes 3217 numbers; these comprise collections made by M. Exell alone, and collections made by A.W. Exell, M. Exell and Gossweiler (but not Mendonça) after 14 June until the Exell couple left Luanda. Gossweiler collected 1416 numbers during the expedition. On the whole, the expedition resulted in collections from most provinces, with the exception of Cabinda, Cuando Cubango, Uíge, and Zaire. Transcripts of the collections recorded during *Missão Botânica II* are at the Univ. of Lisbon and digitised at JSTOR Global Plants (JSTOR, 2021). **Herb.:** BM, COI, LISC. **Ref.:** Exell (A.W,), 1938a,b, 1939; Paiva, 2005; Bernaschina & Ramires, 2007. Figures 84 and 85.

Itinerary of the Missão Botânica in 1937 **(based on Exell, 1937b)**

February – Arthur Exell (AE) and Mildred Exell (ME) depart from London, England; Luis Carrisso (LC), Ana Carrisso (AC), Francisco Mendonça (FM), Francisco Sousa (FS), and Jara de Carvalho (JC) depart from Lisbon, Portugal.

20 March – AE and ME arrive at Luanda, Angola, and meet LC, FM, FS, JC, AC, and John Gossweiler (JG).

Fig. 84. *Missão Botânica* 1937. **A,** Calandula Falls (Malanje); **B,** Xassengue (Lunda Sul); “Solo performances were given by the witch-doctors in their fantastic grass garments, while the women acted as a kind of chorus” (Exell, 1938b); **C,** Desert outside Moçâmedes (Namibe) after the rains; **D,** Ruacaná Falls (Cunene). — Photographer unknown. Private collection of Estrela Figueiredo.

26 March – Luanda to Malanje by train. While at Malanje: excursions to Lucala (28 March), Condo Falls of the Cuanza River (29 March), and Pungo Andongo (31 March).
1 April – Malanje to Cambundi.
2 April – Cambundi to Xassengue.
10 April – Xassengue to Cacolo.
11 April – Cacolo to Saurimo.
16 April – LC, FS, JC, and AC travel to Dundo. AE, FM, JG and ME remain in Saurimo.

AE, FM, JG, and ME:

16 April – Excursion to Chicapa River.
17 April – Excursion to Luachimo River.
19 April – AE, FM, JG travel to Muriege.
20 April – AE, FM, JG return to Saurimo.

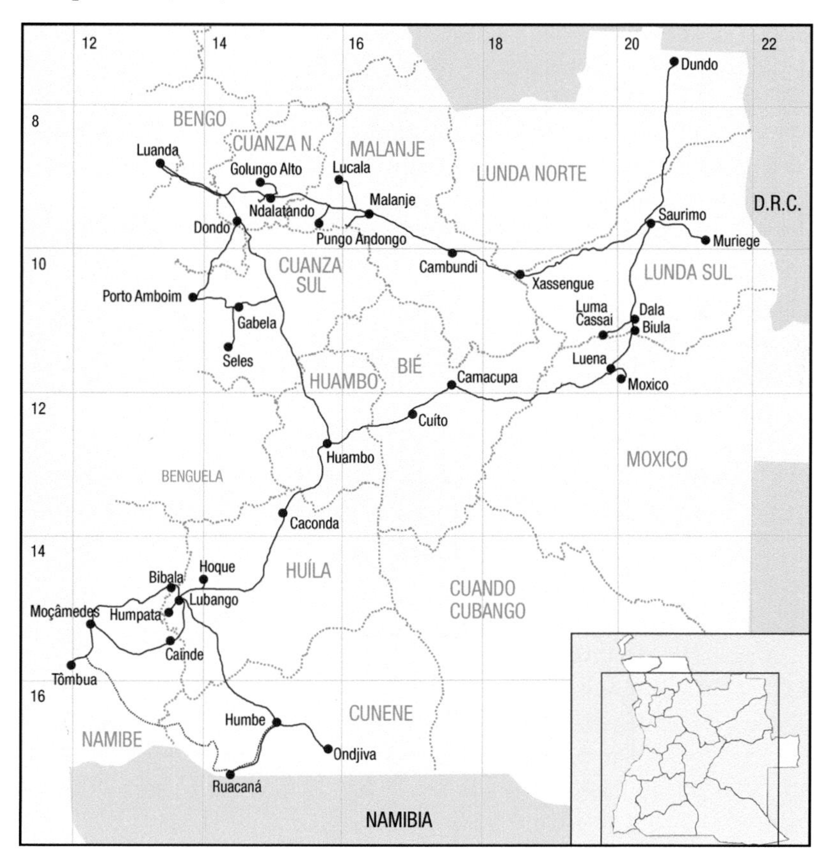

Fig. 85. The itinerary of *Missão Botânica II* (1937).

LC, AC, FS, and JC:
16–20(?) April – Dundo.
21 April – Return to Saurimo.
21 April – The team reconvenes.
24 April – Saurimo to Dala.
26 April – Dala to Biula. ME remains there; others continue to Luma Cassai.
28 April – Return to Biula and all depart for Dala.
1 May – Dala to Luena.
FM and JG:
5 May – Depart with the lorries to Huambo.
9 May – Arrival at Huambo.
LC, AE, ME, and AC:
6 May – At Luena. Excursion to Moxico.
7 May – Overnight train to Camacupa.
8 May – Arrival at Camacupa; by car to Cuíto and Huambo.
9 May – The team reconvenes.
12 May – Huambo to Cuima; car accident; AE, FM, JG, and ME continue to Lubango, LC, FS, JC, and AC return to Huambo.
AE, FM, JG, and ME:
14 May – Cuima to Lubango via Caconda.
17 May – Lubango to Bibala.
18 May – Bibala to Moçâmedes.
20 May – Arrival at Moçâmedes.
22 May – Moçâmedes to Tômbua.
26 May – Tômbua to Moçâmedes.
LC, FS, JC, and AC:
12 May – Huambo.
22–23 May – Collect at Caconda, Quipungo, and between Quipungo and Capelongo, while en route to Lubango.
24 May – At Hoque.
27 May – Arrival at Moçâmedes.
27 May – The team reconvenes.
LC, FS, JC, and AC:
27 May – 14 June – Excursions in Namibe and Cunene as far as Humbe.
14 June – LC dies in the Namibe desert.
AE, FM, JG, and ME:
28 May – Leave Moçâmedes towards Lubango.
29 May – Pass Bero River.
31 May – Arrival at Tampa River region.

2 June – Tampa to Ungueria and Lubango.
6 June – AE, FM, JG, and Moreno Júnior travel to Cunene and Ruacaná Falls.
14 June – AE, FM, and JG return to Lubango; travel to Moçâmedes. ME remains in Lubango.
18 June – AE, FM, and JG return to Lubango and meet ME; others remain in Moçâmedes.

AE, FM, JG, and ME:

22 June – Lubango to Huambo.
23 June – Trip to Chicala.
24 June – Huambo to Quibala.
25 June – Quibala to Dondo.
26 June – Dondo to Luanda.
7 July – AC arrives in Luanda after sea journey from Moçâmedes; the team reconvenes.
8 July – FM, FS, JC, and AC depart from Luanda for Lisbon; AE, JG, and ME travel to Dondo.

AE, JG, and ME:

9 July – Dondo to Gabela.
13 July – Gabela to Porto Amboim to Dondo.
14 July – Dondo to Ndalatando.
15 July – Ndalatando to Luanda.
17 July – AE and ME depart from Luanda and return to Europe.

Mission Rohan-Chabot
(1912–1914)

French expedition under the auspices of the *Ministère de l'Instruction Publique* and the *Société de Géographie de Paris*, and led by the 23-year-old Count Jacques de Rohan-Chabot (1889–1958). The expedition lasted 20 months, from 1912 to 1914, during which time southern Angola and western Zambia were explored. Sailing from Lisbon on 1 April 1912, the expeditionary team arrived at Moçâmedes, Angola, twenty days later. On 28 April, Rohan-Chabot took the train from Moçâmedes to the end of the line, 169 km distant. From there he travelled on horseback to Chibia (Huíla) to procure transportation and guides. He was later met by the other expedition participants, and at the beginning of July 1912, they were at Lubango where the group was stationed for a while. Rohan wrote from Huíla to his father on 2 August, and at the beginning of September 1912, he telegraphed from Cahama (Cunene) on the Caculovar River. From there, the expedition went south to the Cunene Falls (Ruacaná) and then up north through

Humbe to Capelongo (Huíla). In December 1912, they were at Menongue (Cuando Cubango) where they remained until the peak of the rainy season had passed; later they continued eastwards. In January 1913, the expedition was at Cutato and thereafter explored the course of several rivers in Cuando Cubango, going as far east as Lealui and Livingstone in Zambia. The group returned to France in 1914. The expedition focused on geodesy, geography, and ethnology. Only a few plant collections were made, with these consisting mostly of ferns. The itinerary was published by Rohan-Chabot (1914). World War I started that year, and the publication of the results of the expedition was delayed. Rohan-Chabot served in the war and was decorated for his service. After the war, from 1921 to 1937, five volumes (tomes) of results were published: t. 2 (1930), t. 3, fasc. 1 (1925), t. 4, fasc. 1 (1924), t. 4, fasc. 3 (1928) ["1925"], and t. 5 (1937). The ferns were treated in t. 4, fasc. 3. Rohan-Chabot later became director general of the Red Cross, and during World War II was arrested by the Germans in 1944, but later released. **Angola:** 1912–1914; Cuando Cubango, Cunene. **Herb.:** K, P, PC, US. **Ref.:** Froidevaux, 1912; Hulot, 1912; Rohan-Chabot, 1914; Mission Rohan-Chabot, 1928; Bossard, 1993; Romeiras, 1999; Figueiredo & al., 2008; Le Crom, 2009.

Mocquerys, Albert
(1860–1926)

Bio.: b. either Rouen or Évreux in Normandy, France, 1860; d. likely Tunisia, 1926. A commercial natural history collector who was active in several parts of the world, including in West Africa from 1889 to 1891 and in Madagascar from 1897 to 1898. In 1899 and 1900, he made some plant collections in São Tomé and Príncipe and in Angola. **Angola:** 1899; Luanda; a single plant collection from Angola is known. **Herb.:** P. **Ref.:** Vegter, 1976; Dorr, 1997; Dorr & al., 2017.

Monard, Albert
(1886–1952)

Bio.: b. Les Ponts de Martel, Switzerland, 2 September 1886; d. La Chaux-de-Fonds, Switzerland, 27 September 1952. Zoologist. He started his career as a primary school teacher. In 1919, he obtained a doctorate in natural sciences from the Univ. of Neuchâtel. From 1921 to 1952, he was curator at the *Musée d'histoire naturelle* in La Chaux-de-Fonds. Between 1928 and 1947, he undertook four expeditions to Africa, with two of these having been to Angola. The first (July 1928–February 1929) was a hunting expedition led by the medical doctor W. Hertig. It resulted in numerous zoological collections. The second expedition, the *IIème Mission scientifique suisse en Angola*, took place from April 1932 to October 1933, with Monard as the leader, and Théodore Delachaux (1879–1949) and Thiébaud (q.v.) participating. It was only during this second Angolan expedition that plant collections were made, with c. 125 specimens being deposited at Herb. Z+ZT. Afterwards, Monard published several papers on the zoological material collected during the expeditions. **Angola:** April 1932–October 1933; Cunene, Huambo, Huíla, Lunda Sul; collected also with Thiébaud (q.v.). **Herb.:** Z+ZT. **Ref.:** Monard, 1930, 1935; Romeiras, 1999; Almaça, 2005; Figueiredo & al., 2008; Jacquat, 2008.

Monod, Théodore
(1902–2000)
Bio.: b. Rouen, France, 9 April 1902; d. Versailles, France, 22 November 2000. Biologist, environmentalist, explorer, and collector extraordinaire. After finishing school, he obtained *certificats d'études supérieures préparatoires* at the Univ. of Paris, in zoology and botany in 1920 and in geology the following year. He worked on oceanographic cruises for two years. In April 1922, he graduated with a first degree in Science at the Univ. of Paris, and he joined the *Muséum national d'Histoire naturelle* as an assistant; he remained with the *Muséum* his entire career. That same year (1922) he undertook his first

Fig. 86. Théodore Monod at Adrar de Mauritanie (Oued Akerdil) in December 1998. By Bruno Lecoquierre. Reproduced under licence CC BY-SA 3.0.

overseas expedition and in Mauritania conducted the first of many desert crossings. During his lifetime, he covered several thousands of kilometres in deserts on foot or camel. His interests were eclectic, and he made discoveries in areas ranging from biology to archaeology and authored numerous writings (more than 700 works), including twelve books. He obtained his doctorate in natural sciences in January 1927 with a thesis on the Gnathidae (Crustacea). From 1928 to 1930, he was in Algeria for military service. In 1938, Monod was appointed *Secrétaire-général* of the *Institut Français d'Afrique Noire* (IFAN) of which he became director in 1962. He remained attached to the *Institut* until 1965. Simultaneously, from 1942 to 1975, he succeeded Gruvel (q.v.) as holder of the chair *Pêches et productions coloniales d'origine animale* (after 1958 entitled *Pêches Outre-Mer*) at the *Muséum*. Monod was a vegetarian, pacifist, advocate of animal rights, and environmentalist, and he is well-known as a pioneer of the environmental movement in France. His extensive archives, which are at the *Muséum national d'Histoire naturelle*, include his five-metre-long personal archives. His plant collection numbers exceed 26,000. Monod received several decorations and is commemorated in many species names, such as *Maytenus monodii* Exell (Celastraceae) and *Selaginella monodii* Alston (Selaginellaceae), two species he collected when he briefly visited the island of São Tomé in 1956 for an IFAN Conference. His Angolan collections are at Herb. IFAN. **Angola:** 1955; it is not known how many collections were made. **Herb.:** AL, BM, COI, E, IFAN, K, MPA, MPU, NY, P, PC. **Ref.:** Vegter, 1976; Romeiras, 1999; Figueiredo & al., 2008; Daire & al., 2013, Desmond, 2017. Figure 86.

Monteiro, António da Silva
(fl. 1937–1938)

Bio.: Cavalry lieutenant (*Tenente-picador*) and farmer. He owned the farm *Quinta do Ribatejo* in Humpata, Angola, and collected in the region. In 1938, he sent the plants to the herbarium of the Univ. of Coimbra (Herb. COI) via Moreno Júnior (q.v.). **Angola:** 1937–1938; Humpata. **Herb.:** COI. **Ref.:** Gossweiler, 1939; Vegter, 1976; Bossard, 1993; Figueiredo & al., 2008.

Monteiro, Joachim John
(1833–1878)

Bio.: b. London, England, 1833; d. Maputo, Mozambique, 6 January 1878. Mining engineer. He studied at the Royal School of Mines (that later became the Imperial College London), and at the Royal College of Chemistry, obtaining first-class honours. In 1858, he moved to Angola to work on the copper carbonate deposits at Bembe as a mine engineer for the Western Africa

Malachite Copper Mining Co. The mines had been acquired from the former slave trafficker Francisco António Flores from Brazil. The company had built a loading pier at Ambriz (Bengo) and a mining camp in the interior at Bembe (Uíge). Monteiro travelled between Ambriz and Bembe several times. To get to Bembe, the track went via Quibala (Zaire) and the distance travelled was "not less than 130 miles" (Monteiro, 1875). The journey was done on hammock and it normally took Monteiro four days, with eight porters and light luggage. At Bembe, the conditions were harsh. Of the initial technical staff of twelve Cornish engineers, six soon died of malaria and four fell ill and were sent home. Only Monteiro and another engineer remained in the company, with Monteiro eventually rising to the position of chief engineer. During his time in Angola, he travelled much and explored places he visited with his wife Rose Monteiro (q.v.). The couple made many collections of plants, insects, and animals that they sent to England, to the Royal Botanic Gardens, Kew, and the British Museum. Several specimens were labelled "Coll. Mr. & Mrs. Monteiro". One of their trips, in August 1872, took them up the Cuanza River to Dondo and the Cambambe cataracts (Cuanza Norte). Below the cataracts they collected specimens of a strange plant that were sent to Kew. It was later described as *Angolaea fluitans* Wedd., and was a new genus of Podostemaceae. In February 1873, they travelled up the Congo River on a steamer to Boma (D.R. Congo). There they also collected

Fig. 87. Joachim Monteiro with Rose Monteiro. Photographer unknown. Reproduced from Gomes e Sousa (1943).

specimens of undescribed plants, such as *Mussaenda monteiroi* Wernham (Rubiaceae). The mining company at Bembe eventually went bankrupt. In 1863, it had 24 white employees and 400–600 black men working under Monteiro's direction, but the costs of the business were enormous, and the mine was not profitable. While working there, Monteiro discovered that fibres extracted from the baobab (*Adansonia digitata* L., Malvaceae) could be used to make paper. He worked on that discovery for some years and, in 1865, finally decided to develop it, establishing houses at Ambriz for bartering baobab fibre, and pressing and shipping it to England. In April 1873, Monteiro and his wife made one last trip to Bembe for a one-month stay. This time, thirty porters were needed as they took a tent, provisions, bedding, clothes, a Madeira-cane chair that was used to cross streams, and even an inflatable, rubber bathtub. Additionally, a chef and his helper, with pots and pans, a laundry man with soap and irons, and one man with drying papers and boxes for collecting plants and insects were part of the group. At Bembe, the couple explored the surroundings and went up Bembe hill. Afterwards they returned to Ambriz and later to England. In 1875, Monteiro published the book *Angola and the River Congo*. The following year, with his wife, he moved to Mozambique to take up a position of labour agent and

Fig. 88. Camp of Joachim and Rose Monteiro at Quilumbo. Artist unknown. Reproduced from Monteiro (1875).

emigration agent to the government of the then Cape Colony (now part of South Africa). He died two years later at the age of 45. He is commemorated in the names of several plants and animals of which he collected the types. **Angola:** 1858–1872; Bengo (Ambriz), Benguela (Benguela), Cuanza Norte (Cambambe, Massangano), Uíge (Bembe). **Herb.:** B, BM, K (main, 346 specimens), NY, P, US, W. **Ref.:** Monteiro, 1875; Anonymous, 1878; Jackson, 1901; Gomes e Sousa, 1939, 1943; Exell, 1960; Vegter, 1976; Liberato, 1994; Romeiras, 1999; Figueiredo & al., 2008; Birmingham, 2015; Ferreira (F.A.), 2015. Figures 87 and 88.

Monteiro, Rose Bassett
(1840–1897)
Bio.: b. Hampstead, Middlesex, England, 19 April 1840; d. London, England, February 1897. Naturalist and collector. Née Bassett. She was daughter of Henry Bassett (1803–1846), an architect, and Maria Bassett (1803–1880). She had one sibling, Henry Bassett (1837–1920), who was a chemist and to whom she dedicated her book on Delagoa Bay. Her brother, Henry, had three children, the oldest being Prof. Henry Bassett (b. 1882), a chemist at the Univ. of Reading, England, who provided information on his aunt to Gomes e Sousa (1943). In 1858, Rose accompanied her husband J. Monteiro (q.v) to Angola where he accepted an appointment with the Western Africa Malachite Copper Mining Co. at Ambriz (Bengo), which had operating mines in the interior at Bembe. Husband and wife were both keen collectors of plants and animals and enjoyed exploring the country. During their time in Angola, Rose often accompanied her husband on excursions. In 1872, a trip up the Cuanza River to Dondo and the Cambambe cataracts (Cuanza Norte) resulted in collections that included material of a new genus, *Angolaea* Wedd. (Podostemaceae). The following year, the couple travelled together to Bembe for the last time and made several collections. They also travelled up the Congo River as far as Boma (D.R. Congo), where she was the first white woman to visit. This caused a sensation, with crowds gathering to cheer at the arrival of the couple. The Monteiros remained in Angola until 1873, then returned to England. In 1876, they moved to Maputo, Mozambique, where J. Monteiro died two years later. Rose returned to England but kept their property in Maputo, a cottage in the area that is now occupied by the city market. A few years later, she returned to Mozambique where she earned an income from selling zoological and botanical specimens to collectors in Europe. She remained in Mozambique for five years, living in her cottage in Maputo. She also continued studying natural history and collecting plants that she sent to the Royal Botanic Gardens, Kew (Herb. K); between 1883 and 1893, at least 38 collections made by her

in Mozambique were received at Herb. K. She also collected insects that she sent to other collectors and authors such as the naturalist Roland Trimen (1840–1916). The specimens she sent to Trimen, many of which are cited in his work on South African butterflies, included drawings and notes. After her final return to England, she published the book *Delagoa Bay: Its natives and natural history* in 1891. She is commemorated in some names of which she collected the type, such as *Aloe monteiroae* Baker (Asphodelaceae). **Angola:** 1858–1872; Bengo (Ambriz), Benguela (Benguela), Cuanza Norte (Cambambe, Massangano), Uíge (Bembe); collected with (J.) Monteiro. **Herb.:** BM, K, NY, US. **Ref.:** Monteiro, 1875; Trimen, 1887–1889; Monteiro, 1891; Gomes e Sousa, 1939, 1943; Gossweiler, 1939; Alice Gomes e Sousa, 1949; Exell, 1960; Vegter, 1976; Bossard, 1993; Liberato, 1994; Romeiras, 1999; Bassett Family Association, 2006–2021; Harries, 2007; Figueiredo & al., 2008; Crouch & al., 2015. Figures 87 and 88.

Monteiro, Rui Fernando Romero
(1922–unknown)
Bio.: b. 1922. Forester. He was initially with the *Instituto de Investigação Científica de Angola*, where he was in charge of the Forestry Division in 1962. In 1964, he was attached to the *Instituto de Investigação Agronómica de Angola*, which he directed later, in the 1970s. He published several papers on the forests of Maiombe (Cabinda), Dembos (Bengo), and Moxico. **Angola:** 1958–1966; Bié, Cabinda, Bengo (Dembos), Luanda; collected with (R.) Santos (q.v.) & Murta (q.v.). **Herb.:** BM, BR, COI, LISC (1237 specimens), LISU, LUA, LUAI, LUBA, PRE. **Ref.:** Monteiro, 1957, 1962, 1965, 1970; Monteiro & Frade, 1960; Vegter, 1983 (& sub "Mendes, E.J.S.M."); Bossard, 1993; Liberato, 1994; Romeiras, 1999; Figueiredo & al., 2008.

Morais, Artur Augusto Taborda de
(1900–1959)
Bio.: b. Avidagos, Mirandela, Portugal, 4 June 1900; d. Luanda, Angola, 20 July 1959. Botanist. In 1934, he graduated with a B.Sc. from the Univ. of Coimbra, where he was appointed as a lecturer in botany and remained until 1940, when he was suspended for disciplinary reasons. He then taught in secondary schools in Lisbon and in Torres Vedras and, after moving to Angola, he taught in high schools in Lubango and Luanda. In 1958, he was contracted by the *Instituto de Investigação Científica de Angola* where he worked until his untimely death about a year later. Early in his career, he published on the flora of Portugal and later, in Angola, he published some popular articles on Angolan themes, including one on *Welwitschia mirabilis*

Hook.f. (Welwitschiaceae). According to Fernandes (1959), he collected over 1000 numbers in Angola. **Angola:** 1958–1959. **Herb.:** LUAI, LUBA. **Ref.:** Morais, 1958; Fernandes, 1959; Figueiredo & al., 2008.

Moreno, Francisco J. Tomás
(fl. 1959–1972)
Bio.: Technician with the *Instituto de Investigação Científica de Angola* in the 1960s. He jointly collected with several collectors and was in the team of the *Campanhas de Angola* (q.v.) in 1959. Not to be confused with Moreno Júnior (q.v.). **Angola:** 1960–1972; Bié, Cuando Cubango, Cuanza Norte, Cuanza Sul, Cunene, Huambo, Huíla, Malanje, Namibe; collected also with Baptista de Sousa (q.v.) in 1970; Barbosa (q.v.) in 1961–1972; (C.A.) Henriques (q.v.) in 1962 in Namibe; and Lopes (q.v.) in 1971 in Bié. **Herb.:** BM, COI, LISC (242 specimens), LUA, LUAI, LUBA. **Ref.:** Bossard, 1993; Romeiras, 1999; Figueiredo & al., 2008.

Moreno Júnior, Mateus Martins
(1892–1970)
Bio.: b. Faro, Portugal, 27 September 1892; d. Lisbon, Portugal, 19 May 1970. Army officer and author. He was an active author from an early age, having started periodicals while still attending high school, and he continued to write throughout his life, contributing to many national periodicals and newspapers. Moreno, F.A. Mendonça (q.v.) and two other Algarvians founded a local periodical that was issued from 1911 to 1913. After attending high school in Faro, Moreno Júnior went to Lisbon at the end of 1914 to study mathematics at the Univ. of Lisbon. However, in 1917 he was mobilised to serve in World War I, as militia ensign (*alferes miliciano*) of the army artillery and was sent to France. After the war ended, he decided to continue in the army and studied at the *Escola de Guerra* (military academy) and graduated with the *Curso Superior Colonial.* He was stationed in Angola for some years. While stationed in Lubango, he also taught at the local high school, created a military museum, as well as a periodical. When he returned to Lisbon, he lectured at the *Colégio Militar* (military secondary school) in Lisbon from 1942 to 1944 and at the *Escola Superior Colonial.* He retired with the rank of major, having received several decorations. As a member of the *Sociedade Broteriana*, he collected plant specimens in Huíla and sent them to the herbarium of the Univ. of Coimbra (Herb. COI). **Angola:** 1934–1936; Huíla. **Herb.:** COI. **Ref.:** Gossweiler, 1939; Vegter, 1976; Bossard, 1993; Romeiras, 1999; Figueiredo & al., 2008; Mendonça & Martins, 2014.

Moura, F. Alberto
(fl. 1953)
Angola: 1953; Cabinda. **Herb.:** COI, LISC (7 specimens), LUA. **Ref.:** Bossard, 1993; Romeiras, 1999; Figueiredo & al., 2008.

Muatchilende
(fl. 1946–1954)
Note: He collected for the Diamang Collections (q.v.). Likely Fernando Muatxilende (note spelling). who worked at the *Museu do Dundo, Companhia de Diamantes de Angola*, Lunda, from 1946 and was still listed as a member of the staff in 1953. **Angola:** 1954; Lunda Sul. **Herb.:** DIA†?, LISC (1 specimen). **Ref.:** Porto & al., 1999; Romeiras, 1999; Figueiredo & al., 2008.

[Mumbala, R.]
Note: Recorded in JSTOR Global Plants (JSTOR, 2021). This refers to the Mumbala River.

Murta, Frederico
(fl. 1958–1974)
Bio.: Technician with Univ. of Luanda, stationed in Huambo, Angola. He assisted R.F. Romero Monteiro (q.v.) in some expeditions to collect material of *Brachystegia* Benth. (Fabaceae). **Angola:** 1958–1974; Benguela, Bié, Cabinda, Cuando Cubango, Cuanza Norte, Cuanza Sul, Huíla, Malanje, Namibe, Uíge; collected with Dechamps (q.v.), (Romero) Monteiro, (R.) Santos (q.v.), and (Manuel) Silva (q.v.). **Herb.:** BM, BR, COI, LISC (369 specimens), LISU, LUA, LUAI, LUBA. **Ref.:** Bossard, 1993; Romeiras, 1999; Figueiredo & al., 2008; G. Cardoso de Matos, pers. comm., 2021.

[Mussengue]
Note: Recorded in JSTOR Global Plants (JSTOR, 2021). This may refer to a locality or vernacular name.

N

Naumann, Ferdinand Christian [also as "Friedrich Carl"]
(1841–1902)
Bio.: b. Ehrenbreitstein, Germany, 6 February 1841; d. Klosterlausnitz, Germany, 26 July 1902. Naval surgeon. He studied natural sciences and

medicine in Berlin and Heidelberg and graduated as a medical doctor in 1865. Two years later, he became a naval surgeon. From 1868 to 1871, he participated in an expedition on the ship *Medusa* to South America and the Pacific, and from 1874 to 1876 he was on the ship *Gazelle* under the command of Capt. Georg Freiherr von Schleinitz (1834–c. 1910) on a trip around the world. The biologist Théophile Rudolphe Studer (1845–1922) was part of the team on the latter voyage. This journey was strenuous due to the workload, the many changes of climate, overexertion, shortages, and other deprivations. The two surgeons on board were overloaded with many patients to be treated. Sixteen deaths occurred during the trip. Afterwards, Naumann settled in Gera, Germany, where he lived with his mother, had a medical practice, and botanised. He married after the death of his mother in 1881. During the *Gazelle* expedition he collected numerous specimens. One of the collecting localities is the Congo estuary, where he made collections on the northern bank at Ponta da Lenha (name no longer in use, corresponding to a locality between Banana and Boma, in the D.R. Congo) and at Boma, in September 1874. He landed at least in one location on the southern bank of the river, in Angola, at Pedra do Feitiço (Zaire Province, Angola) but it is not known if he made any collections there. His collections were sent to the herbarium of the Botanic Garden and Botanical Museum Berlin-Dahlem (Herb. B) and many were destroyed during World War II. Duplicates are found in other herbaria. His private herbarium that was deposited at the herbarium of the Friedrich Schiller Univ. Jena (Herb. JE) included some likely duplicates. He is commemorated in many taxa described from his collections, including the genus *Naumannia* Warb. (Zingiberaceae) and, for example, *Solanum naumannii* Engl. (Solanaceae), of which the type was destroyed. **Angola:** 1874, if collections were made, none has been located. **Herb.:** B, BM, E, F, FH, G-BOIS, HBG, JE (main), K, KIEL, LE, M, MO. **Ref.:** Anonymous, 1889; Prahl, 1904; Urban, 1916; Vegter, 1983 (as "Friedrich Carl"); Bossard, 1993; Liberato, 1994; Romeiras, 1999; Frahm & Eggers, 2001; Figueiredo & al., 2008, 2020; Glen & Germishuizen, 2010; Pusch & al., 2015; J. Müller, pers. comm., November 2019.

Neto, Georgina António
(1960–)
Bio.: b. 1960. Herbarium assistant at the herbarium of the former *Centro Nacional de Investigação Científica* (Herb. LUAI) in 1999. She was one of the trainees of the SABONET project. In 2003, she also trained at the herbarium of the *Centro de Botânica* (Herb. LISC) of the *Instituto de Investigação Científica Tropical* in Lisbon. **Angola:** 1999. **Herb.:** LUAI. **Ref.:** Smith & Willis, 1999. Figure 89.

Fig. 89. Georgina Neto. Photographer unknown. South African National Biodiversity Institute. Reproduced with permission.

Newman, Mark Fleming
(1959–)

Bio.: b. U.K., 28 February 1959. Botanist. In 1991, he was with the Seed Bank, Royal Botanic Gardens, Kew. Afterwards he was at the Royal Botanic Garden Edinburgh. **Angola:** 1991; Cunene, Huíla, Namibe; collected with Alcochete (q.v.), Gerrard Reis (q.v.), and E. Matos (q.v.), mainly for seeds, with herbarium voucher specimens for identification. **Herb.:** K, LISC (1 specimen). **Ref.:** Figueiredo & al., 2008.

Newton, Francisco ("Francis", "Frank")
(1864–1909)

Bio.: b. Porto, Portugal, 18 May 1864; d. Matosinhos, Portugal, 9 December 1909. Naturalist. He is often listed as "Francisco Xavier Oakley (sic, O'Kelly) de Aguiar Newton". This is due to a 19th century Portuguese custom of adding the surnames of ancestors to a person's official name. Regardless, at baptism he was registered as Francisco Newton. He was the son of the amateur botanist Isaac Newton (1840–1906), who was born in Porto, of British and Irish ascendance. It is likely that the older Newton influenced his son's interest in natural history. In 1880, at the age of 16, Francisco went to Africa. From 1881 to 1884 or 1885, he was in Angola and then started to make plant collections. It is likely that he was based at Namibe as he started by exploring the area surrounding the town up to the Giraul River and the Namibe desert. He soon went further inland towards the Chela Mountains, Humpata, and Huíla. In January 1882, he accompanied Duparquet (q.v.) along the Caculovar River down to Humbe. There he joined the hunting party of Axel Wilhelm Eriksson (1846–1901), a trader, farmer, and explorer who travelled widely collecting natural history specimens. They travelled up the Cunene River towards Mulondo. Returning to the Humbe area, in October he met the Earl of Mayo (Dermot Robert Wyndham-Bourke, 1851–1927) with whom Johnston (q.v.) had been travelling. Shortly afterwards, in December 1882, Newton wrote to Kew from the Cunene River noting that he had collected specimens of 300 species of flowering plants and 25 ferns and had been in Huíla during the rainy season (i.e., October to April) and had collected orchids there. He returned to Namibe with the Mayo party. In a letter dated August 1885 that Newton sent to Júlio Henriques (1838–1928) at the herbarium of the Univ. of

Fig. 90. Francisco Newton. Photographer unknown. © Arquivo Histórico dos Museus da Universidade de Lisboa. Reproduced with permission.

Coimbra (Herb. COI), Newton commented on a previous document that he had sent and attached a list of the localities he had visited with the correct spelling checked by an "Africanist". Neither of these documents appears to be extant (M. Dias da Silva, pers. comm., 2018). It might be based on these documents as well as on the collections labels that Henriques (1885) described Newton's travels. In the same letter, Newton mentioned that he had sent a box of plants to the Earl of Mayo. These specimens were integrated in Johnston's collection and deposited at the herbarium of the Royal Botanic Gardens, Kew (Herb. K). Newton returned to Portugal and was appointed to the position of naturalist with the government to explore the islands of São Tomé and Príncipe in the Gulf of Guinea. From October 1885 to January 1892, he was in São Tomé and Príncipe with a short visit to Benin in 1886. After a stay in Porto, he returned to São Tomé from 1892 to 1895, also visiting Príncipe, Anobom, and Bioko. From 1896 to 1897, he collected in East Timor, and from 1898 to 1902 he was in the Cape Verde Islands with a visit to Guinea-Bissau in March 1900. He returned to Angola in 1903 on a zoological expedition during which he collected animal specimens in several provinces but apparently did not then collect plants. Overall, he collected extensively, including specimens that became types of over 50 plant names. He is commemorated in at least 22 vascular plant names and the names of seven vertebrates. **Angola:** 1881–1884; Huíla (Humpata, Huíla, Serra da Chela), Namibe (Moçâmedes). **Herb.:** B, BM, C, COI, G, H, JE, K, LISC (2 fragments), LISU, PO, US (grass type fragments), W, Z+ZT. **Ref.:** Henriques, 1885; Gossweiler, 1939; Silva, 1940; Vegter, 1983; Bossard, 1993; Liberato, 1994; Romeiras, 1999: Figueiredo & al., 2008, 2019a. Figure 90.

Nogueira, Rui
(fl. 1971)
Angola: Huíla. **Herb.:** LUAI, LUBA.

Nolde, Ilse (Baronin von)
(1889–1970)
Bio.: b. Groß-Schwülper, Niedersachsen, Germany, 22 May 1889; d. Reinbek, Germany, 27 September 1970. Naturalist and artist. Née von Marenholtz. She was the daughter of Gebhard von Marenholtz and Margarete, who was born Gräfin v. d. Schulenburg-Wolfsburg. Nolde became interested in plants at an early age. She spent some time in Florence, Italy, where she was exposed to art. Later she was mentored and encouraged by the German botanist Gottfried Wilhelm Johannes Mildbraed (1879–1954), who at the time was the keeper of the herbarium of the Botanic Garden and Botanical Museum Berlin-Dahlem

(Herb. B). She became an accomplished plant collector and exceptional botanical artist. On 27 September 1918, she married Harald Baron von Nolde, and in 1928 the couple moved to Quela (Malanje), Angola, to run a coffee plantation. While there, she collected plants from at least January 1932 to December 1938. At the end of 1938, after the death of her husband, she returned to Germany, and afterwards worked in the *Reichsinstitut für ausländische und koloniale Forstwirtschaft* at Reinbek near Hamburg for several years. Her botanical estate (field notebooks and illustrations but excluding specimens) is held at the herbarium of the Univ. of Hamburg (Herb. HBG); it includes numerous well-executed botanical drawings and handwritten descriptions. She is commemorated in several names, such as that of the genus *Noldeanthus* Knobl. (Oleaceae) and the species *Ritchiea noldeae* Exell & Mendonça (Capparaceae) and *Cochlospermum noldeae* Poppend. (Bixaceae); the last endemic to the Quela highlands where Nolde focussed her collecting activities. From her late-40s onwards she published on ecology, inter alia of the Quela highlands, where her research focussed on the "Fazenda Camacol" at 1219 m above sea level. The main set of her collection of 887 numbers was held at Herb. B; it was mostly destroyed during the bombing of Berlin in World

Fig. 91. Ilse von Nolde. Photographer unknown. © Botanischer Garten und Botanisches Museum Berlin, Freie Universität Berlin. Reproduced with permission.

War II, but duplicates are held elsewhere. The numbering is not chronological. The whereabouts and fate of a 477-page manuscript entitled "*Angola—Land, Menschen und Schicksal einer afrikanischen Kolonie*" unfortunately remains unknown. **Angola:** 1932–1938; Malanje (Quela). **Herb.:** B (main), BM, COI, LISC (22 specimens), MO. **Ref.:** Gossweiler, 1939; Esdorn, 1972; Vegter, 1983; Bossard, 1993; Romeiras, 1999; Poppendieck, 2004; Figueiredo & al., 2008, 2020. Figure 91.

Noronha
(fl. 1953–1960)
Angola: 1953–1960. **Herb.:** COI, LUA. **Ref.:** Figueiredo & al., 2008.

Nunes, Pereira
(fl. 1959–1972)
Note: Referred as "Dr Pereira Nunes" in herbarium labels; this could be an individual who published on fishing in Angola. **Angola:** 1959, 1972; Malanje (1959), Cunene (1972). **Herb.:** LISC (2 specimens). **Ref.:** Nunes, 1959.

O'Donnell, Henrique Figueiredo
(fl. 1921–1933)
Bio.: Engineer. Graduated from the *Instituto Superior Técnico* in Lisbon, Portugal. He was part of the team of the *Missão Geológica de Angola* (1921–1926). The *Missão* left Lisbon in January 1922 and arrived in Lobito (Benguela) the following month. A base camp was subsequently established at Huambo. In 1925, three groups were formed to survey different areas of the country; O'Donnell was the leader of one of the groups. In 1926, the *Missão* was replaced by the new *Serviço da Carta Geológica de Angola* of which O'Donnell became the leader. The *Serviço* lasted until the early 1930s and resulted in the production of a geological map of Angola (Mouta & O'Donnell, 1933) that was presented at the 16th International Geology Congress in Washington, D.C., U.S.A. in 1933. Later in the 1930s, O'Donnell was involved in surveys for the new port in Luanda and also worked for the port of Lisbon. While in Angola. he collected a few hundred plants in the Benguela region. **Angola:** 1929–1933; Benguela. **Herb.:** COI. **Ref.:** Mouta & O'Donnell, 1933; Gossweiler, 1939; Vegter, 1983; Bossard, 1993; Romeiras, 1999; Brandão, 2008; Figueiredo & al., 2008. Figure 92.

Fig. 92. Henrique O'Donnell (extreme left) and the team of the *Missão Geológica de Angola* in Huambo in 1922. Photographer unknown. From University of Lisbon-IICT Photography Collection - UL/IICT/MGG 21756. Reproduced with permission.

Oliveira, António Lopes Branquinho d' – see **Branquinho d'Oliveira**

Oliveira, Ermelinda de
(fl. 1959)
Angola: August–September 1959; Benguela (Quileba), Huambo (Cachiungo, Bailundo). **Herb.:** COI. **Ref.:** Romeiras, 1999; Figueiredo & al., 2008.

P

Palminha, Francisco Prudêncio
(fl. 1955–after 1976)
Bio.: b. [23 September 1913?]; d. after 1976. Algologist. He was associated with the *Missão de Biologia Marítima*, an entity that in 1966 became the *Centro de Biologia Aquática Tropical* of the *Junta de Investigações do*

Ultramar. In 1976, he was still active in Lisbon. He published a few papers on algae. **Angola:** 1958–1959; no herbarium specimens have been located. **Herb.:** M. **Ref.:** Vegter, 1983; Figueiredo & al., 2008.

Paulian, Renaud
(1913–2003)
Bio.: b. Neuilly-sur-Seine, near Paris, France, 28 May 1913; d. Bordeaux, France, 16 August 2003. Zoologist and entomologist. He obtained his Ph.D. at the Univ. of Paris. Most of his career was dedicated to administration. In 1947, he was appointed as the deputy director of the *Institut de recherche scientifique de Madagascar* in Tananarive, Madagascar, where he remained until 1961. Later, he was attached to the *Office de la Recherche scientifique et Technique Outre-mer* as *Inspecteur général de recherches.* Around 1966, he was director of the *Institut d'Études Centrafricaines* in Brazzaville, Rep. of the Congo, and later chancellor (*Recteur*) of the Univ. of Abidjan, Côte d'Ivoire. He collected zoological and botanical specimens in several regions such as on the Atlas Mts. (1938), Mt. Cameroon (1939), in the Côte d'Ivoire (1945, 1966–1969), Central and West Africa (1961–1969), Namibia (1957), and the Mascarenes. After retiring in October 1981, he was a regular visitor at the *Laboratoire d'Entomologie* of the *Muséum national d'Histoire Naturelle*, Paris, and then pursued his scientific interests. His memoirs (Paulian, 2004) were published posthumously. **Angola:** 1961–1969. **Herb.:** MO, P, TAN. **Ref.:** Vegter, 1983; Dorr, 1997; Cambefort, 2008; Figueiredo & al., 2008.

Pearson, Henry Harold Welch
(1870–1916)
Bio.: b. Long Sutton, Lincolnshire, England, 28 January 1870; d. Wynberg, Cape Town, South Africa, 3 November 1916. Botanist. He apprenticed as a chemist's assistant and worked for a while as a teacher before entering the Univ. of Cambridge in 1893. There he graduated with first classes in both parts of the natural sciences tripos (examinations to qualify for a degree) in 1896, also winning the Darwin Prize of his college. In 1898, he travelled to Ceylon (now Sri Lanka) to study grasslands and on his return to Cambridge was appointed as assistant curator at the herbarium of the university (Herb. CGE). In 1899, he moved to the Royal Botanic Gardens, Kew (Herb. K) as an assistant on Indian botany, and later as assistant to Sir William Turner Thiselton-Dyer (1843–1928), the director, who eventually encouraged Pearson to establish a botanical career at the Cape. In 1900, Pearson obtained an M.A. from Cambridge Univ. He immigrated to South Africa in 1903 to be the first incumbent of the Harry Bolus professorial chair in botany at the South African College in Cape

Town. With a particular interest in *Welwitschia* Hook.f. (Welwitschiaceae), he initiated a series of expeditions to Namibia shortly after arriving in the country. In 1907, he obtained a doctorate from the Univ. of Cambridge for his work on *Welwitschia*. From November 1908 to June 1909, he undertook a first expedition to the arid western parts of southern Africa funded by the Percy Sladen Memorial Trust. After exploring Namibia, he sailed from Lüderitz Bay (Namibia) to Lobito in Angola and travelled in the country from March to June 1909. His main aim was to collect material of *Gnetum* L. (Gnetaceae), which he found in Montebelo, in Cazengo (Cuanza Norte), but many other plants were collected during the expedition and he recorded observations on the vegetation that were later published (Pearson, 1910, 1911). While in Cazengo he visited the Granja de S. Luis, where the Cazengo colonial garden, later to be known as *Estação Experimental do Cazengo*, had recently been established and where Gossweiler (q.v.) was based. After returning to the coast and sailing to Moçâmedes, Pearson explored the interior as far as Xangongo (Cunene) via Lubango (Huíla). At the time, the railway from Moçâmedes to the interior was built up to km 107 (the railhead). From there to the Huíla plateau the journey was difficult and slow. The wagon road to the Boer settlement of Humpata (see page 160) took seven days to travel. The alternative, which Pearson took, was a more gradual climb that took three days or less, but could only be done on foot, on horseback or by *machila* (hammock). After exploring the plateau, he went south along the Caculovar River reaching Xangongo (Cunene). Pearson sailed from Angola in mid-June and called at Cabinda, São Tomé (São Tomé and Príncipe), and São Vicente (Cape Verde), but apparently did not collect at either of these ports of call. He arrived in Portugal a month later, likely while en route to England. Pearson's collections were sent to Herb. K. A record of collections received at Herb. K between 1907 and 1912, lists 912 specimens from south Angola and Great Namaqualand; these were obtained after conclusion of the Percy Sladen Memorial Expedition 1908–1909. The specimens were sent to Kew undetermined. Over the next years, as the material was being studied, several specimens became types of new taxon names, but some of these were later synonymised. For example, *Kalanchoe pearsonii* N.E.Br. (Crassulaceae) is a synonym of the earlier *K. lindmanii* Raym.-Hamet, which is based on a collection made by Fritzsche (q.v.). The Angolan specimens collected during the expedition fall within Pearson's collecting number 2015 and number 2965, however the numbering is not chronological. The first set of the collection is at Herb. K, with a second set at the herbarium of the South African Museum (Herb. SAM, now held at Herb. NBG). Pearson (1911) published an overview of the expedition with the list of collecting localities and a map. The second Percy Sladen Memorial Expedition took place from

November 1909 to January 1910. This expedition did not result in collections from Angola. Pearson was an industrious collector and amassed nearly 10,000 numbers during his short lifetime. He was an advocate for the creation of a botanical garden in Cape Town and finally in 1913 his (and others') efforts succeeded, with the establishment of the Kirstenbosch National Botanic Garden, of which he became honorary director. He was the first editor of the *Annals of the Bolus Herbarium*, founded in 1914. Developing research and teaching at a high level, he was an influential and respected professor and an inspiration for his students. His early death at the age of 46, from acute pneumonia following a minor operation, was a great loss for South African botany. He is buried in the Kirstenbosch National Botanical Garden and is commemorated in several names, including the genus *Pearsonia* Dümmer (Fabaceae) and, for example, the species *Sansevieria pearsonii* N.E.Br. (Asparagaceae), which he collected in Angola and recognised as new but did not describe. **Angola:** March–June 1909; Cuanza Norte, Cunene, Huíla, Moçâmedes. **Herb.:** BM, BOL, COI, K, LISC, NBG (SAM), PRE. **Ref.:** Pearson, 1910, 1911; Anonymous, 1912, 1985; Gossweiler, 1939; Compton, 1965; Rycroft, 1975, 1980; Gunn & Codd, 1981; Stafleu & Cowan, 1983: 131–132; Vegter, 1983; Bossard, 1993; Romeiras, 1999; Figueiredo & al., 2008; Huntley, 2012. Figures 93 and 94.

Fig. 93. Harold Pearson. Photographer unknown. South African National Biodiversity Institute. Reproduced with permission.

Itinerary of the Percy Sladen Memorial Expedition 1908–1909 in Angola

24 March – Lobito (Benguela).
25 March – Luanda (Luanda).
31 March – Luanda to Cassoalala (Cuanza Norte).
1 April – Cassoalala to Ndalatando (Cuanza Norte) to Granja de S. Luis (Cazengo, Cuanza Norte).
2–5 April – Granja de S. Luis.
6–8 April – Montebelo (Cazengo, Cuanza Norte).
9–12 April – Granja de S. Luis.

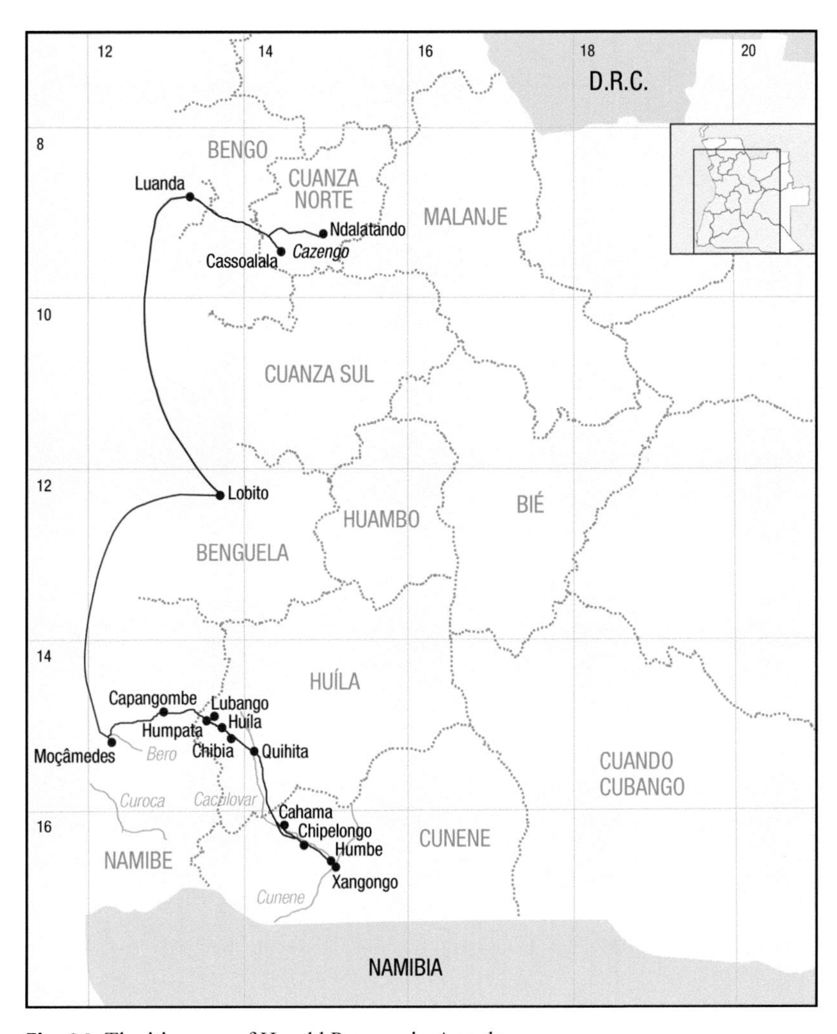

Fig. 94. The itinerary of Harold Pearson in Angola.

19 April – Lobito (Benguela).
22 April – Moçâmedes (Namibe). Went c. 13 km southeast towards Curoca River, and to the mouth of Bero River.
27 April – Moçâmedes to the railhead at km 107.
28 April – Railhead (km 107).
29 April – Camp north of railway, 1–5 km east of railhead.
30 April – West of previous and south of railway.
1 May – South of railway.
2–4 May – From the point they were on 29 April to Humpata (Huíla).
5–7 May – Humpata.
8 May – Humpata to Lubango (Huíla).
10 May – Humpata towards Huíla (Huíla).
11 May – Huíla to *Missão da Huíla.*
12 May – To Chibia (Huíla).
12–13 May – To Quihita (Huíla).
14–15 May – To Gambos fort (Huíla) and on to *Missão do Tchiapepe* (Huíla).
16–17 May – On to Cahama (Cunene).
18 May – Cahama and then left bank of Caculovar River.
19 May – To *Missão do Chipelongo* (Cunene).
20 May – To Calculovar River opposite Humbe (Cunene).
21 May – To Humbe and through Cunene River marshes to Xangongo (Cunene).
22 May – To *Missão do Chipelongo* (Cunene).
23 May – Chipelongo and on towards Ediva.
24 May – To Cahama (Cunene).
25–26 May – Cahama (Cunene) and towards Gambos (Huíla).
27 May – *Missão do Tchiapepe* (Huíla) and Gambos fort.
28–29 May – To Quihita (Huíla).
30 May – To Chibia (Huíla).
4 June – To Huíla (Huíla).
5 June – To Tchivinguiro (Huíla).
6 June – To Munhino River (Namibe).
7 June – Moçâmedes (Namibe).
11 June – Benguela (Benguela) to km 82.5 on the Benguela railway.
12 June – Lobito (Benguela).

Pechuël-Loesche [also as "Pechuël-Lösche"], Moritz Eduard (1840–1913)

Bio.: b. Zöschen near Merseburg, Germany, 26 July 1840; d. Munich, Germany, 29 May 1913. Traveller and geographer. After attending school at Halle, he joined the merchant navy and travelled widely in the 1860s, including to the West Indies, North America, the coast and islands of the Atlantic and Pacific Oceans, and both the Southern Ocean (surrounding Antarctica) and the Arctic Ocean. He returned to Germany and continued his studies at Leipzig, graduating with a doctorate in natural sciences in 1872. From 19 August 1874 to 5 May 1876, he joined the expedition of the *Deutsche Gesellschaft zur Erforschung Aequatorial-Afrikas* to the Kingdom of Loango (now the Rep. the of Congo and Cabinda, Angola). The leader of this expedition was Güssfeldt (q.v.) and the team included, among others, Soyaux (q.v.) as botanist, and, for a short period, Mechow (q.v). Pechuël-Loesche was in charge of the meteorological surveys, but he also conducted ethnographic studies, and investigated the fauna and flora. He was mostly based at Chinchoxo (Cabinda). He undertook a trip to the Kouilou River (Rep. of the Congo) in 1875 and to Mayumba Bay (Gabon) in 1876. In that year, as the expedition was cancelled and the team was recalled to Germany, he left with Soyaux and arrived in Berlin on 30 June 1876. In 1881, he was appointed by Leopold II of Belgium to join the Congo venture with the purpose of opening a route from Stanley Pool (now Malebo Pool, D.R. Congo) to the Indian Ocean and to establish stations along the way.

Fig. 95. Eduard Pechuël-Loesche c. 1882. Photographer unknown. From Wikimedia Commons. Public domain.

The expedition team included others such as Teusz (q.v.). The main objective of the venture was not scientific as purported, but rather commercial, especially to sign contracts with the chiefs of the peoples encountered to grant the rights of commerce to Leopold II. Realising this, Pechuël-Loesche did not remain long with the project, departing in 1883. In 1884, he travelled to Namibia with his wife and made collections there. In 1886, he worked as a lecturer at the Univ. of Jena, Germany, and obtained his habilitation. He joined the Univ. of Erlangen in 1895. He retired in 1902 and then moved to Munich where he died in 1913. He published numerous journal articles and several books. His diaries, together with his watercolours, reveal the quotidian aspects of the expeditions, including the relationship between Europeans and Africans. He was strongly opinionated about the way German expeditions were organised and the observations he made along the Loango coast (including on botany) were rigorous, detailed, and sound. According to Heintze (2010), he was the only one of the many 19th century German explorers in Angola to establish a deeper relationship with, and have a better understanding of, the Africans. Much of his knowledge of African culture and languages was obtained from women, which he acknowledged in his diaries. During his stay in Chinchoxo he had an African *Zeitfrau* (temporary wife). He is commemorated in the genus *Pechuel-loeschea* O.Hoffm. (Asteraceae) and some species names. *Pechuel-loeschea leubnitziae* (Kuntze) O.Hoffm. commemorates both Pechuël-Loesche and his wife Elsbeth (née von Leubnitz). Urban (1916) recorded that he collected mosses from "Westafrika (a. 1876, 1882, 1884)". According to Baker (1903: 517), Pechuël-Loesche collected in the Lower Congo, but we could not locate any specimens nor other reference to collections he made in Angola. **Angola:** 19 August 1874–5 May 1876; Cabinda; apparently, he did not collect. **Herb.:** B, G (lichens). **Ref.:** Pechuël-Loesche, 1887, 1907; Güssfeldt & al., 1888; Baker, 1903; Urban, 1916; Gunn & Codd, 1981; Vegter, 1983; Romeiras, 1999; Figueiredo & al., 2008, 2020; Heintze, 2010, 2011. Figure 95.

Pedro
(fl. 1964)
Angola: 1964; Moxico; collected with (Brito) Teixeira (q.v.). **Herb.:** LISC, LUA. **Ref.:** Figueiredo & al., 2008.

Peles, S.A.
(fl. 1970)
Bio.: Portuguese man who worked at *Companhia de Diamantes de Angola* in Lunda. **Angola:** 1970; Lunda Norte. **Herb.:** DIA†?, LISC (6 specimens). **Ref.:** G. Valente, pers. comm., January 2019.

Pereira, A. Gomes
(fl. 1959–1967)
Angola: 1959–1967; Cuango Cubango, Zaire. **Herb.:** LISC (8 specimens), LUA. **Ref.:** Bossard, 1993; Figueiredo & al., 2008.

Pereira, Alfredo Martiniano
(fl. 1900s–1916)
Bio.: Portuguese agronomist who directed the *Posto Algodoeiro do Quilombo*, in Angola in the 1900s. He published a few articles on the cultivation of cotton and rice. He married Sara Futscher and the couple had ten children. Some descendants of the family Futscher Pereira became public figures. He collected with Gossweiler (q.v.). **Angola:** 1916; with Gossweiler. **Herb.:** COI, LISC (1 specimen). **Ref.:** Pereira, 1908a, 1908b, 1908c; Vera Futscher Pereira, pers. comm., August 2019.

Pereira, Brás
(fl. 1960–1967)
Note: Likely G. Brás Pereira, who published some reports on animal husbandry in Angola between 1964 and 1967. **Angola:** 1960. **Herb.:** LUAI. **Ref.:** Aguiar, 1984; Romeiras, 1999; Figueiredo & al., 2008.

Pereira, Jesus
(fl. 1953)
Angola: 1953; Bengo. **Herb.:** LISC (3 specimens).

Pereira, João Alexandrino
(1913–1984)
Bio.: b. 1913; d. 1984. Agrarian technician with the *Instituto de Investigação Agronómica de Angola*; he specialised in pasture management. He took Angolan nationality in 1977. Not to be confused with J. Alves Pereira (q.v.). **Angola:** 1933–1984; Cuando Cubango, Huambo, Malanje. **Herb.:** COI, LUA. **Ref.:** Bossard, 1993; Romeiras, 1999; Figueiredo & al., 2008.

Pereira, Joaquim Lima
(fl. 1951–1988)
Bio.: Veterinarian. He graduated with a B.Sc. and Ph.D. from the Technical Univ. of Lisbon, Portugal. He was appointed as professor at the Univ. of Luanda, Angola, and was funded by the *Instituto de Investigação Agronómica de Angola*. He also practised as a veterinarian at Quipungo, Angola. Later, from 1976 to 1980, while back in Portugal, he was attached to the *Instituto Politécnico de Vila*

Real and tasked with directing the commission responsible for establishing the *Instituto Politécnico*, and as a professor of zootechnics. He became vice-chancellor of the *Instituto Universitário de Trás-os-Montes e Alto Douro* in 1981, and chancellor from 1985 to 1987. From 1983 to 1988, he also presided over the commission for the establishment of the *Instituto Politécnico de Bragança*. **Angola:** 1951–1967; Benguela, Cunene, Huíla, Namibe. **Herb.:** LISC (52 specimens), LUA. **Ref.:** Bossard, 1993; Romeiras, 1999; Figueiredo & al., 2008.

Pereira, José Alves
(1912–unknown)
Bio.: b. 1912. He also collected in Moçambique. Not to be confused with J. Alexandrino Pereira (q.v.). **Angola:** 1959; Bengo. **Herb.:** COI, LUA. **Ref.:** Figueiredo & al., 2008.

Pereira, Manuel dos Santos
(fl. 1940s–1950s)
Bio.: Veterinarian. In the 1940s, he was the director of the *Estação Zootécnica da Humpata* in Huíla, Angola. It was on his advice that the *Posto Experimental do Caraculo* was created in 1948 at Caraculo, about 78 km from the coast of Namibe at the foothills of the Chela Mountains, with the purpose of breeding caracul sheep for their pelts (astrakhan). Santos Pereira became the director of this experimental station. The *Posto* covered 16,000 ha and was devised as a model for further farms in the area, with Pereira hoping to attract 1600 white settler families to run farms with herds amounting to two million sheep. An area of nine million hectares (the Caracul Reservation) was established for these settlers. The workers were to be accommodated in villages. A model, purpose-built experimental village with modernised *rondavels* (traditional circular dwellings) where thatched roofs were replaced with bricks still exists in Caraculo. Although the venture attracted attention and praise at the time, it was based on forced labour, it displaced populations, and contributed to the separation of communities. It would eventually fail to appeal to prospective resident settlers, as the farm tenants remained in cities and delegated the farm operations to foremen. **Angola:** 1951; Namibe. **Herb.:** LISC (13 specimens). **Ref.:** Bossard, 1993; Saraiva, 2014.

Pessanha, Martim Vaz Taborda
(fl. 1960)
Bio.: Agronomist. He published on plant cultivation in Angola. **Angola:** 1960; Luanda (Calumbo). **Herb.:** LISC (1 specimen), LUA. **Ref.:** Pessanha, 1963; Bossard, 1993; Figueiredo & al., 2008.

Peter, Gustav Albert
(1853–1937)
Bio.: b. Gumbinnen, East Prussia (now Russia), 21 August 1853; d. Göttingen, Germany, 4 October 1937. Botanist. He studied at the Univ. of Königsberg (now Kaliningrad) and obtained a doctorate in 1874. Afterwards, he was appointed to the Univ. of Munich. From 1888 to 1923, he was professor of botany at the Univ. of Göttingen. In 1913, he undertook a first collecting expedition to Africa that lasted almost six years. A second expedition took place in 1925. On that second trip he collected in Angola, at Lobito, which was a port of call during his sea journey. He collected at several other ports of call in Namibia and in South Africa en route to his destination, which was German East Africa (now Tanzania). He amassed a large herbarium of about 50,000 numbers that was acquired by the Botanic Garden and Botanical Museum Berlin-Dahlem (Herb. B) in 1936. Duplicates were distributed to several herbaria. **Angola:** 1925; Benguela (Lobito); one collection known. **Herb.:** B (main), A, BERN, BM, BP, BR, C, CGE, CORD, DAO, E, E-GL, G, GOET (main), H, K, L, LE, M, MANCH, P, PRC, S, US, W, WRSL. **Ref.:** Stafleu & Cowan, 1983: 189–191; Vegter, 1983; Codd & Gunn, 1985; Plug, 2020.

Phillips, Richard Cobden
(fl. 1874–1890s)
Bio.: From Manchester, England. Trader. He worked for the trading house Hatton & Cookson, a Liverpool-based company founded in 1838 that traded with West Africa. From at least 1874 to 1877, he was in Landana and in Molembo (Malembo, south of Landana), in Cabinda, Angola, and posted letters to the Royal Botanic Gardens, Kew, regarding exchanging plant material. In one of these letters (Phillips, 1877), he sent a photo of the country as seen from his house, likely at Malembo where he wrote the letter. From 1883 to 1885, he was at the Lower Congo as evidenced by his archives deposited at the Royal Geographical Society (RGS). His papers include letters written from Banana (D.R. Congo) and from Ponta da Lenha (name no longer in use, corresponding to a locality between Banana and Boma, in the D.R. Congo). According to information in the RGS archives, among his papers there are also 18 photographs taken along the Congo River, including of the explorer Henry Morton Stanley (1841–1904) and his party. In Stanley's account of his voyages in Africa (Stanley, 1878, vol. 2: 468), there are engravings that reproduce photographs by "Mr Phillips". After travelling through the "Dark Continent", and before embarking for Zanzibar in August 1877, Stanley was stationed at the bay of Cabinda for a while, where "on the southern point of the bay stands a third factory of the enterprising firm of Messrs Hatton

and Cookson, under the immediate charge of their principal agent, Mr. John Phillips" (Stanley, 1878, vol. 2: 468). A factory (in Portuguese, *feitoria*) was the term then used for a trading post. The other two factories mentioned by Stanley were at Boma and Ponta da Lenha. A map of Cabinda (Child, 1745) shows that as early as by the mid-18th century there were three factories on the southern side of the bay. Stanley mentioned that he was received by John Phillips and stayed in a cottage overlooking the sea. A depiction of his quarters and of the expedition group, also rendered from photographs by Mr Phillips, are included in the account. There is no mention of R.C. Phillips in Stanley's account, but it is known (Phillips, 1876) that Phillips had a brother named John Searle Ragland Phillips (1850–1919), a newspaper editor, who was likely Stanley's host in Cabinda. The photographs mentioned were registered in November 1877 as owned by J. Phillips resident in Cheshire and taken by R.C. Phillips resident in Cabinda. Pechuël-Loesche (q.v.) mentioned both John Phillips (as Hatton and Cookson's chief agent) and R.C. Phillips in his diaries written in 1875 (Heintze, 2011). Pechuël-Loesche was close to the latter and stayed at his house in Landana. At the time, R.C. Phillips had an African wife and a child. According to Pechuël-Loesche, he had a good relationship with the local people. Based on his experience in the Congo area, R.C. Phillips published a paper on the sociology of the people of the Lower Congo (Phillips, 1888). In the 1890s, he was one of the voices that manifested opposition to the atrocities committed under the rule of Leopold II of Belgium in the Congo Free State (Pavlakis, 2016). He is listed by Jackson (1901) as having sent material from South Africa, which was received at the

Fig. 96. Richard Phillips's quarters in Cabinda. Engraver unknown. Reproduced from Stanley (1878).

herbarium of the Royal Botanic Gardens, Kew (Herb. K) in 1878, but there is no reference to him in Gunn & Codd (1981) and no material has been seen. There is, however, at least one specimen from Cabinda at Herb. K that was received from Phillips in 1878 (K001008842). **Angola:** Cabinda. **Herb.:** B, K. **Ref.:** Child, 1745; Phillips, 1876, 1877, 1888; Stanley, 1878; Jackson, 1901; Heintze, 2011; Pavlakis, 2016; Anonymous, s.d.(f). Figure 96.

Pinto, Alexandre Alberto da Rocha Serpa – see **Serpa Pinto**

Pinto, Henrique Vieira
(fl. 1949–1953)
Bio.: Forester. He was also active in Mozambique. **Angola:** June 1949, 1953; Cabinda. **Herb.:** COI, K, LISC (27 specimens), LUA, US (ex Yale School of Forestry). **Ref.:** Pinto, 1961; Bossard, 1993; Romeiras, 1999; Figueiredo & al., 2008.

Pinto, Óscar Rodrigues
(fl. 1952–1973)
Bio.: Agronomist. In 1952, he graduated from the *Instituto Superior de Agronomia* in Lisbon. He later joined the *Instituto do Algodão de Angola*, an institute that was created by the Portuguese government in 1961. **Angola:** 1973; Cunene. **Herb.:** LISC (1 specimen), LUA. **Ref.:** Pinto, 1984; Bossard, 1993; Ferrão, 1993; Figueiredo & al., 2008.

Pinto Basto da Costa Ferreira, Maria Fernanda Duarte
(1938–)
Bio.: b. São Filipe de Benguela, Angola, 16 July 1938. Herbarium technician. Educated in Angola, she attended some courses at the Univ. of Coimbra, Portugal, and at the *Estudos Gerais Universitários de Angola*, in Chianga, Angola. From March 1962 to October 1974, she worked as a technician at the herbarium of the *Divisão de Botânica, Instituto de Investigação Agronómica de Angola* (IIAA; Herb. LUA) in Huambo. After the revolution in Portugal in 1974, the inevitable independence of Angola (that followed in 1975), and associated political unrest, and the general departure of Portuguese nationals, remaining in Huambo became a liability. Much to her regret, she left her country. She was one of the last refugees to leave taking the last airplane out of Angola. In Portugal, in August 1976, she became a technician at the herbarium of the *Centro de Botânica, Instituto de Investigação Científica Tropical* (Herb. LISC) in Lisbon. She curated the collections and later was (informally) in charge of their arrangement. Although she retired in August

2003, she actively continued as a volunteer at the herbarium up to March 2020, when restrictions due to COVID-19 put a stop to voluntary activities. She is an experienced plant identifier with a good general knowledge of the Angolan flora, and her help was much requested from both the staff of Herb. LISC and the academic staff and students of the nearby *Instituto Superior de Agronomia*. She provided identification training to dozens of Angolan and Guinean students who sojourned to Lisbon since the mid-1970s. She was not only a tutor to the trainees, but also a friend, being always willing to help them. She additionally provided determinations of material collected by others during expeditions to African countries. Under the name M.F. Pinto Basto, she co-authored a few books, published papers on the African flora, and contributed several family treatments to *Flora de Cabo Verde*. She collected c. 500 numbers in Angola, Guinea-Bissau, and São Tomé. Most of her collections from Angola were integrated into (Brito) Teixeira's collection ("Teixeira & al." in sched.). She is commemorated in *Maerua pintobastoae* J.A.Abreu & al. (Capparaceae). **Angola:** 1962–1974; Huambo (Chipipa, Chianga), Namibe (Parque Nacional do Iona); mostly integrated into the collection of (Brito) Teixeira (q.v.). **Herb.:** BR, LISC, LUA, K, WAG. **Ref.:** Pinto Basto, 1970; Abreu & al., 2014; M.F. Pinto Basto, pers. comm., 2021. Figure 97.

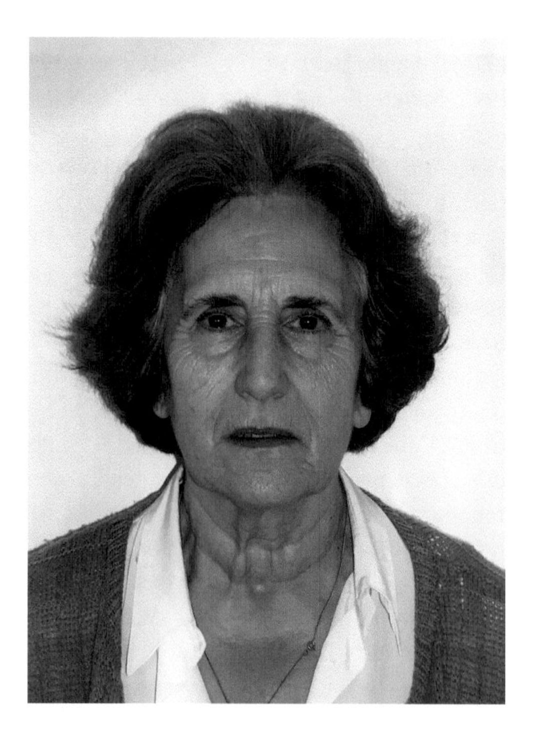

Fig. 97. Maria Fernanda Pinto Basto. Photographer unknown. Private collection of Maria Fernanda Pinto Basto. Reproduced with permission.

Pittard, Priscilla
(fl. 1937–1938)
Note: Her identity is unknown. She collected a few hundred specimens in Benguela, including the specimen that is the type of the endemic *Vangueria pachyantha* (Robyns) Lantz (Rubiaceae). There is a 1916 embarkation record for a Priscilla Pittard, 20 years old, leaving Liverpool with her father Peers R. Pittard (a merchant clerk), mother Violet, and sister Daisy for the Cape Verde Islands (L. Dorr, pers. comm. fide Ancestry.com). **Angola:** 1937–1938; Benguela (Ganda, Lobito). **Herb.:** BM, COI, LISC (1 specimen), MO. **Ref.:** Gossweiler, 1939; Vegter, 1983; Bossard, 1993; Romeiras, 1999; Figueiredo & al., 2008.

Ploeg, Douwe Taeke Engelbertus van der
(1919–2006)
Bio.: b. Bergum, Netherlands, 28 March 1919; d. Sneek, Netherlands, 20 September 2006. Botanist. He trained as a schoolteacher, and was also a self-taught botanist and floristic specialist, especially on the flora of Friesland, Netherlands, where he was a central figure in Frisian floristics for several decades. He had a special interest and expertise in wetland and aquatic plants, such as the pondweeds, *Potamogeton* L. (Potamogetonaceae), and sedges (Cyperaceae), both taxonomically challenging groups. Some of his best-known works are *De flora fan de Fryske sangrounen*, *Atlas fan de floara fan Fryslân*, and *List fan Fryske plantenammen*. In 2009, a posthumous commemorative Van der Ploeg Symposium was held in Eernewoude, in the province of Friesland; Ploeg would have celebrated his 90th birthday then. **Angola:** 1991; Huíla. **Herb.:** L. **Ref.:** Papenburg, 2010.

Pocock, Mary Agard
(1886–1977)
Bio.: b. Cape Town, South Africa, 31 December 1886; d. Grahamstown, Eastern Cape, South Africa, 10 July 1977. Algologist, botanist, and artist. She started as a schoolteacher in England, followed by a stint teaching in Cape Town, South Africa (1913–1917), and afterwards continued her academic studies at the Univ. of Cambridge, England (1919–1921) and later at the Univ. of Cape Town, obtaining her Ph.D. in 1932. At various times over the next 30 years, she lectured at Rhodes Univ. in Grahamstown, South Africa, and in 1942 was instrumental in establishing the university's herbarium (Herb. RUH), which was incorporated into the herbarium of the Albany Museum in Grahamstown (Herb. GRA) in 1993. During World War II, she served in the South African Women's Auxiliary Services. She was

Fig. 98. Mary Pocock. Photographer unknown. South African National Biodiversity Institute. Reproduced with permission.

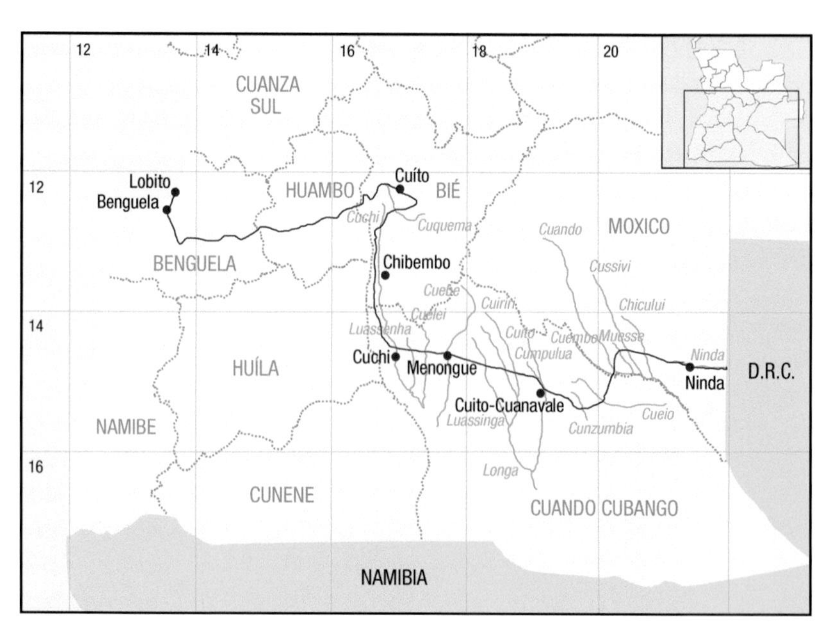

Fig. 99. The itinerary of Mary Pocock in Angola.

awarded a D.Sc. (*honoris causa*) at Rhodes Univ. in 1967. She undertook many expeditions and made a large collection of over 30,000 specimens. In 1925, Pocock and the anthropologist Dorothea Frances Bleek (1873–1948) undertook their famous two-women-on-foot (and in *machila*, i.e., hammock) expedition from the Victoria Falls (Zimbabwe), across Angola to Lobito (Benguela). On 11 May 1925, they entered Angola through the eastern border, at c. 15°S, and five months later they left the country embarking at Lobito on 17 October 1925. They explored a region in the extreme east of Angola, around 15°S, which at the time was virtually unknown. Serpa Pinto (q.v.) and Capelo (q.v.) and Ivens (q.v.) in their *Expedição portuguesa ao interior da África austral* (q.v.) had passed through it on their trans-African journeys, but only Baum (q.v.) in 1900 and Gossweiler (q.v.) in 1907 had made significant collections in the area. The Cuito-Cuanavale region remained undercollected for many years (see *Campanhas de Angola*). The expedition was mostly done on foot or in *machila*, with porters for food and equipment. Pocock collected plants, took photographs, painted watercolours, and kept a journal with observations and details of the route taken and the settlements visited. The itinerary was published by Balarin & al. (1999). Pocock collected 990 numbers during the expedition, and she spent 1927 working on her collections at the Royal Botanic Gardens, Kew (Herb. K) and the Natural History Museum, London (Herb. BM) in England. Her botanical estate is kept at Herb. GRA. She is commemorated in several names, such as *Lampranthus pocockiae* (L.Bolus) N.E.Br. (Aizoaceae). **Angola:** 1925; numbers 264–989. **Herb.:** B, BM, BOL, COI, GRA (incl. RUH), K, KMG, NBG (SAM), PRE, STE, others. **Ref.:** Gossweiler, 1939 (also listed as "M.A. Pococh"); Jacot Guillarmod, 1978; Gunn & Codd, 1981; Stafleu & Cowan, 1983: 305; Vegter, 1983; Bossard, 1993; Romeiras, 1999; Smith & Willis, 1999; Balarin & al., 1999; Figueiredo & al., 2008; Dold & Kelly, 2018. Figures 98 and 99.

Pogge, Paul
(1839–1884)

Bio.: b. Ziersdorf, Mecklenburg, Germany, 27 December 1839; d. Luanda, Angola, 17 March 1884. Explorer. He studied at Berlin, Heidelberg, and Munich from 1858 to 1860, graduating as a lawyer. For a while he ran his father's estate, but after undertaking a hunting trip to South Africa in 1865, he became interested in the exploration of the continent. He offered his services to the *Deutsche Gesellschaft zur Erforschung Aequatorial-Afrikas* as a self-employed hunter and was accepted as a member of an expedition to Angola led by Alexander von Homeyer (1834–1903). Pogge arrived at

Luanda in February 1875 and proceeded inland with four companions. At Pungo Andongo (Malanje), von Homeyer, the botanist Soyaux (q.v.), and a third member of the expedition took ill and had to return to Germany. On 11 June 1875, Pogge and Lux (q.v.) left Pungo Andongo and arrived at Malanje two days later, on 13 June. The following day, they continued towards the interior in an easterly direction, reaching Mona Quimbundo on 26 August. By then Lux had also taken ill and Pogge continued alone. On 15 September 1875, Pogge set off towards the Cassai River and on 30 October he was on the banks of the Cassai, north of Dilolo. He continued in a northeasterly direction and finally arrived at the capital (the Musumb) of the Lunda Kingdom. The Musumb was the locality where the Mwat Yamv resided; the location changed whenever a new king was on the throne. The Musumb, where Pogge arrived on 5 December 1875, was beyond the Lulua and Luisa rivers. Pogge was not authorised to go further north or east. On 17 April 1876, he had run out of provisions and gifts, so he headed back to Luanda. On 27 June 1876, he crossed the Cassai back to what is now Angolan territory and by 21 July he was camped between the Chicapa and Chiumbe rivers. He reached Mona Quimbundo on 30 July 1876. He was back in Malanje in October 1876 and later at the coast. During this expedition he made several collections, mostly of biological specimens. A narrative was published as a book in 1880. Afterwards Pogge approached the *Afrikanische Gesellschaft in Deutschland* with his own project: the establishment of a station in Lunda, following the model of Leopold II's *Association internationale Africaine.* This was approved and in November 1880, Pogge and Wissmann (q.v.) left Hamburg. They disembarked at Luanda on 7 January 1881 and proceeded to Malanje, where they arrived on 25 January 1881. On 8 February, they met Buchner (q.v.), who was returning from Lunda, and later, on 20 February, they met Mechow (q.v.) and Teusz (q.v.), who were on their way back from Cuango. On 22 March, they jointly celebrated the birthday of the German emperor Wilhelm I. After this meeting, Pogge and Wissmann decided to follow a northeasterly route to the country of the Luluwa (D.R. Congo). With the German flag in front of the caravan and with Buchner guiding them for a day, they left Malanje on 3 June 1881. After Sanza they headed in a southeasterly direction, reaching Mona Quimbundo on 20 July. On 1 August 1881, with 69 porters, they travelled north from there along the Luele and Chicapa rivers. They reached the Cassai River on 2 October 1881 and after crossing it they proceeded to what is now territory of the D.R. Congo. On 23 October, they parted ways for a while but reunited shortly afterwards. By mid-April 1882, they were at the Lualaba River near Nyangwe (D.R. Congo). From there they took different routes. Pogge turned back on

5 May 1882, heading to the Lulua River where he arrived on 21 July 1882 and settled in order to establish a station. More than a year later, his means were exhausted, and no replacement had been sent so on 9 November 1883 he began the journey to return home. On 16 December 1883, he crossed the Cassai back into present-day Angola. Finally, back at Malanje he met his companion Wissmann, who was by then on his second expedition in Angola. Pogge was almost unrecognisable due to illness and Wissmann made efforts to help him return to the coast. Finally, Pogge arrived at Luanda on 28 February 1884, but two weeks later he died of pneumonia. He was buried at the protestant cemetery. His diaries survived in spite of his wish to have them destroyed. Together with letters and an account by Wissmann (1889), they provide information on this second expedition. Pogge is commemorated with a statue in Rostock, as well as several plant names of which he collected the type specimens. During the two expeditions he collected 1648 numbers. Collection numbers 1–547 refer to Pogge's first expedition (1875–1876); numbers 548–1648 refer to the second expedition (1881–1884), with 70 numbers missing. The collections were mostly destroyed in Berlin during the bombing of the herbarium of the Botanic Garden and Botanical Museum Berlin-Dahlem (Herb. B) in World War II, but some duplicates, mostly fragments, exist at other herbaria. The numbering of the collections does not follow a chronological sequence. Apparently, all numbers from

Fig. 100. Paul Pogge in 1878. Artist unknown. From Wikimedia Commons. Public domain.

the second expedition, i.e., above number 547, refer to localities in the D.R. Congo. **Angola:** February 1875–end of 1876, January 1881–March 1884; Cuanza Norte, Luanda, Lunda Norte, Lunda Sul, Malanje. **Herb.:** B (main), BM, BR, GOET, HBG, K, P. **Ref.:** Pogge, 1880; Wissmann, 1889; Urban, 1916; Gossweiler, 1939; Mendonça, 1962a; Vegter, 1983; Bossard, 1993; Liberato, 1994; Romeiras, 1999; Lindgren, 2001; Figueiredo & al., 2008, 2020; Heintze, 2010. Figures 100, 101, and 102.

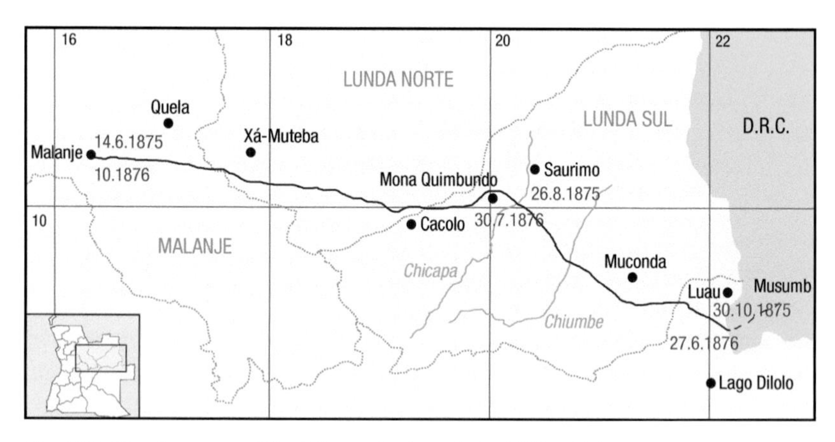

Fig. 101. The itinerary of Paul Pogge in Angola in 1875–1876.

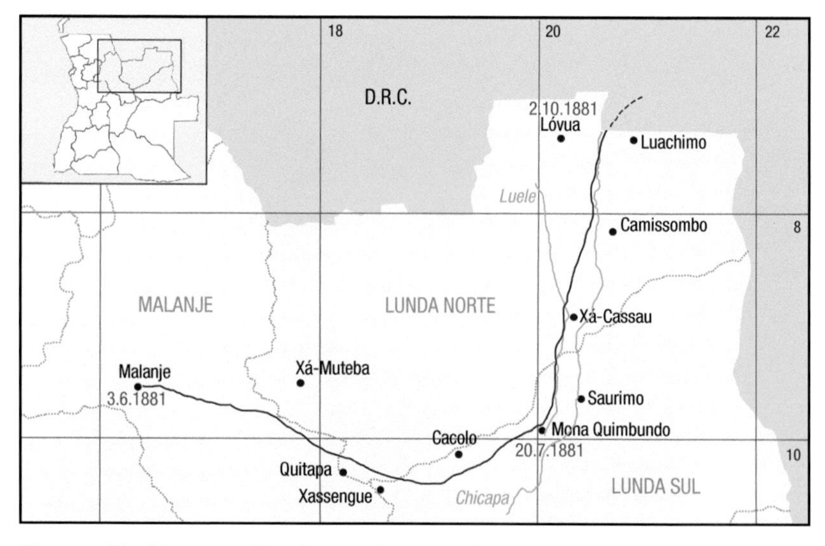

Fig. 102. The itinerary of Paul Pogge in Angola in 1881–1884.

Ponte, António Mendes da (1924–2009)

Bio.: b. Portugal, 22 June 1924; d. Caldas da Rainha, Portugal, 2 March 2009. Agronomist. In September 1948, he graduated as an agronomist at the *Instituto Superior de Agronomia* in Lisbon. In 1949, he accepted an appointment with the *Junta de Exportação do Café*, an institution created in Angola in 1940, which became the *Instituto do Café de Angola* in 1962. From 1949 to 1959, Ponte developed the *Estação Regional da Ganda*, of which he became the first director. In 1959, he joined the *Instituto de Investigação Científica de Angola* as chief of the *Centro de Estudos da Cela*. He later became head of the agronomy department (1962–1964), then adjunct director (1964), and finally director of the Institute (1965–1969). Afterwards he was a provincial inspector with the *Serviços de Agricultura e Florestas* (1970) and with the *Serviços de Planeamento e Integração Económica* (1972–1975), both in Angola, and he worked for a business venture from 1970 to 1972. With the independence of Angola, he immigrated to Brazil where he worked at first with an agronomical research company (1976–1980) and afterwards as a consultant. In the 1990s, he was involved in projects to recover coffee plantations in Angola. He published several articles on agricultural issues, mostly related to the cultivation of coffee. **Angola:** 1968–1973; Benguela, Cuando Cubango; collected with (Manuel) Silva (q.v.). **Herb.:** LISC (40 specimens), LUA. **Ref.:** Bossard, 1993; Romeiras, 1999; Figueiredo & al., 2008; Madalena Ponte (daughter) and Bruno de Araújo (grandson), pers. comm., December 2019. Figure 103.

Fig. 103. António Ponte. Photographer unknown. Private collection of Madalena Ponte. Reproduced with permission.

Powell-Cotton, Antoinette ("Tony")
(1913–1997)

Bio.: b. [Thanet, Kent?], England, [20 July?] 1913; d. [Thanet, Kent?], England, [July?] 1997. Nurse. Daughter of the explorer and naturalist Percy Horace Gordon Powell-Cotton (1866–1940), who was the founder of the Quex Museum at Birchington, England. In 1936 and 1937, she travelled with her older sister, Diana Powell-Cotton (1908–1986), on two expeditions to Angola. The sisters acquired an old lorry and went to the area of the Donguena, Cuamato, and Cuanhama peoples, with the purpose of recording indigenous culture. They carried salt that they exchanged for permission to make films and take photographs. During the expeditions they acquired ethnographic artefacts that amount to nearly 3000 objects. Some of these are on display at the Quex Museum. A few photographs were published in the *Illustrated London News* (11 September 1937). Antoinette also made a few hundred plant collections. During World War II she trained as a nurse and afterwards, apart from nursing activities, she assisted in running the Museum, and participated in local archaeological excavations. The papers relating to the expeditions to Angola are held at the British Museum. **Angola:** 1936–1937; Cunene. **Herb.:** BM, COI, MO. **Ref.:** Gossweiler, 1939; Vegter, 1983; Bossard, 1993; Romeiras, 1999; Figueiredo & al., 2008; Glen & Germishuizen, 2010; Anonymous, s.d.(g).

Poynton, Richard James
(1925–2013)

Bio.: b. South Africa, 23 June 1925; d. Pretoria, South Africa, 11 April 2013. Forest scientist. Although according to Gunn & Codd (1981) he was born in Pretoria, Kruger (2013) stated he was from KwaZulu-Natal. Poynton matriculated in 1942 at Michaelhouse, the Anglican Diocesan College in the Balgowan Valley in the midlands of KwaZulu-Natal. He attended the Univ. of Stellenbosch, in the Western Cape, graduating with a B.Sc. in forestry in 1948. In 1968, he obtained an M.Sc. degree, also from the Univ. of Stellenbosch, and in 1979, he graduated with a Ph.D. from the Univ. of the Witwatersrand. He spent his career at the South African Forestry Research Institute (SAFRI) where he became deputy director of research and was in charge of the herbarium. In 1960, SAFRI assigned him to the production of what would become his *magnum opus*, the work *Tree planting in southern Africa*. It consists of three encyclopaedic volumes: the first volume on pines, published in c. 1977; the second on eucalypts, published in 1979; and the third on other genera, published in 2010, crowning his career at the age of 86. His work constitutes the most comprehensive analysis of exotic tree-planting for any place in the world (Kruger, 2013). Apart from these three monumental volumes, Poynton also published several other very useful works on silvi- and arboriculture, such as Poynton (1984). He also amassed an invaluable herbarium (Herb. PRF), which is now housed in the National Herbarium in Pretoria (Herb. PRE). He collected c. 1500 numbers in southern Africa. **Angola:** 1964. **Herb.:** PRE (PRF). **Ref.:** Poynton, [1977?], 1979, 1984, 2010; Gunn & Codd, 1981; Vegter, 1983; Kruger, 2013.

Price, James ("Jim") Henry
(1932–2007)
Bio.: b. London, England, 6 February 1932; d. 25 October 2007. Algologist. After working for an insurance company and following conscription to serve with the Royal Air Force, he trained as a teacher at Loughborough Teachers Training College. In 1959, he initiated his studies at Liverpool Univ. After graduating, he became a lecturer in botany and marine sciences at Portsmouth Polytechnic. In 1963, he joined the Natural History Museum, London, where he worked until he retired in 1992. He conducted fieldwork in several regions, including in Angola in 1974, as part of a coastal survey undertaken with John (q.v.) and Lawson (q.v.). **Angola:** January–February 1974; with John and Lawson. **Herb.:** BM (main), BRAD, US. **Ref.:** Lawson & al., 1975; Tittley, 2008.

Pritchard, Noel Marshall
(1933–2004)
Bio.: b. 1933; d. [5 April?] 2004. Bryologist. Brought up in London, England, he went to the Univ. of Oxford in 1951 and graduated with a B.Sc. in botany. He undertook an expedition to Angola and afterwards continued his studies for a D.Phil. degree. In 1957, he joined the Univ. of Aberdeen, Scotland, as assistant lecturer. He lectured there until his retirement in 1985. He was also curator of the Cruickshank Botanical Garden, Aberdeen, Scotland from 1964 to 1985. **Angola:** June–September 1954; Huíla, Namibe; c. 100 numbers. **Herb.:** BM, COI, LISC (79 specimens). **Ref.:** Vegter, 1983; Bossard, 1993; Romeiras, 1999; Gimingham, 2005; Figueiredo & al., 2008.

Queiroz
(fl. 1940)
Angola: November 1940; Lunda Norte (Dundo). **Herb.:** COI, LISC (11 specimens). **Ref.:** Figueiredo & al., 2008.

R

Raimundo, António Rodrigues da Fonseca
(1926–2014)
Bio.: b. Portugal, 4 July 1926; d. Portugal, 22 February 2014. Agronomist with *Instituto de Investigação Agronómica de Angola* where he succeeded (Brito) Teixeira (q.v.) as head of the *Divisão de Botânica e Ecologia.* He participated in surveys for the project *Zonagem agro-ecológica de Angola* in the 1970s, with Castanheira Diniz (q.v.) and others. With the independence of Angola, c. 1975 he returned to Portugal. He collected over 1500 numbers. **Angola:** 1969–1974; Bengo, Benguela, Bié, Cabinda, Cuanza Norte, Cuanza Sul, Cunene, Huambo, Huíla, Luanda, Malanje, Uíge, Zaire; collected with Bamps (q.v.), Baptista de Sousa (q.v.), Cardoso de Matos (q.v.), and Maia Figueira (q.v.). **Herb.:** BM, BR, COI, ELVE, K, LISC (1086 specimens), LUA. **Ref.:** Vegter, 1983; Raimundo, 1985; Bossard, 1993; Liberato, 1994; Romeiras, 1999; Figueiredo & al., 2008; G. Cardoso de Matos, pers. comm., 2021.

Ramalho, Paulo Amado de Melo – see **Amado de Melo Ramalho**

Rattray, John
(1858–1900)
Bio.: b. Caputh, Perthshire, Scotland, 29 June 1858; d. Perth, Scotland, 9 December 1900. Botanist. He graduated in 1880 with an M.A. from the Univ. of Aberdeen and in 1883 with a B.Sc. (natural sciences) from the Univ. of Edinburgh. He became a fellow of the Royal Society of Edinburgh in 1885, and later, in 1892, a fellow of the Linnean Society. He collected plants and animals during an expedition on the cable-laying steamship *Buccaneer* to West Africa in 1885–1886. The plant specimens collected in the several ports of call of this expedition (Canary Islands, Senegal, Guinea, Sierra Leone, Ghana, São Tomé, Príncipe, and Angola) have labels printed with "Collected by John Rattray, H.M. Challenger Commission, Edinburgh". The *Challenger* expedition that was undertaken in the wooden steam corvette H.M.S. *Challenger* took place from 1872 to 1876 and it was the first great voyage of oceanographical exploration. With the exception of the Arctic, all oceans were surveyed. However, the itinerary of the expedition did not include the Angolan coast and Rattray was not part of its scientific team. He worked at, and later was in charge of, the Marine Station created by John Murray in 1884, until he joined the editorial staff of the Challenger Office, which Murray directed. In a paper published in 1886 with the results of the expedition to West Africa,

Rattray wrote that it was by the kindness of John Murray that he could join the *Buccaneer* expedition. Rattray was working for the Challenger Commission and was allowed time off to travel on the *Buccaneer*, and for this reason the Challenger Commission appears on the labels. Although Stafleu & Cowan (1983) stated that this collector is not the Scottish algologist and diatomist John Rattray (1858–1900), Figueiredo & al. (2019c) demonstrated otherwise. Rattray published several papers and described many species of diatoms and is commemorated in the algal genus *Rattrayella* De Toni (Eupodiscaceae). His last known publication dates from 1894, so his fruitful publishing career lasted only ten years. It is not known what occupied him for the next, and final, six years of his short life. He is buried in the St Anne's church graveyard at Dowally, Scotland. Two other people with the surname Rattray collected in southern Africa, but not in Angola: George Rattray (1872–1941) in South Africa and Zimbabwe, and James McFarlane Rattray (1907–1974) in Zimbabwe. According to his detailed diary of the expedition to West Africa, while stationed at Luanda harbour, John Rattray procured specimens of most of the species that occurred on the "island" (Ilha de Luanda, a land spit off Luanda). **Angola:** 10–17 February 1886; Luanda. **Herb.:** BM, E, K. **Ref.:** Rattray, 1886, 1894; Hedgpeth, 1946; Gunn & Codd, 1981; Stafleu & Cowan, 1983: 586; Vegter, 1983; Gardiner, 1989; Figueiredo & al., 2008, 2019c.

Reis, A.
(fl. before 1999)
Herb.: LUBA. **Ref.:** Smith & Willis, 1999.

Rey, H.
(fl. 1974)
Angola: 1974; Huíla; collected with (A.) Borges (q.v.) and Constantino (q.v.). **Herb.:** LISC (1 specimen). **Ref.:** Figueiredo & al., 2008.

Reynolds, Gilbert Westacott
(1895–1967)
Bio.: b. Bendigo, Victoria, Australia, 10 October 1895; d. Mbabane, Swaziland (now Eswatini), 7 April 1967. Optometrist and amateur botanist. He went to South Africa in 1902, where he was educated. In 1921, he joined the optician business of his father in Johannesburg. In 1930, he started his own practice that allowed him to travel extensively in the country. As a keen plant grower, he started collecting and studying plants, especially aloes. Some of the living material of aloes that he collected during his travels is still in cultivation in the Mlilwane Game (now Wildlife) Sanctuary, Eswatini (formerly Swaziland).

He collected extensively in several countries in the *Flora of Southern Africa* and *Flora Zambesiaca* regions, as well as in especially eastern Africa and Madagascar. On the whole, he travelled over 320,000 km [200,000 mi.] hunting aloes in Africa. On his travels in Angola, he was accompanied by Smuts (q.v.), to whom his first book (Reynolds, 1950) is dedicated. This Smuts is not to be confused with General J.C. Smuts (1870–1950), who wrote the foreword to Reynolds (1950). The 1959 expedition to Angola was undertaken with a car towing a caravan; 8000 km [5000 mi.] were covered. Reynolds and Smuts entered the country at the Dilolo border (D.R. Congo) and arrived at Luau (Moxico). Travelling in a car and towing a caravan was not easy and soon they realised that their vehicle could not get through the deep sand in the tracks of the region. They travelled westwards and after getting stuck in sand repeatedly, they arrived at Luena (Moxico) where they were told that taking the car and caravan to Munhango (Bié) was impossible. Rather, the car and caravan had to be transported by railway. Again, on arrival at Munhango they were told that "Caminho muito mau, muita areia" (road very bad, much sand) and so they took the car and caravan by train to Camacupa (Bié). From there on, they finally travelled in their vehicle, but in some instances, they had to leave the caravan and continue by car only, returning later to fetch the caravan. It was near Camacupa

Fig. 104. Gilbert Reynolds next to *Aloe duckeri* Christian in Malawi. Photographer unknown. South African National Biodiversity Institute. Reproduced with permission.

that Reynolds saw the first new species of *Aloe* L. (Asphodelaceae) that he would later describe as *A. guerrae* Reynolds, naming it for Guerra (q.v.). On the way to Cuíto (Bié), two further new species were found near Chinguar, these were described by Reynolds as *A. grata* Reynolds and *A. rupicola* Reynolds. At Huambo (Huambo) they were stranded while the cracked rear end of the car chassis was fixed. They proceeded south to Cuvango (Huíla) and west to Matala (Huíla) where they crossed the Cunene River. At the pont, the caravan got stuck both on the way in and the way out, and "in all Africa [Reynolds had] not found any pont quite as bad as that one". From there they continued to Lubango (Huíla). Leaving the caravan there, they went to Moçâmedes (Namibe) and explored the desert. Returning to Lubango, again with the caravan, they continued to Catengue, Benguela, and Lobito (Benguela) where another new species of aloe was found, *A. catengiana* Reynolds. They continued to Luimbale

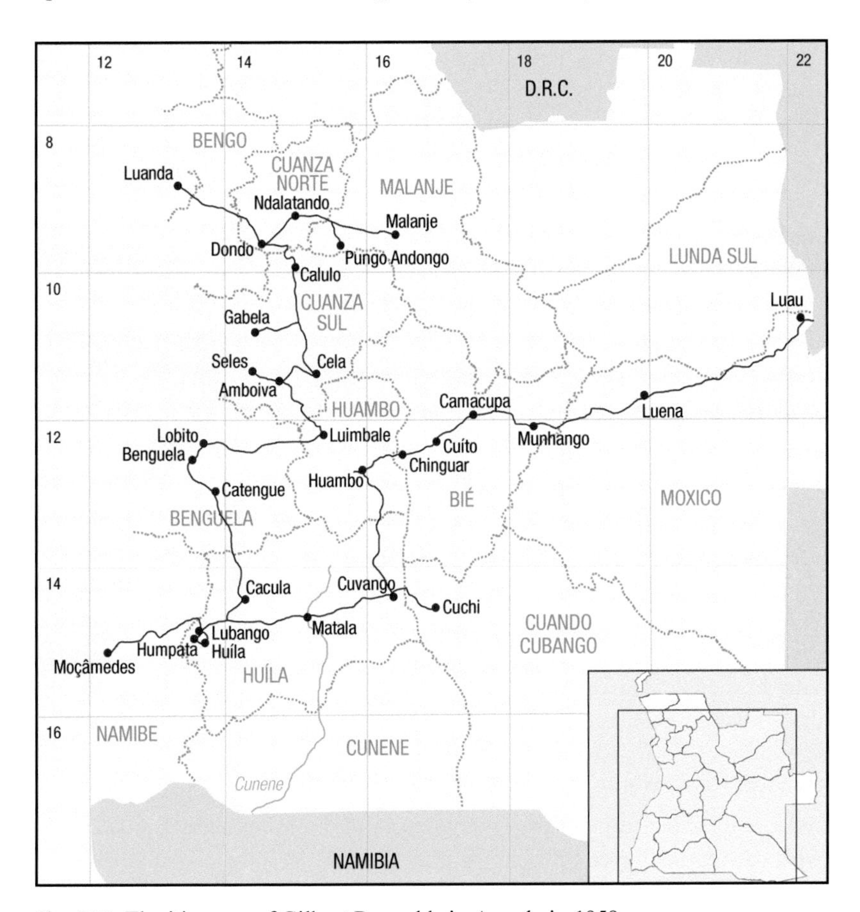

Fig. 105. The itinerary of Gilbert Reynolds in Angola in 1959.

(Huambo), which only resulted in the enigmatic comment "the less said about Luimbale the better", Amboiva, and Seles (Cuanza Sul) where the fifth new aloe was found, *A. gossweileri* Reynolds. They headed to Malanje (Malanje) and finally Luanda where the expedition ended; they returned to South Africa by sea. Reynolds lived in Johannesburg until 1960, at which time he retired to Mbabane, Swaziland (now Eswatini). In 1952, he was awarded an honorary doctorate by the Univ. of Cape Town. He is commemorated in five plant names, including *A. reynoldsii* Letty. **Angola:** 1959–1960, 1964, 1965; at least Bengo, Benguela, Bié, Cuando Cubango, Cuanza Norte, Cuanza Sul, Huambo, Huíla, Luanda, Malanje, Moxico, and Namibe. **Herb.:** BM, BOL, EA, K, LISC (6 specimens), LUA, PRE, S, SRGH. **Ref.:** Reynolds, 1950, 1960, 1966; Kimberley, 1971; Gunn & Codd, 1981; Stafleu & Cowan, 1983: 750; Vegter, 1983; Dorr, 1997; Figueiredo & al., 2008; Walker, 2010; Polhill & Polhill, 2015. Figures 104 and 105; see also maps in Reynolds (1960, 1966: frontispiece).

Ribeiro, A.N.
(fl. 1939)
Angola: 1939; Lunda Sul (Saurimo). **Herb.:** COI. **Ref.:** Figueiredo & al., 2008.

Richards, Mary Alice Eleanor Stokes
(1885–1977)
Bio.: b. Dolgellau, Wales, 3 August 1885; d. Dolserau, near Dolgellau, Wales, 12 April 1977. Née Stokes. Amateur botanist and collector extraordinaire. Since an early age, she had an interest in botany, but her parents did not allow her to pursue a career in science. A very energetic woman, she was active with volunteer public duties and involved in several organisations. During World War I, she turned her house into a Red Cross hospital and was later honoured with a Royal Red Cross Medal for her work for the organisation. Her interest in botany and natural history continued throughout her life, particularly in field botany, and she amassed c. 27,000 magnificent collections that are deposited in numerous herbaria; those from Wales are at the Welsh National Herbarium (Herb. NMW) in Cardiff. In 1907, she married Henry M. Richards (1870–1942). After he died during World War II, she continued living in Wales, but in 1950 she visited Zambia, and having returned to that country a few more times in the following years, she eventually moved there in 1955, at 70 years of age. She remained in Africa until 1974, first in Zambia, then in Tanzania, until she had to return to Europe in 1974 at the age of 88 due to her failing eyesight. In 1964, she was conferred the honorary degree of M.Sc. from the Univ. of Wales. She is commemorated in numerous species names and the genus *Richardsiella* Elffers & Kenn.-O'Byrne (Poaceae). On labels of Angolan specimens, she is

named "Mrs H.M. Richards" or "Mrs M. Richards". Her notebooks are held by the Library and Archives of the Royal Botanic Gardens, Kew. **Angola:** 1962; between Caianda (Moxico) and Zambia. **Herb.:** A, B, BM, BR, DSM, E, EA, K, LISC (1 specimen), LMA, NDO, NMW, NU, NY, P, PRE, S, SRGH, UCNW, UZL. **Ref.:** Exell, 1960; Milne-Redhead, 1978; Vegter, 1983; Glen & Germishuizen, 2010; Polhill & Polhill, 2015.

[Ritchie, A.H.]
Note: Entomologist who was initially with the agriculture department in Jamaica and in 1916 with the Sugar Planters' Association of Jamaica. In the late-1920s–1930s, he was Government Entomologist with the agriculture department in Tanzania (then Tanganyika). Both in Jamaica and in Tanzania he made collections and published several reports. He died sometime before 1944. A specimen he collected in Tanzania (K000240690) is wrongly databased at Herb. K as originating from Angola, the information taken into a database of collectors (JSTOR, 2021). Ritchie did not collect in Angola. **Ref.:** Anonymous, 1916; Wallace & Wallace, 1944; Polhill & Polhill, 2015.

Roberts
(fl. 1973)
Angola: 1973; collected with Huntley (q.v.) and J.D. Ward (q.v.). **Herb.:** PRE.

Rocha, Arnaldo V.
(fl. 1963)
Bio.: He worked at *Posto Zootécnico do Cafu* in Cunene, Angola. Not to be confused with Albano Rocha da Torre (q.v.) or V. Rocha (q.v.). **Angola:** 1963; Cunene, Huíla, Namibe. **Herb.:** LISC (98 specimens). **Ref.:** Romeiras, 1999; Figueiredo & al., 2008.

Rocha, Vicente
(unknown–1993)
Bio.: d. 20 February 1993. Medical doctor with the *Serviços de Portos, Caminhos de Ferro e Transportes de Angola*. He retired in 1975. Not to be confused with A.V. Rocha (q.v.) or Albano Rocha da Torre (q.v.). **Angola:** 1963; Huambo. **Herb.:** LISC (8 specimens). **Ref.:** Rocha, 1965; Figueiredo & al., 2008.

Rocha da Torre, Albano
(1909–1993)
Bio.: b. Meadela, Viana do Castelo, Portugal, 5 February 1909; d. Meadela, Viana do Castelo, Portugal, 19 April 1993. Agrarian technician. He studied

Fig. 106. Albano Rocha da Torre. Photographer unknown. Private collection of the Rocha family. Reproduced with permission.

at the *Escola de Regentes Agrícolas* in Santarém, Portugal, graduating c. 1938. Most of his working career was spent in Angola, in Malanje, where he collected. After the independence of Angola, he returned to Portugal and worked in Loures, Lisbon, at the dairy factory *União das Cooperativas Abastecedoras de Leite de Lisboa*. After retirement, he returned to the family farm that he shared with his brother (António R. Torre, q.v.) and occupied himself with modernising it. His collections are labelled as originating from "Albano Rocha". Not to be confused with A.V. Rocha (q.v.) nor V. Rocha (q.v.). **Angola:** 1948–1953; Malanje. **Herb.:** COI, LISC, LUA. **Ref.:** Bossard, 1993; Romeiras, 1999; Figueiredo & al., 2008; J. Paiva, pers. comm., October 2018. Figure 106.

Rocha da Torre, António – see **Torre**

Rodin, Robert Joseph
(1922–1978)

Bio.: b. Sacramento, California, U.S.A., 15 July 1922; d. San Luis Obispo, California, U.S.A., 27 June 1978. Botanist. He graduated with a B.A. degree from the Univ. of California in 1943, during World War II, and joined the

U.S. Marine Corps serving from 1943 to 1945 in Guam and China. After the war ended, he was the botanist on the 1947–1948 Univ. of California African Expedition assisting Edwin Meyer Loeb (1894–1966) and Ella-Marie Loeb (q.v.) in November and December 1947. In 1951, he was awarded a Ph.D. in botany from the Univ. of California at Berkeley and became a professor of biology at Forman Christian College, Lahore, West Pakistan (now Pakistan). Returning to the U.S.A., he took the position of professor of biology at California Polytechnic Univ. in San Luis Obispo, which he occupied from 1953 to 1976, when he retired. He worked on pteridophytes, Gnetales, and ethnobotany. He returned to Africa in 1973 to continue the work he had initiated 26 years previously in the southwestern region. As in the previous expedition, he was based at Oshikango, "within sight of the Angolan border" (Rodin, 1985). He visited Angola for a few days to consult the leading authority on Ovambo ethnology, Estermann (q.v.). Rodin then made a few collections, some with Estermann. **Angola:** 1973; Huíla. **Herb.:** BH, BM, BOL, BR, C, E, ECON, F, ISU, K, LAH, M, MO, OBI, P, PRE, RAW, SRGH, UC, US, W, WIND. **Ref.:** Lellinger, 1979; Vegter, 1983; Rodin, 1985.

Rodrigues, G.S.
(fl. 1934)
Angola: 1934; Bié. **Herb.:** LISC (5 specimens). **Ref.:** Figueiredo & al., 2008.

Rodrigues, J.
(fl. 1951)
Angola: 1951; Moxico; collected also with Araújo (q.v.). **Herb.:** LISC (1 specimen), LUA. **Ref.:** Bossard, 1993; Romeiras, 1999; Figueiredo & al., 2008.

Rodrigues, Luís
(unknown)
Angola: Huíla. **Herb.:** LUAI. **Ref.:** Figueiredo & al., 2008.

Rodrigues, Maria Zulmira
(unknown)
Angola: Namibe. **Herb.:** LUAI. **Ref.:** Figueiredo & al., 2008.

Rodrigues, Maximino
(unknown)
Angola: Namibe. **Herb.:** COI. **Ref.:** Figueiredo & al., 2008.

Rohan-Chabot – see ***Mission Rohan-Chabot***

Roseira, Arnaldo Deodato da Fonseca
(1912–1984)
Bio.: b. Nossa Senhora das Neves, São Tomé, São Tomé and Príncipe, 29 April 1912; d. Porto, Portugal, 8 March 1984. Botanist. After attending high school in Porto, he enrolled as a student at the Univ. of Porto in 1929. Graduating in 1934, he was appointed as lecturer, remaining with the university for 48 years, from 1934 to 1982. He obtained his Ph.D. in 1944 and in 1957 was appointed as a full professor holding the chair of botany. He directed the Faculty of Science from 1972 to 1974, the *Instituto de Antropologia Dr. Mendes Correia* from 1958 to 1959 and again from 1969 to 1973, and the *Instituto Botânico Dr. Gonçalo Sampaio* from 1960 to 1974 and again from 1981 to 1982, being also in charge of the botanical garden. In 1975, during the revolutionary period (the Carnation Revolution in Portugal), he was removed from his position for political reasons and made to retire on a minimum wage, in a process then called *saneamento* (sanitising) of the state. He was reintegrated in 1976 and finally in 1980 his professorship was reinstated. He retired in 1982. Roseira worked mostly with the flora of Portugal, but he also collected in Africa, mainly in São Tomé, and published on the flora of this island. From 1961 to 1962, he travelled to Angola and Mozambique to lecture in summer courses, and from 1964 to 1966 he was posted in Angola as a professor and vice-chancellor of the *Estudos Gerais Universitários de Angola* (precursor of the Univ. of Luanda, created in 1968, which became Univ. of Angola in 1976, and since 1985 is the Univ. Agostinho Neto). At that time, he made some collections. He is commemorated in *Lasiodiscus rozeirae* Exell (Rhamnaceae). **Angola:** 1964; Luanda (Foz do Cuanza); a few hundred specimens; collected with Barbosa (q.v.) and Wild (q.v.). **Herb.:** BM, COI, LISC (4 specimens), [LUAU†?], PO (main). **Ref.:** Vegter, 1983; Caldas, 1984; Figueiredo & al., 2018a; C. Vieira, pers. comm., 2018.

Roux, Edward ("Eddie") Rudolph
(1903–1966)
Bio.: b. Johannesburg, South Africa, 24 April 1903; d. Johannesburg, South Africa, 2 March 1966. Plant physiologist. He graduated in botany and zoology from the Univ. of the Witwatersrand, Johannesburg, with a B.Sc. in 1924, B.Sc. (Hons.) in 1925, and M.Sc. in 1926. Afterwards he obtained a scholarship to study at Cambridge Univ., England, where he obtained a Ph.D. in 1929. He was briefly with the chemistry department of the Univ. of Cape Town and after World War II, in 1946, he became a senior lecturer in plant physiology at Univ. of the Witwatersrand and a professor of botany in 1962. He became involved in political activities very early in his youth, helping to found the

Young Communist League in 1921. In 1924, he joined the Communist Party of South Africa (CPSA) where he was active until 1936. In 1957, he resumed political activity joining the Liberal Party. His political activities eventually affected his career. Although he was listed as a former member of the CPSA since 1950, it was only in the early 1960s that he experienced persecution by the Nationalist government. In 1964, he was banned from teaching, publishing, attending gatherings, being quoted or leaving Johannesburg. He died two years later in 1966. During his career he published several scientific papers and political articles, and in the 1930s he was in charge of the CPSA weekly newspaper *Umsebenzi*. He is better known for his political history books: *S. P. Bunting, a political biography* (Roux, 1944) and *Time longer than rope* (Roux, 1948). He was honoured with a memorial issue of *The Rationalist* (vol. 11, issue 3). Roux's archives are at the library of the Univ. of the Witwatersrand. **Angola:** 1950; Namibe (Moçâmedes). **Herb.:** J, PRE. **Ref.:** Roux, 1944, 1948; Anonymous, 2006, 2018b; Roux & Roux, 1970; Vegter, 1983; Figueiredo & al., 2008; Glen & Germishuizen, 2010.

Rycroft, Hedley Brian
(1918–1990)
Bio.: b. Pietermaritzburg, KwaZulu-Natal, South Africa, 26 July 1918; d. East London, Eastern Cape, South Africa, 1 December 1990. Botanist and administrator. He graduated from the Natal Univ. College (now the Univ. of KwaZulu-Natal) with a B.Sc. in 1939 and an M.Sc. in 1941. He then continued studying at the Univ. of Stellenbosch, Western Cape, graduating with a B.Sc. in forestry in 1944, during World War II. In order to earn some income while studying, he escorted nurses to the hospital, drove the ambulance, attended post-mortems, and assisted during surgery. After graduating, he took the position of forest research officer and district forest officer at Jonkershoek, near Stellenbosch. While based there he studied the mountain vegetation around Stellenbosch, with this research resulting in a Ph.D. being conferred on him by the Univ. of Cape Town in 1951. In 1954, he became the director of the National Botanic Gardens of South Africa, based at Kirstenbosch, Cape Town. He was the third director after Pearson (q.v.) and Robert Harold Compton (1886–1979). During Rycroft's directorship, a number of regional botanical gardens were developed. The regional expansion of the national botanical gardens network stimulated interest in the South African flora and indigenous gardening, and through several national and international initiatives attention was focussed on the horticultural value of South African plants. He authored at least two books on the world-renowned Kirstenbosch National Botanical Garden in Cape Town (Rycroft, 1975, 1980). Rycroft also held the chair of Harold Pearson Professor

of Botany at the Univ. of Cape Town, lecturing on ecology, until he retired in 1983 and moved to KwaZulu-Natal. He is commemorated in *Aspalathus rycroftii* R.Dahlgren (Fabaceae). **Angola:** April 1962; Cunene (Ruacaná Falls). **Herb.:** K, MO, NBG (SAM), NH, NU, PRE, STE. **Ref.:** Rycroft, 1975: 11, 1980: 28–29; Gunn & Codd, 1981; Vegter, 1983; McCracken & McCracken, 1988: 106; Oliver, 1991; Figueiredo & al., 2008.

S

Salbany, Armando
(unknown–1965)
Bio.: d. 1965. Botanist. In 1939, he was an employee of the *Junta de Exportação do Algodão Colonial* in Mozambique. He obtained an M.Sc. in botany from the Univ. of the Witwatersrand in Johannesburg, South Africa, in 1946. During the 1950s, he published on pastures in Angola and soil conservation in Mozambique. He collected also in Mozambique. He was in Lubango (Huíla) in 1955 and joined the *Campanhas de Angola 1955–1956* (q.v.) for a while. **Angola:** 1955–1956; Huíla. **Herb.:** LISC, LUA. **Ref.:** Salbany, 1947, 1953, 1956; Exell & Hayes, 1967; Vegter, 1986; Bossard, 1993; Romeiras, 1999; Figueiredo & al., 2008.

Sales, Eugénio Eduardo de Melo
(fl. 1960–1963)
Angola: 1960–1963; Bié, Cuanza Sul; collected with (Brito) Teixeira (q.v.). **Herb.:** LISC, PRE; **Ref.:** Bossard, 1993; Romeiras, 1999; Figueiredo & al., 2008.

Samariamba
(fl. 1955)
Note: Likely a member of the staff at *Companhia de Diamantes de Angola* in Lunda. **Angola:** 1955; Moxico. **Herb.:** DIA†?, LISC (4 specimens). **Ref.:** Romeiras, 1999; Figueiredo & al., 2008.

Sampaio Martins, Eurico
(1944–)
Bio.: b. Monte Perobolso, Almeida, Portugal, 18 January 1944. Biologist. He lived in Angola from 1972 to 1975. After graduating from the Univ. of Lisbon in 1971, he worked for *Junta de Investigações do Ultramar* for a while. In 1972,

Fig. 107. Eurico Sampaio Martins at the mouth of the Cuanza River in 2009. Photographer and private collection of Luis Catarino. Reproduced with permission.

he went to Angola to work as a scientific officer at *Instituto de Investigação Agronómica de Angola* in Huambo. Returning to Portugal after the independence of Angola, he then joined the *Instituto de Investigação Científica Tropical* in Lisbon where he worked at *Centro de Botânica* as a researcher until retirement in 2010. Since 2002, he is deputy editor of *Flora Zambesiaca*, for which he also authored several treatments. He also published family accounts for *Flora de Moçambique*. His collection consists of c. 1700 numbers. **Angola:** 1973–1975; Benguela (1973), Bié (1974), Cuanza Norte (1974), Cuanza Sul (1974), Huambo (1973–1975), Huíla (1974), Malanje (1974), Namibe (1974), Uíge (1974); collected with Bamps (q.v.), Cardoso de Matos (q.v.), and Maia Figueira (q.v.); c. 400 numbers. **Herb.:** BR, LISC, LISU (62 specimens), LUA, MA. **Ref.:** Sampaio Martins, 1990, 2009; Bossard, 1993; Liberato, 1994; Romeiras, 1999; Sampaio Martins & Martins, 2002; Figueiredo & al., 2008; E. Sampaio Martins, pers. comm., October 2018. Figures 107 and 78.

Sanjinje, Matende
(fl. 1945–1972)
Bio.: Tchokwe *soba* (native chief) who worked for *Museu do Dundo, Companhia de Diamantes de Angola* in Lunda Norte, Angola, since 1945. In 1946, he participated in the *Campanha etnográfica ao Tchiboco*, a five

month-long expedition organised by the *Museu do Dundo* with the purpose of acquiring artefacts. José Redinha (1905–1983), curator of the *Museu*, was the leader of the expedition that included c. 50 native people. Sanjinje was described by Redinha as a friend and collaborator of the museum who was the *soba* in charge of indigenous protocol in the campaign. He was the interlocutor when permission was required from local chiefs to camp on their lands. He also assisted when transacting with the local peoples, mediated conflicts, and promoted the museum image. It has been suggested that he might have been appointed *soba* under the influence of the *Museu* itself. Well-connected in villages of Lunda and across the border in the D.R. Congo, he continued working for the *Museu* as a guide, interpreter, and facilitator during field work, becoming the main guide and chief of the ancillary staff of the *Laboratório de Investigação Biológica* directed by Barros Machado (q.v.). He was still working there in 1972. Sanjinje was also an experienced hunter and had a good field knowledge of natural history. He collected for the Diamang Collections (q.v.), up to at least 1962. **Angola:** 1953–1962; Lunda Norte, Lunda Sul, Moxico. **Herb.:** DIA†?, LISC (80 specimens). **Ref.:** Redinha, 1953; Porto & al., 1999; Romeiras, 1999; Figueiredo & al., 2008; Bevilacqua, 2016; Ceríaco & al., 2020; J.A.Q. Barcelos and G. Valente, pers. comm., January 2019. Figure 108; see also Figure 52.

Fig. 108. Matende Sanjinje in 1958. By Eduardo Luna de Carvalho. Reproduced under licence CC BY-NC-ND 2.5 PT. https://www.diamang.com

[Santos, C.]
Note: Listed in Figueiredo & al. (2008) based on a record in the Herb. PRE database that has since been removed.

Santos, Pedro José Vieira dos
(fl. 1959)
Angola: 1959; Namibe, Uíge; collected on his own in Uíge, and with (Brito) Teixeira (q.v.) in Namibe. **Herb.:** COI, LISC (1 specimen), LUA. **Ref.:** Bossard, 1993; Romeiras, 1999; Figueiredo & al., 2008.

Santos, Romeu Mendes dos
(fl. 1955–1989)
Bio.: Technician with *Serviços de Agricultura* and *Instituto de Investigação Científica de Angola*. He participated in the two *Campanhas de Angola* (q.v.) expeditions as a technician (*preparador*). He collected over 4000 numbers in almost all provinces of Angola, on his own or with others. Transcripts of his collection records are at the Univ. of Lisbon and digitised at JSTOR Global Plants (JSTOR, 2021). The transcripts refer to numbers 1–186 (September 1955–November 1956), 188–416 (November 1959–January 1961), 424–696 (August 1961–1 December 1961), 697–838 (December 1961), 839–1197 (January 1962–May 1964), and 1997–2515 (1966). He published on useful plants, and on the flora and vegetation of Cuando Cubango, including an article on common names and a book with a vegetation map. **Angola:** 1955–1972; all provinces except Cabinda; collected also with others such as Barbosa (q.v.), Barroso Mendonça (q.v.), Filipe (q.v.), (Romero) Monteiro (q.v.), Murta (q.v.), and Schultz (q.v.). **Herb.:** BM, COI, K, LISC (2366 specimens), LISU, LUA, LUAI, LUBA, PRE. **Ref.:** Santos, 1967, 1972, 1982, 1989; Vegter, 1986 (as "Santos, R."); Bossard, 1993; Liberato, 1994; Romeiras, 1999; Figueiredo & al., 2008.

Santos Júnior, Joaquim Rodrigues
(1901–1990)
Bio.: b. Barcelos, Portugal, 1901; d. Águas Santas, Maia, Portugal, 1990. Anthropologist. He earned three degrees from the Univ. of Porto (B.Sc. in 1923, medicine in 1932, and Ph.D. in anthropology in 1944). He was employed by the Univ. of Porto in 1923, becoming a professor in 1954. During his stay at the Univ. of Porto he was also curator of the *Museu e Laboratório Antropológico*, director of the *Instituto de Zoologia* and *Estação de Zoologia Marítima Dr. Augusto Nobre*, and medical practitioner. During this time, he undertook several anthropological expeditions to Mozambique, with the purpose of gathering (mostly biometric) data on local populations. From 1968 to

1971, he was a visiting professor at the Univ. of Luanda, Angola. He retired in 1971. Santos Júnior's academic career was mostly based on studies of the vast amounts of physical anthropology data acquired during the expeditions, which he used for his Ph.D. dissertation and several papers. In addition to these now reproachable texts, he also published on other subjects such as zoology and prehistory, these being later exalted in a *Festschrift* published after his death. **Angola:** 1969–1970. **Herb.:** K, LUAI, PO (main), US. **Ref.:** Santos Júnior, 1950; Vegter, 1986; Rodrigues, 1990–1993; Pereira, 2005; Figueiredo & al., 2008; C. Vieira, pers. comm., December 2018.

Saraiva, António Leite
(fl. 1963–1975)
Bio.: Technician. He graduated from the Agricultural School of Tchivinguiro (Huíla, Angola) and trained for two years at the *Divisão de Botânica e Ecologia* of the *Instituto de Investigação Agronómica de Angola* at Chianga, Huambo (Huambo). Afterwards he worked at the *Centro de Estudos da Ganda* and *Centro de Estudos da Humpata*. After the independence of the country, he moved to Portugal, settling at Vendas Novas (Évora). **Angola:** 1963; Benguela (Ganda); collected with (Manuel) Silva (q.v.). **Herb.:** COI, LISC. **Ref.:** Figueiredo & al., 2008; G. Cardoso de Matos, pers. comm., 2021.

Schinz, Hans
(1858–1941)
Bio.: b. Zürich, Switzerland, 6 December 1858; d. Zürich, Switzerland, 30 October 1941. Botanist. He obtained a doctorate at the Univ. of Zürich in 1883. The following year he joined an expedition to German South West Africa (Namibia) organised by Adolf Lüderitz (1834–1886). The expedition arrived in Cape Town, South Africa, where they remained for four weeks, until 20 October, when they sailed to Angra Pequena (now called Lüderitz, in Namibia). In March 1885, Schinz left Lüderitz's expedition and pursued his own journey, departing from Angra Pequena in a northerly direction on a long journey. From the Ondonga region, north of Etosha, he went in a northwestern direction into Angola. He crossed the Cunene River in August 1885. For a few weeks he was based at "Onkumbi" on the banks of the Cunene River. This refers to Humbe (Cunene), the southernmost Portuguese post at the time. Gomes e Sousa (1939) incorrectly stated that "Schinz went to a Portuguese Catholic mission at Oncumbi and at that time the natives burned it down, killing two missionaries, but Schinz managed to escape safely". In fact, Schinz was with the missionary Martti Rautanen (1845–1926) in Namibia when he learned from a traveller of the attack, which took place in June 1885, on the mission.

The mission that was destroyed was near the Cuvelai River, the *Missão de São Miguel de Cuanhama*, and not the *Missão do Humbe*; the *Missão de São Miguel de Cuanhama* is further east from Humbe. Schinz travelled and collected in present-day Angolan territory until 24 September 1885, when he returned to present-day Namibia. According to his itinerary map (Schinz, 1891), he did not go further than Humbe and did not go to the *Missão de São Miguel de Cuanhama.* After returning to Zürich, Schinz spent two years in Berlin, studying his collections and writing up the results of the expedition. He became a professor at the Univ. of Zürich in 1892 and director of the botanical garden of Zürich in 1893. He retired in 1928; in the same year he was awarded an honorary doctorate by the Univ. of Berne. In addition to several papers on the African flora, he published the *Conspectus florae Africae* with Théophile Durand (1855–1912); only two parts (out of the several planned) of this *Conspectus* were issued. He is well-known for co-authoring a Flora of Switzerland (*Flora der Schweiz*) in 1928. He is commemorated in several plant names. **Angola:** August–September 1885; Cunene. **Herb.:** B, BM, BREM, C, COI, CORD, E, FR, G, GE, GRA, H, K, L, LE, P, SGO, W, Z+ZT (main). **Ref.:** Schinz, 1891; Gomes e Sousa, 1939, 1949; Gossweiler, 1939; Gunn &

Fig. 109. Hans Schinz. Photographer unknown. South African National Biodiversity Institute. Reproduced with permission.

Codd, 1981; Stafleu & Cowan, 1985: 175–181; Vegter, 1986; Bossard, 1993; Liberato, 1994; Dorr, 1997; Romeiras, 1999; Anonymous, 2007; Figueiredo & al., 2008. Figure 109.

[Schmidt, Johann Anton]
(1823–1905)
Note: Botanist who studied at the Univ. of Heidelberg and Univ. of Göttingen and in 1850 obtained a doctorate in botany. After an expedition to the Cape Verde Islands from 1851 to 1852, he became a lecturer at Heidelberg until 1863, when he moved to Hamburg as a private scholar. Schmidt was listed by Romeiras (1999) as having collected in Angola in 1851, but there is no indication that his expedition to Cape Verde extended to Angola. **Ref.:** Frahm & Eggers, 2001; Figueiredo & al., 2020.

Schmitz, André
(1920–2018)
Bio.: b. La Roche-en-Ardenne, Belgium, 30 July 1920; d. Arlon, Belgium, 31 August 2018. Agronomist and botanist. He graduated in agronomy (*Ing. Eaux et Forêts*) from the *Institut agronomique de l'État* at Gembloux, Belgium. From 1946 to 1963, he was assistant and later head of the section (*chef de groupe*) at the forest division of the *Institut National pour l'Étude*

Fig. 110. André Schmitz. Photographer unknown. From Leteinturier & Malaisse (2001). Reproduced with the permission of Botanic Garden Meise, Belgium.

Agronomique du Congo Belge. In 1961, he graduated with a B.Sc. in botany from the Univ. of Liège, Belgium and in 1967 with a Ph.D. from the same university. From 1963 to 1972, he was a lecturer and participated in some expeditions to Katanga (D.R. Congo). In 1972, he became attached to the *Fondation Universitaire Luxembourgeoise* and later, from 1977 to 1988, he undertook several expeditions to Africa and Asia. He published on various subjects, including phytosociology, and collected c. 8500 numbers, mostly in Katanga. **Angola:** 1950, 1953; Benguela; a few collections. **Herb.:** BM, BR, K, P. **Ref.:** Schmitz, 1963; Bamps, 1973; Vegter, 1986; Leteinturier & Malaisse, 2001. Figure 110.

Schultz
(fl. 1968)
Angola: 1968; Cunene. **Herb.:** COI, LISC, PRE; collected with Filipe (q.v.) and R. Santos (q.v.). **Ref.:** Figueiredo & al., 2008.

[Schultze, W.Z.]
Note: Listed in Figueiredo & al. (2008) based on a record at Herb. PRE database that has since been removed.

[Schulze, E.]
Note: Listed in Figueiredo & al. (2008) based on a record at Herb. PRE database that has since been removed.

[Schumann, Karl Moritz]
(1851–1904)
Note: Curator of the *Botanisches Museum* in Berlin-Dahlem, Germany. He is listed in the literature as a collector in Angola based on a specimen (fragment) that he sent to Herb. K in 1900 and that was wrongly databased as collected by Schumann himself. **Ref.:** Vegter, 1986; Figueiredo & al., 2020.

Schütt, Benedictus Ludwig Heinrich Otto
(1843–1888)
Bio.: b. Husum, Germany, 6 January 1843; d. Istanbul, Turkey (then Constantinople, Ottoman Empire), 1888. Topographer and explorer. He studied in Görlitz and graduated in engineering from the Polytechnic Institute of Berlin. Afterwards, he worked as a topographer in the construction of railways in the Ottoman Empire. After an interruption due to the Franco-Prussian War (1870–1871), he continued his topographical work in Syria and Mesopotamia (a region including parts of present-day Iraq, Iran, Syria, Kuwait, and Turkey)

until 1877. The quality of his work as a topographer made him known to the *Afrikanische Gesellschaft in Deutschland* and he was invited to lead an expedition to Africa. The aim was to continue Pogge's (q.v.) exploration of Angola. Accompanied by the architect Paul Gierow, Schütt arrived in Luanda on 10 or 12 December 1877. On 4 January 1878, they followed the usual route up the Cuanza River to Dondo, then overland to Malanje, where they arrived on 22 February 1878. From there, the expedition departed in a northwesterly direction on 4 July 1878. They crossed the Lui River and passed Cassanje towards the Cuango River. In the country of the Mbangala, they were not allowed to cross the river and had to return to the country of the Mbondo on the other bank of the Lui. They followed a more southerly route to Mona Quimbundo, approximately the same route that Lux and Pogge had taken. From there, they travelled north along the Luele River and then along the Chicapa River, towards Lunda near the Luachimo River. They reached as far north as 10–11 km from the confluence of the Cassai and Luachimo rivers and then turned back. From the right bank of the Luachimo, they went northwest, crossing the Chicapa, Lovua, and Luxico rivers, then in a southwesterly direction towards the Cuango River that they crossed on 14 April 1879. On 12 May, they were back in Malanje. They met Buchner (q.v.) and gave him advice on travelling itineraries. On their way back to the

Fig. 111. Otto Schütt in 1878. Artist unknown. From Wikimedia Commons. Public domain.

coast, in Pungo Andongo, they met Mechow (q.v.), who was on his way to the Cuango River. Schütt and Gierow arrived in Luanda on 21 June and departed on 24 June 1879. Afterwards, Schütt did some topographical work in Japan until 1882. Information on his Angolan expedition can be found in a report and travel diaries. The diaries were edited into a book in 1881. Schütt made ethnographical, ornithological, and mineralogical collections that were sent to Berlin. His name appears in publications as Otto H. Schütt but is given by Heintze (2010) as Benedictus Ludwig Heinrich Otto Schütt. According to Bossard (1993), Schütt made collections from Malanje to Cassai. However, he was not listed by Urban (1892, 1916) as one of the collectors who sent specimens to the Botanic Garden and Botanical Museum Berlin-Dahlem, Germany (Herb. B). No plant collections have been located in databases nor were references to such found in the literature. **Angola:** 10 or 12 December 1877–24 June 1879; Cuanza Norte, Lunda Norte, Malanje; apparently he did not collect. **Herb.:** B. **Ref.:** Schütt, 1881; Bossard, 1993; Romeiras, 1999; Figueiredo & al., 2008, 2020; Heintze, 2010. Figure 111.

Seely, Mary Kathryn
(1939–)
Bio.: b. San Mateo, California, U.S.A., 13 September 1939. Zoologist. She graduated from the Univ. of California with a B.A. in 1961 and a Ph.D. in 1965. In May 1967, she went to Namibia with her then husband, the ornithologist Rolf Jensen. In 1970, she became director of the Desert Ecological Research Unit in Gobabeb, a position she held until 1998. It was under her leadership that in 1990 the Desert Research Foundation of Namibia was founded. An expert in desert ecology, she was also a member of the National Planning Commission (1995–1997) and a member of the "Vision 2030 Core Team" of the President of Namibia (from 2001). She retired in 2006. Seely published several books on the Namib desert and received three honorary professorships and several honorary doctorates. **Angola:** 1974. **Herb.:** K, PRE, WIND. **Ref.:** Gunn & Codd, 1981; Vegter, 1986; Figueiredo & al., 2008; Anonymous, 2019.

Semedo, Claúdio Manuel Bugalho
(unknown–2003)
Bio.: d. Portugal, 2003. Agronomist. He led the third expedition of the *Missão de Estudos Florestais a Angola* (M.E.F.A., q.v.) in 1959, and part of the fourth and final expedition in 1960. After D'Orey (q.v.) retired from the directorship of the *Jardim-Museu Agrícola Tropical* in Lisbon in 1975, Semedo took over that position in which he remained until his death. **Angola:** 1959–1960;

Cabinda (under M.E.F.A.), Luanda (1960, a few collections). **Herb.:** COI, K, LISC (10 specimens), LUAI. **Ref.:** Semedo, 1982; Liberato, 1994; Romeiras, 1999; Figueiredo & al., 2008.

Serpa Pinto, Alexandre Alberto da Rocha (1846–1900)

Bio.: b. Santa Catarina de Tendais, Cinfães, Portugal, 20 April 1846; d. Lisbon, Portugal, 28 December 1900. Army officer and explorer. He studied at the military school in Lisbon and joined the infantry in 1863. The following year he graduated as an ensign. In 1869, he served in the Massangano military campaign in Mozambique. He was subsequently promoted to the rank of lieutenant. In 1875, he was deployed to Madeira. In 1876, he returned to Lisbon and offered his services for a planned geographical expedition to Africa but was first deployed to the Algarve. He was eventually appointed to the expedition team, together with Capelo (q.v.), the leader, and Ivens (q.v.). On 5 July 1877, Serpa Pinto left Lisbon with Capelo and initiated the *Expedição portuguesa ao interior da África austral* (q.v.). Shortly afterwards, he was promoted to the rank of major. Eight months later, in March 1878, while very ill, he arrived in Belmonte (later known as Silva Porto, now Cuíto, in Bié, Angola) to meet Capelo and Ivens. At this time a definitive break-up of the expedition team took place. Capelo and Ivens continued with the official expedition, while Serpa Pinto followed his intention of crossing the continent as an independent traveller. On 23 May, he left Belmonte and on 6 June started off towards the Cuanza River, which he crossed on 14 June. He then travelled in a southeasterly direction and by 19 July was already near Muié (in Moxico), and on 13 August he was travelling along the Ninda River, soon afterwards passing to the territory that is now in the D.R. Congo. Although he originally planned to reach Mozambique overland, he turned south and six months later, on 12 February 1879, entered Pretoria (South Africa), which was then the capital of the Transvaal Republic (the Zuid-Afrikaansche Republiek). The Transvaal Province covered most of the present-day northern provinces of South Africa (North-West, Limpopo, Mpumalanga, and Gauteng). By then, Serpa Pinto's expedition team was reduced to just a few men. Finally, on 19 April, he embarked in Durban (Kwazulu-Natal, South Africa) for Maputo (Mozambique) and from there returned to Europe via Aden (Yemen) and Cairo (Egypt). He arrived in Lisbon on 9 June 1879, where he was reunited with the small group of African men who had accompanied him to the end of the expedition and who had arrived in Lisbon the previous day, together with his pet parrot. He was given a hero's reception and subsequently travelled in Europe doing presentations about the expedition. In 1884, he was appointed to lead another expedition, this time to survey Lake Nyassa (now Lake Malawi). Later, as Portuguese consul to Zanzibar, he undertook an expedition to explore the Shire valley (Malawi). The aims of this expedition included the signing of treaties with the local chiefs to secure territories for Portugal, in line with the claims made on the "Rose-coloured map" (see page 260) Against the advice of Johnston (q.v.), then the British consul in Mozambique, the bellicose Serpa Pinto crossed the Shire River to the highlands of Malawi, dropped the British flag and precipitated a diplomatic crisis that resulted in the 1890 British Ultimatum. Serpa Pinto returned to Lisbon on 20 April 1890 and was promoted to lieutenant-colonel. In 1894, he became a general and was appointed as governor of Cape Verde. He returned to Portugal in 1897. He was aide-de-camp to King D. Luis I and later to King D. Carlos I and was made a viscount. His biography was written by his daughter (Pinto, 1937). Serpa Pinto published the account of his 1877 expedition (Serpa Pinto, 1881a) in London, in Portuguese. It was simultaneously published in English (Serpa Pinto, 1881b). Later in 1881, Capelo and Ivens published their account of the expedition in Lisbon (Capelo & Ivens, 1881, 1882), to which they added a note contesting the grievances exposed by Serpa Pinto in his book. The

Fig. 112. Alexandre Serpa Pinto and his remaining companions at the end of the expedition, in Pretoria, South Africa. By Mr Gross. Public domain.

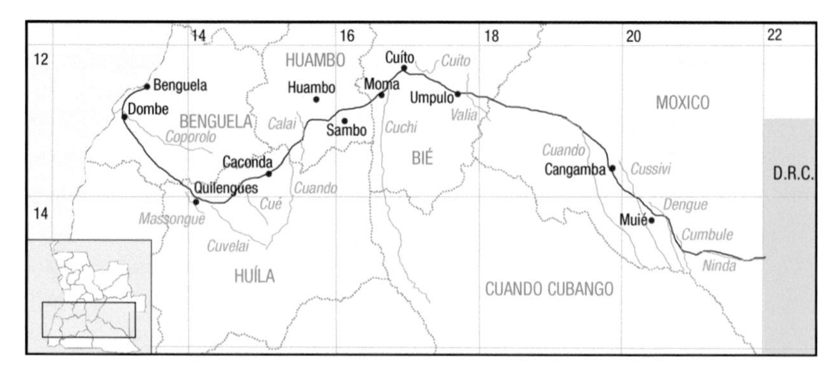

Fig. 113. The itinerary of Alexandre Serpa Pinto in Angola.

plants collected by Serpa Pinto during the 1877 expedition (in August 1878) were studied by Ficalho & Hiern (1881), who published two new plant names commemorating the explorer, *Bauhinia serpae* Ficalho & Hiern (Fabaceae) and *Dianthus serpae* Ficalho & Hiern (Caryophyllaceae). **Angola:** 1877–1878; Benguela, Bié, Huambo, Huíla, Moxico. **Herb.:** COI, K, LISU (70 specimens). **Ref.:** Capelo & Ivens, 1881, 1882; Ficalho & Hiern, 1881; Serpa Pinto, 1881a, 1881b; Pinto, 1937; Gossweiler, 1939; Romariz, 1952; Nowell, 1982; Vegter, 1983 (sub "Pinto, Serpa"), 1986; Santos, 1988; Bossard, 1993; Liberato, 1994; Romeiras, 1999; Figueiredo & al., 2008. Figures 112 and 113.

The "Rose-coloured map"

The "Rose-coloured map" refers to a map showing Portugal's claim to a large area of the African territory as a pink strip from Angola on the west coast to Mozambique in the east. Two such maps were produced, the first of which was issued in 1886. The two maps differed in the border claimed along the Cunene River in southern Angola. The Portuguese claims went against the "Cape-to-Cairo" expansionist plans of Great Britain and Cecil Rhodes and led to conflict between the two countries. This resulted in the "1890 British Ultimatum", in which Britain demanded Portugal's retreat from territories they claimed as occupied in present-day Zimbabwe and Malawi. The King of Portugal, D. Carlos I, had no choice but to concede to British wishes. This had repercussions not only in Africa, but also in Portugal where it contributed to the unpopularity of the King, the strengthening of the Republican movement, and eventually the fall of the monarchy. **Ref.:** Nowell, 1982.

Serra, Manuel Campos de Magalhães
(c. 1943–2021)
Bio.: b. c. 1943; d. Elvas, Portalegre, Portugal, 2021. Technician. He was with the *Instituto de Investigação Agronómica de Angola* at Chianga, Huambo, as an assistant technician (*assistente técnico 1ª classe*) in 1964. After the independence of Angola, in 1975 he returned to Portugal and settled in Elvas (Portalegre), where he died in 2021 of COVID-19. **Angola:** 1960–1968; Huambo; collected with Andrade (q.v.) and (Brito) Teixeira (q.v.). **Herb.:** BM, COI, LISC, LUA, LUAI. **Ref.:** Bossard, 1993; Figueiredo & al., 2008; G. Cardoso de Matos, pers. comm., 2021.

[Shantz, Homer LeRoy]
(1876–1958)
Note: American plant physiologist, phytogeographer, and ecologist. He obtained a doctorate at the Univ. of Nebraska in 1905 and joined the U.S. Dept. of Agriculture in 1908, remaining there until 1926. In 1919, he was appointed to the American Commission to Negotiate Peace to advise on natural vegetation and crop potentialities in Africa. He then participated in the Smithsonian expedition of 1919–1920 from Cape Town (South Africa) to Cairo (Egypt). A second trip to East Africa followed in 1924. After two years as a professor at the Univ. of Illinois, he served as president of the Univ. of Arizona from 1928 to 1936. He then became chief of the Division of Wildlife Management in the United States Forest Service, a position he held until his retirement. From 1955 to 1956, he undertook a third and final expedition to Africa with B.L. Turner (q.v.) to resurvey the localities where he had worked earlier and to assess progress since 1919. Many of the original study sites were rephotographed, resulting in a publication, with Turner (Shantz & Turner, 1958). Shantz visited several countries between South Africa and Egypt, including Mozambique, and present-day Zimbabwe, Zambia, the D.R. Congo, Malawi, and Tanzania. Shantz's archives, including maps and photographic documents, are at the Univ. of Arizona. The 1919–1920 journal and photographs are accessible online (https://uair.library.arizona.edu). The main sets of the collections made during these expeditions are at the herbaria of the Smithsonian Institution, Washington, D.C. (Herb. US) (1919–1920) and Univ. of Arizona, Tucson (Herb. ARIZ) (1955–1956); seed collected likely went to the U.S. National Plant Germplasm System. As a result of the map of the itinerary of the 1955–1956 expedition published by Turner (2016) it has been wrongly assumed that collections could have been made in Angola. **Ref.:** Shantz, 1919–1920; Shantz & al., 1923; Shantz & Turner, 1958; Sauer, 1959; Phillips, 1963; Gunn & Codd, 1981; Wagner, 1992; Turner, 2016; G. Yatskievych, pers. comm., December 2018.

Sieiro, Delmiro Modesto
(fl. 1959–1960)
Bio.: He studied the giant sable antelope (*palanca-negra-gigante*) and authored and co-authored a few works on the subject. He made some plant collections in the giant sable antelope reserve, the *Reserva da Palanca Negra Gigante* in Malanje, Angola. **Angola:** 1959 (Malanje), 1960. **Herb.:** LISC (2 specimens), LUBA. **Ref.:** Frade & Sieiro, 1960; Sieiro, 1974a, 1974b; Figueiredo & al., 2008.

Silva, Cândido da
(fl. 1960)
Angola: 1960; Luanda, Malanje. **Herb.:** COI, LISC (2 specimens). **Ref.:** Bossard, 1993; Romeiras, 1999; Figueiredo & al., 2008.

[Silva, Costa e]
Note: Listed in Figueiredo & al. (2008). We could not locate a source for this information. It is likely an error from a database.

Silva, Francisco de Azevedo e
(fl. 1953–1954)
Bio.: Engineer (according to Bossard, 1993), likely a forester. **Angola:** 1953–1954; Cabinda. **Herb.:** COI, LISC (70 specimens), LUA. **Ref.:** Bossard, 1993; Romeiras, 1999; Figueiredo & al., 2008.

Silva, Hélder José Lains e – see **Lains e Silva**

Silva, João Augusto
(1910–1990)
Bio.: b. Brava, Cape Verde Islands, 1910; d. Paço d'Arcos, Portugal, 1990. Artist, writer, and naturalist. He worked in the colonial administrations of Guinea-Bissau, Angola, and Mozambique. From the 1940s to 1960s, he lived in Mozambique and was attached to the *Instituto de Investigação Científica de Moçambique*. He eventually became administrator of the Gorongosa Park in Mozambique and published a photographic and written record of wildlife in Gorongosa. Later, in Portugal, he was curator of the Zoological Garden in Lisbon. From 1969 to 1970, he was in Angola with an expedition funded by the National Geographic Society and *Direcção dos Serviços de Veterinária de Angola* to study the hippotragine antelope, with R.D. Estes and R.E. von K. Estes (q.v.). During that expedition he collected plants and took photographs, and later authored a book with the results of the study; he authored several

other titles on fauna and wildlife, as well as novels. **Angola:** 1970; Malanje. **Herb.:** LISC (24 specimens). **Ref.:** Silva, 1964, 1965, 1972; Figueiredo & al., 2008.

Silva, Joaquim José da
(c. 1755?–1810)
Bio.: b. Rio de Janeiro, Brazil, or Angola, c. 1755; d. Angola, 2 April 1810. Civil servant. He studied mathematics and medicine at the Univ. of Coimbra, Portugal, and graduated in 1778. He worked as a naturalist at the *Real Museu e Jardim Botânico da Ajuda* in Lisbon, assisting Domingos Vandelli (1735–1816) until 1783. He was appointed by royal decree to collect natural history specimens in Africa and, simultaneously, as a civil servant with the administrative position of government secretary in Angola. In 1783, he embarked to Angola with the Portuguese artist José António (unknown–1784) and the Italian naturalist and artist, Angelo Donati (unknown–1783). In October 1783, they arrived in Luanda after a journey of 146 days, 19 of which were spent at Benguela where Silva collected some plants. Shortly after their arrival in Luanda, Donati died. António died the following year. Silva was sent on a few trips that had purposes other than natural history studies. In October 1783, he was sent to Cabinda to investigate the iron content of the rocks used to build the fort. On the way there he stopped at Dande, north of Luanda, to deal with the shipping of lime for the construction of a fort. In 1784, he travelled to Massangano on the banks of the Cuanza River, searching for manatees and hippopotami; he was *Capitão-Mor* of that settlement for four years. On these trips he took the opportunity to collect plants and other objects. On 25 May 1785, he finally sailed from Luanda to Benguela, with the aim of joining a series of expeditions ordered by José de Almeida e Vasconcelos, the Baron of Moçâmedes and governor of Angola from 1784 to 1790. These expeditions to the Cunene River region were to explore the region and to engage in punitive action against some rebel *sobas* (chiefs). The sea voyage took 28 days; 13 days from Luanda to Sumbe, where they arrived on 7 June and remained anchored for five days before sailing to Benguela, which took a further ten days, arriving there on 22 June. Two missions were conducted under the leadership of Lieutenant-colonel engineer Luis Cândido Cordeiro Pinheiro Furtado (1749–1822). On the first mission, the coast was surveyed from sea and from land. The sea expedition was undertaken in boats (*embarcações redondas*) following the coast. Furtado arranged for the local trader Gregório José Mendes to lead the overland expedition. The boats surveyed the coast and arrived at Angra do Negro to meet the frigate that had arrived there on 3 August. The

boats continued to Cabo Negro, the southernmost point before returning to Benguela. On his arrival back at Benguela, in September, Furtado found out that the overland expedition still had not left. After exerting pressure, this expedition finally departed with 1050 men on 30 September or on 4 October. This expedition followed a coastal route. They arrived at Angra do Negro on 3 November and returned along an inland route via Quilengues and were back in Benguela on 29 December 1785. The second mission had as aim to take punitive measures against disobedient local *sobas* and the discovery of the course of the Cunene River. In August 1785, under the command of António José da Costa, this mission left Benguela, journeying to the interior; Silva joined this group. After 20 days of travel, they reached Quilengues in mid-September 1785. They were based there until November 1785 and only arrived at Cabo Negro nine months later on 10 August 1786. The return journey took about eight months. Silva was in Benguela in April 1787 and back in Luanda in July 1787. This second mission took the extraordinary length of over two years. The reason for this was later exposed by the Secretary of State for Navy and Overseas, Martinho de Melo e Castro (1716–1795), in an official letter to the Governor of Angola in 1791. Costa's *modus operandi* was to establish camp and make the local people sustain

Fig. 114. Joaquim Silva (left) and José António surveying the mouth of Dande River in 1783. By José António. Detail from Silva (s.d.: fig. 84) © Arquivo Histórico dos Museus da Universidade de Lisboa. Reproduced with permission.

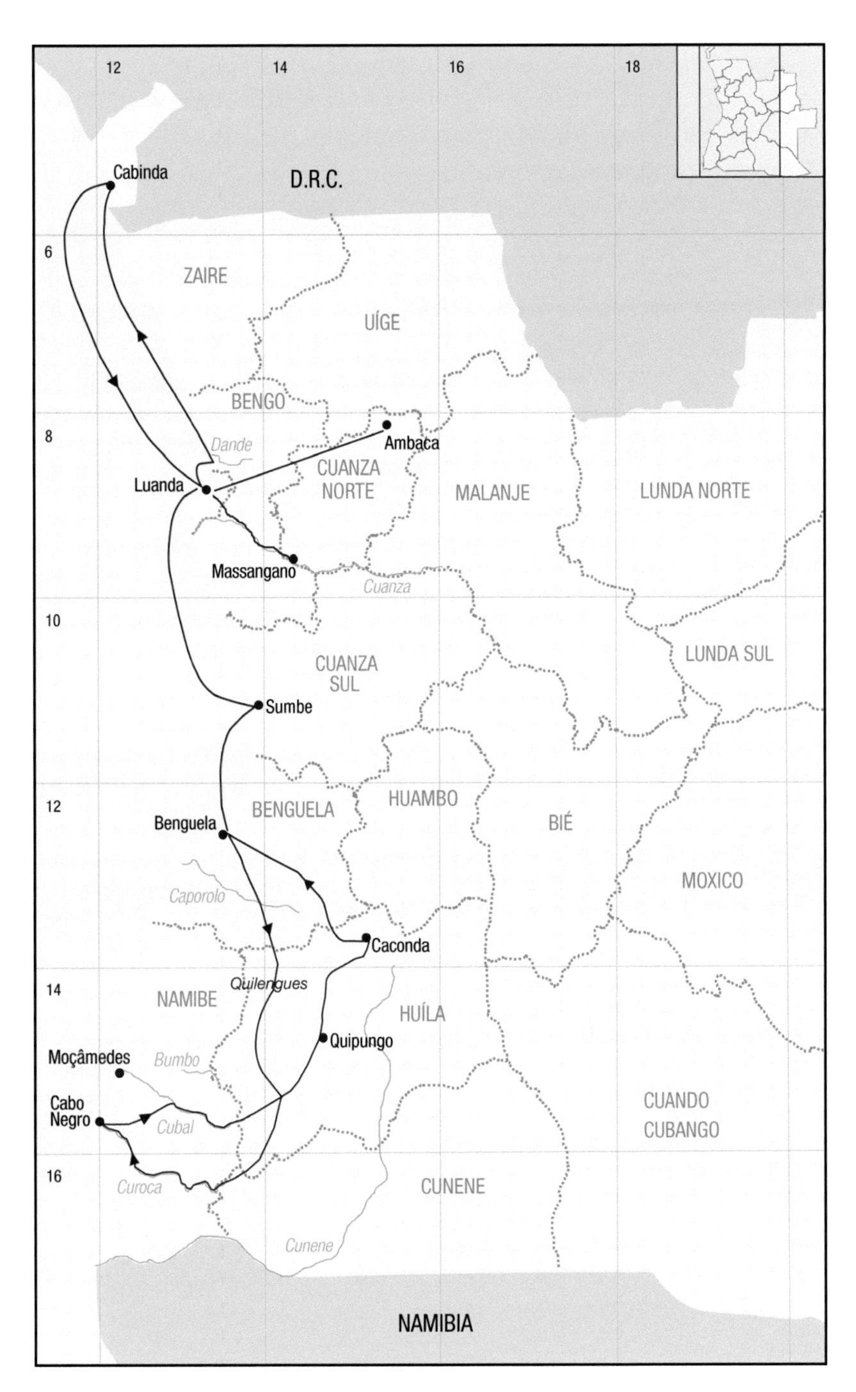

Fig. 115. The itinerary of Joaquim Silva in Angola.

the troops and his entourage. Additionally, he summoned other *sobas* of surrounding chiefdoms and their people and kept them retained until they paid a ransom in ivory and slaves to be allowed to return to their settlements. These extorted goods and people were then sent to Benguela to be sold. The *sobas* who did not obey to his summons were attacked by Costa, who then invaded their lands and made slaves of their subjects and confiscated their goods. Once an area was exhausted, he would move on to another chiefdom. This went on until he was finally ordered back to Benguela. Even so, he managed to take 80 more slaves that he sold there. In spite of all this, and of not having accomplished any of the objectives he had been instructed to achieve, Costa was promoted and given the command of a fort. Silva wrote a report of the journey, which has not been located and of which only a few excerpts were published posthumously (Silva, 1813a, 1813b). The published excerpts do not mention any of these events. Silva, who wrote letters to Melo e Castro, reporting on his activities in Angola, apparently fell out of grace with him for not collecting a sufficient number of specimens and when in October 1787 Silva requested a transfer to Brazil, Castro did not reply to the request. The itinerary of the expedition that Silva undertook in Angola is not known with certainty due to the lack of information. Silva's collections that were at the Royal Ajuda Museum in Lisbon, were taken to Paris by the French naturalist Étienne Geoffroy St. Hilaire (1772–1844) during the Peninsular War and the Napoleonic invasion of Portugal (1807–1811). The collection consisted of 216 specimens. The specimens lack collector name and number, and only a few have locality information. Some lists of material collected by Silva and sent to Lisbon are extant and some were published. **Angola:** 1783–1808; Bengo (Dande), Benguela, Cabinda, Cunene, Cuanza Norte (Ambaca, Massangano), Huíla (Caconda, Quipungo), Luanda, Namibe. **Herb.:** P. **Ref.:** Silva, 1813a, 1813b, s.d.; Hamy, 1908; Pombo, 1935; Exell & Mendonça, 1956; Teixeira, 1962; Simon, 1983; Vegter, 1986; Santos, 1988; Bossard, 1993; Liberato, 1994; Romeiras, 1999; Vieira (C.C.), 2006; Figueiredo & al., 2008; Figueiredo & Smith, 2021b. Figures 114 and 115.

Silva, Manuel da
(1916–1994)
Bio.: b. Portugal, 1916; d. Portugal, 1994. Portuguese technician. He left Angola after 1975, when the country became independent, and joined the *Estação Agronómica Nacional* in Oeiras, Portugal. **Angola:** 1963–1975; mostly in Cuanza Norte, Cunene, and Huambo; collected with many others, such as Dechamps (q.v.), Lopes (q.v.), Murta (q.v.), and Ponte (q.v.); his own collection stretches to over 4000 numbers. **Herb.:** BM, BR, COI, DAO,

DBN, K, LISC (1450 specimens), LISE, LISU, LUA, PRE. **Ref.:** Vegter, 1986; Bossard, 1993; Romeiras, 1999; Figueiredo & al., 2008, 2018a.

Silva, Maria da Graça Domingos
(unknown–2018)
Bio.: b. Mozambique; d. Portugal, 3 January 2018. Biologist. For a couple of years, she attended the Univ. of Coimbra, Portugal, and then moved to Lisbon, graduating in biology from the Univ. of Lisbon. She accepted an appointment at the *Instituto de Investigação Agronómica de Angola* in Huambo. Afterwards she returned to Mozambique and in c. 1974 moved to Portugal. She settled in the Algarve and worked for the *Parque Natural da Ria Formosa.* After retiring, she moved to the Odivelas region where she died. **Angola:** 1970; Benguela; collected with Pinto Basto (q.v.). **Ref.:** M.F. Pinto Basto, pers. comm., October 2021.

Silva, Pinho
(fl. 1957)
Note: Likely Manuel Pinho da Silva, who in 1960 was a member of the staff at *Companhia de Diamantes de Angola* in Lunda in charge of Tchokwe folklore and music collecting. **Angola:** 1957; Lunda Norte. **Herb.:** DIA†?, LISC (1 specimen). **Ref.:** Romeiras, 1999; Figueiredo & al., 2008; N. Porto and G. Valente, pers. comm., January 2019.

Simão, Joaquim Santos
(fl. 1944–1955)
Bio.: Engineer. He worked in Mozambique at *Repartição Técnica de Agricultura* in Maputo. He collected a few specimens in Angola. His field notes referring only to Mozambican collections (numbers 1–1577, collected from September 1944 to October 1947) are at Univ. of Lisbon, with transcripts of the collection records digitised at JSTOR Global Plants (JSTOR, 2021). **Angola:** 1953, 1955; Cabinda, Moxico. **Herb.:** BM, EA, FHO, LISC (1 specimen), LMA, LUA, SRGH. **Ref.:** Exell & Hayes, 1967; Vegter, 1986; Bossard, 1993; Romeiras, 1999; Figueiredo & al., 2008.

Simpson, S.
(fl. c. 1926)
Note: At least one collection is cited in *Conspectus florae Angolensis* as collected by Simpson in Humbe (Cunene, Angola). Watt (1926: 200) wrote that "Mr. S. Simpson submitted a sample of what he calls 'Wild Cotton found growing freely near Humbe, Mossamedes, Angola'". The collection

(*Simpson 1*) was determined by Vollesen (1987) as *Gossypium herbaceum* L. subsp. *africanum* (G.Watt) Vollesen (Malvaceae). No further information could be traced on this collector. **Angola:** c. 1926; Cunene. **Herb.:** K. **Ref.:** Bossard, 1993; Romeiras, 1999; Figueiredo & al., 2008.

Smith, Christen
(1785–1816)

Bio.: b. Skoger, Drammen, Norway, 17 October 1785; d. Congo River, 22 September 1816. Botanist. He studied medicine and botany at the Univ. of Copenhagen. After travelling in Europe and undertaking expeditions to Madeira and the Canary Islands, he was invited by the Royal Society of London to join the crew of an expedition to the Congo. The aim was to determine whether the Congo River had any connection to the Niger basins of western and central Africa. The expedition was under the command of Captain James Hingston Tuckey (1776–1816). Smith's assistant was the gardener David Lockhart (d. 1846), who also made some collections. The expedition sailed in February 1816. On 9 April, they were in Santiago (Cape Verde Islands). The mouth of the Congo was reached in July 1816. They travelled upriver in the steamboat H.M.S. *Congo* and in a lighter vessel. The vessels could not go further than what is now Matadi (D.R. Congo) as they

Fig. 116. Christen Smith c. 1810. Artist unknown. From Wikimedia Commons. Public domain.

were unable to pass the Yellala rapids that are just upriver. Nevertheless, the expedition continued on foot with a party of 30 men, including Smith and Tuckey, setting off on 22 August. The next day they were at Inga where they remained until 2 September, when they continued upriver reaching Isangila. Lacking provisions, and because of exhaustion and illness, they had to turn back, reaching Inga on 13 September. By then both Tuckey and Smith were ill, with Smith being unable to move. Four days later, on 17 September they arrived back at the place where the H.M.S. *Congo* and the smaller vessel were moored. Smith and Tuckey were taken downriver in the smaller vessel. Smith died, at the age of 30 years, during this journey. Tuckey died a few days later. The ill-fated expedition ended with the loss of more than half its 56 members, including all the officers and scientists. Lockhart survived and provided additional information to an account of the expedition that was based on the diaries of Tuckey and Smith (1818). The botany was contributed by Robert Brown. The account includes a map of the Congo River with localities shown and also a line that appears to indicate the course taken by the vessels (along the left or right banks). The collections of the expedition that refer to Angola are those made on the southern side (left bank) of the Congo River up to Noqui where the expedition arrived on 10 August. After Smith explored the Noqui hills, they proceeded on the right bank of the river, which is now the

Fig. 117. Zaire River and Pedra do Feitiço as seen during Tuckey's expedition. By James Fittler from a sketch by the Lieut. Hawkey. Public domain.

D.R. Congo. From the coast up to Noqui, Smith landed at several places that are now in Angola: Ponta do Padrão, Soyo, Pedra do Feitiço, and on several islets. As noted by Mendonça (1962a), even though Smith's specimens have no labels with an indication of date or locality, several have original labels with a number and one or two letters, such as "S" (specimen BM000922537), "MS" and "O" (specimen BM000922516) that may indicate localities. Although Tuckey is recorded in the literature as a collector (Romeiras, 1999), there is no indication that he collected any specimens while on this expedition. **Angola:** 1816; Zaire (Nóqui hills, Pedra do Feitiço, Ponta do Padrão, Soyo, Congo islets). **Herb.:** B, BM, BR, C, DBN, G, G-DC, K, LE, LINN, MO, P, STR. **Ref.:** Tuckey & Smith, 1818; Buch, 1826; Mendonça, 1962a; Vegter, 1986; Bossard, 1993; Liberato, 1994; Romeiras, 1999; Figueiredo & al., 2008. Figures 116 and 117.

Smith, G.O.
(fl. 1973)
Note: This is likely Garth Owen-Smith (1944–2020), a Namibian conservationist. He started working in the Kunene (Namibia) region in the 1960s, assisting rural communities to link development with wildlife conservation and improving their income through natural resources enterprises. He founded and co-directed Integrated Rural Development and Nature Conservation from 2000 to 2010. In 2011, he published his memoirs (Smith, 2011). He received several awards for his conservation work. **Angola:** June 1973; Cunene (Iona). **Herb.:** PRE. **Ref.:** Figueiredo & al., 2008; Smith, 2011.

Smith, Peter Alexander
(1931–1999)
Bio.: b. Harare, Zimbabwe, 2 August 1931; d. Maun, Botswana, 20 May 1999. Naturalist and conservationist. From 1950 to 1952, he studied at Rhodes Univ., Grahamstown, South Africa. He joined the Colonial Development Corporation as an assistant accountant and worked in Zimbabwe and Namibia (1963–1958), then worked for the Botswana government on tsetse fly control (1959–1970) and for that country's department of water affairs (1970–1972). From then on, he was with the department of agricultural research studying the Okavango drainage system and was mainly concerned with the ecological zoning of the delta, channel blockage, and aquatic weed management. He collected over 6000 plant specimens in the Okavango Delta. His collection was donated to the Harry Oppenheimer Okavango Research Centre of the Univ. of Botswana and formed the basis of the Peter Smith Herbarium (Herb. PSUB). **Angola:** fide Gunn & Codd, 1981. **Herb.:** BR, GAB, K, PRE, PSUB,

SRGH. **Ref.:** Gunn & Codd, 1981; Vegter, 1986; Setshogo, 2005; Jones, 2006; Figueiredo & al., 2008.

Smuts, Neil Reitz
(1898–1963)
Bio.: b. Johannesburg, South Africa, 23 December 1898; d. Johannesburg, South Africa, 6 August 1963. Medical doctor. In August 1917, during World War I, the 18-year-old Lieutenant Smuts was commissioned to the Royal Flying Corps (later the Royal Air Force). He flew Sopwith Camels and scored several victories. His marked courage and initiative in attacking ground targets were noted and he was awarded a Distinguished Flying Cross (D.F.C.) for his service in the Royal Air Force during World War I. After the war, he graduated as a medical doctor in Edinburgh, Scotland. During World War II, he was again in service as a lieutenant-colonel with the South African forces. Afterwards he was awarded the Order of the British Empire (O.B.E.) for services in Italy, North Africa, and Madagascar. As a medical doctor, he practised in Johannesburg until 1957. He had an interest in plants and was a conservationist. He contributed to the costs of publishing several books on the

Fig. 118. Neil Smuts. Photographer unknown. South African National Biodiversity Institute. Reproduced with permission.

South African flora, including Reynolds's *The aloes of South Africa* (1950), which is dedicated to Smuts. He accompanied Reynolds (q.v.) on his expeditions to Malawi, Tanzania, and Zambia in 1958, and to Angola and Zambia in 1959. **Angola:** 1959. **Herb.:** PRE (c. 200 specimens). **Ref.:** Reynolds, 1950: B, 1966: xi; Gunn & Codd, 1981; Vegter, 1986; Wongtschowski, 2003; Figueiredo & al., 2008. Figure 118.

Soares
(fl. 1960s)
Note: Collected a few plants in the Dundo park (Lunda, Angola). The collections were in the private herbarium of Cavaco (q.v.) and were deposited in the herbarium of the *Muséum national d'Histoire naturelle* in Paris (Herb. P) in the early 1960s. **Angola:** Lunda Norte (Parque in Dundo). **Herb.:** P (c. 14 numbers). **Ref.:** Figueiredo & al., 2008.

Soares, Firmino António
(fl. 1957–1974)
Bio.: Forester. He was a civil servant with the *Serviços de Agricultura e Florestas do Ultramar* (colonial agricultural and forestry services) in Portugal. He was stationed in Angola until September 1974. He then joined the *Instituto de Produtos Florestais* in Portugal. He published a few papers on forestry and agriculture matters relating to Angola, the Cape Verde Islands, and Timor. He collected a few plants in Angola. **Angola:** September 1957; Moxico (Cafungo, Luau). **Herb.:** COI, LISC (8 specimens). **Ref.:** Soares, 1959, 1961; Romeiras, 1999.

Soini, Sylvi Esteri
(1920–2018)
Bio.: b. Tyrvää, Finland, 13 March 1920; d. Sastamalala, Finland, 18 March 2018. Agronomist. After World War II ended, she studied agriculture and forestry at the Univ. of Helsinki. She graduated in 1956 and worked for the Agricultural Research Centre until she was invited by the Finnish Missionary Society to take a position of agriculturist with the Lutheran Church in Namibia. She worked for the Lutheran Church for three years, from May 1965 to June 1968, doing agricultural surveys of Owambo and Kawango. Afterwards she became a teacher at the Finnish Mission High School in Oshigambo and at the Engela Institute for the Finnish Missionaries Society. During her stay in Namibia, she collected c. 1500 numbers, some in Angola. In October 1970, she returned to Finland to join the Bureau for Local Experiments of the Agricultural Research Centre. She published her memoirs at the age of 95.

Fig. 119. Sylvi Soini. Photographer unknown. South African National Biodiversity Institute. Reproduced with permission.

Angola: 1969; Cunene. **Herb.:** H (main), PRE. **Ref.:** Roivainen, 1974; Soini, 1981, 2015; Vegter, 1986; Figueiredo & al., 2008; Glen & Germishuizen, 2010; Ranki, 2018. Figure 119.

Sousa, A. Figueira
(fl. 1962–1970)
Angola: 1962–1970. **Herb.:** COI, LUAI, LUBA. **Ref.:** Figueiredo & al., 2008.

Sousa, António de Figueiredo Gomes e – see **Gomes e Sousa**

Sousa, Francisco de
(1893–unknown)
Bio.: b. Portugal, 22 February 1893. Technician who joined the *Instituto Botânico* of the Univ. of Coimbra as a gardener in 1925, then became a collector (1927–1934), herbarium technician (1934–1935), and finally naturalist assistant from 1936 to his retirement in 1958. He was in charge of the seed collection and production of the *Index seminum*. He participated in the *Missão Botânica a Angola II* (q.v.) in 1937 and collected c. 350 numbers with Carrisso (q.v.). **Angola:** March–June 1937; collected with Carrisso (q.v.). **Herb.:** BM,

COI, LISC, LUAI, P. **Ref.:** Gossweiler, 1939; Fernandes, 1984; Vegter, 1986; Liberato, 1994; Figueiredo & al., 2008.

Sousa, João Bernardo Nobre Baptista de – see **Baptista de Sousa**

Sousa, José António de
(fl. 1961–1973)
Angola: 1961–1973; Cunene, Huíla, Namibe; collected also with Barroso Mendonça (q.v.) and Menezes (q.v.) in 1970 and 1973. **Herb.:** COI, K, LISC (126 specimens), LUAI, LUBA, PRE. **Ref.:** Romeiras, 1999; Figueiredo & al., 2008.

[Souter, R.]
Note: Based on a record in the Herb. PRE database due to a misreading of a label of a collection made by Santos (q.v.).

Soyaux, Herman
(1852–c. 1928)
Bio.: b. Breslau, Silesia (then Prussia), Wrocław, Poland, 4 January 1852; d. Brazil, c. 1928. Botanist. After studying botany in Berlin, in 1873 he followed the advice of the botanist Georg Schweinfurth (1836–1925) and joined an expedition of the *Deutsche Gesellschaft zur Erforschung Aequatorial-Afrikas* to the Kingdom of Loango (now the Rep. of the Congo and Cabinda, Angola). He left Berlin on 24 November 1873 and two months later, on 24 January 1874, he arrived at Landana (Cabinda) and later proceeded to Chinchoxo, a station north of Landana, beyond the Chiloango River. Although he was mostly stationed at Chinchoxo, he undertook a trip to the Chiloango and up the Kouilou River (Rep. of the Congo), and reached the forests of Mayombe. In 1875, he was appointed by the *Deutsche Gesellschaft zur Erforschung Aequatorial-Afrikas* to join the expedition led by Alexander von Homeyer (1834–1903) to Angola. After leaving Loango and passing through Landana, Banana, and Ambriz, Soyaux arrived at Luanda on 21 January 1875. There he met Homeyer and Pogge (q.v.). In February 1875, they proceeded inland up the Cuanza River to Dondo and then overland to Pungo Andongo. In April 1875, Soyaux had to return to Luanda to replenish the expedition's fiscal resources and he remained in the city for a few weeks. At that time, Lux (q.v.) arrived to join the expedition and on 12 May 1875 Soyaux and Lux left Luanda to travel to Pungo Andongo. Once there, Soyaux and Homeyer contracted malaria and had to leave the expedition, returning to Luanda on 27 August 1875. Soyaux finally made it back to Loango, arriving very ill at Landana on 3 October 1875.

He recovered and on 5 May 1876, he returned to Berlin. In 1879, he returned to Africa to run a coffee plantation in Gabon for the C. Woermann Co. of Hamburg, Germany. In 1884, while based in Gabon, Soyaux was visited by Büttner (q.v.), who was en route to Angola. Soyaux returned to Germany and in 1885 joined the *Deutschen Kolonialverein*. He visited Brazil in 1886 and settled in that country in 1888 as the leader of a colony of Germans called *Bom Retiro* in Rio Grande do Sul. In 1904, he was secretary-general of the *Centro Económico do Rio Grande do Sul* and lived in Porto Alegre, and from 1913 to 1928 he was president of the *Verband Deutscher Vereine*. He wrote several articles and published two books on his experiences in Africa. He is commemorated in several plant and animal names, including the genus *Soyauxia* Oliv. (Peridiscaceae) and *Mussaenda soyauxii* Büttner (Rubiaceae) that was described by Büttner. Soyaux collected in Cabinda Province from 24 January 1874 to January 1875, and then in Angola (at least in Malanje Province) until 6 May 1876. There were 270 numbers from Angola received at the herbarium of the Botanic Garden and Botanical Museum Berlin-Dahlem (Herb. B). The collection numbers are chronological, those from Cabinda reaching number 223, while those from Malanje range from 226 to 249. By

Fig. 120. Herman Soyaux. Photographer unknown. Ethnologisches Museum, Staatliche Museen zu Berlin. Reproduced under licence CC BY-NC-SA 3.0 DE.

the time that Soyaux returned to Europe, he had amassed 1038 specimens, a figure that includes duplicates. When he later returned to Africa in 1879, he started a new numbering series, which goes up to at least number 457 for material collected in Gabon. **Angola:** 24 January 1874–5 May 1876; Cabinda (Loango, Chinchoxo), Cuanza Norte, Luanda, Malanje (Pungo Andongo). **Herb.:** B, BM, COI, E, G, JE, K, M, P (273 numbers), PC, S, US, W, Z+ZT. **Ref.:** Soyaux, 1879, 1888; Urban, 1916; Gossweiler, 1939; Mendonça, 1962a; Leeuwenberg, 1965; Stafleu & Cowan, 1985: 764–765; Vegter, 1986; Bossard, 1993; Liberato, 1994; Romeiras, 1999; Figueiredo & al., 2008, 2020; Heintze, 2010; Troelstra, 2017. Figure 120.

Spoerndli, J.
(fl. 1947)
Bio.: Missionary with the Congregation of the Holy Spirit (Spiritans). He was stationed at the Catholic Mission of Quibaxe, Cuanza Norte, Angola, in the Congregation of the Holy Spirit (Spiritans). Earlier, he was in Nigeria and published a paper on the Mbari houses of the Igbo people. **Angola:** 1947, Cuanza Norte (Dembos). **Herb.:** LUA. **Ref.:** Spoerndli, 1945; Bossard, 1993; Figueiredo & al., 2008.

Stanton, Angela
(c. 1935–2009)
Bio.: b. c. 1935; d. Westbury-sub-Mendip, Somerset, England, 29 September 2009. She was a secretary at the Imperial College in London. In 1957, she married William Iredale Stanton (1930–2010), a mining geologist. She accompanied him to Angola where he produced 1 : 250,000 geological maps for the areas of Bembe, Maquela do Zombo (Uíge), and Mbanza Congo (Zaire). Later they were in Portugal. In 1970, they returned to England, settling in Westbury-sub-Mendip where W.I. Stanton worked for the River Authority until 1995. While they were stationed in Angola, she collected in the northern provinces. She committed suicide at home in 2009. **Angola:** 1958–1970; Cuanza Norte, Uíge, Zaire; 131 numbers. **Herb.:** BM (main), BR, COI, DPU, LISC (25 specimens), NY. **Ref.:** Vegter, 1986; Bossard, 1993; Romeiras, 1999; Figueiredo & al., 2008; Mullan, 2010.

Stopp, Klaus Dieter
(1926–2006)
Bio.: b. Kötzschenbroda (now a suburb of Radebeul, near Dresden), Germany, 11 July 1926; d. Mainz, Germany, 6 June 2006. Botanist, pharmacologist, and historian. As a young man straight out of school, he became a soldier in

World War II and was taken prisoner by the Soviet army. He was deported to an uranium mine in the Urals, but escaped and made his way back home on foot. Afterwards he studied at the Johannes Gutenberg Univ. in Mainz and obtained a Ph.D. in botany in 1949. He specialised in ethnobotany and published on fruit morphology. He undertook several expeditions, including to South Africa (1950–1951), Congo (1954; fide Meurer, 2007; likely D.R. Congo), Angola (1959–1960), New Guinea (1961), and Zambia and Kenya (1967, 1973). While appointed as scientific assistant (1949–1954), he was curator of the botanic garden at the Johannes Gutenberg Univ. from 1962 to his retirement in 1988, he was a professor of pharmacology at the *Institut für Spezielle Botanik und Pharmakognosie* at the same university. He was also a cartographical historian and collector of maps and antiquarian ephemera, and published several catalogues, including a six-volume catalogue of the birth certificates of German immigrants to the U.S.A., which is now a standard reference for genealogical studies. The plant specimens Stopp collected in Angola are labelled "Itinera africana. IV Angola centralis & austro-occidentalis 1959–60 (Dist. Huambo, Huila et Moçamedes)" and have a sketched map attached indicating the collecting locality. The numbering is not chronological. Although Stopp undertook journeys to South Africa and collected in that country, he was not included in Gunn & Codd (1981),

Fig. 121. Klaus Stopp. Photographer unknown. Reproduced from Earnest & Earnest (2006) with the permission of the Mid-Atlantic Germanic Society.

but his East African collecting activities are described in Polhill & Polhill (2015: 441). Stopp also deposited a collection made by Damann (q.v.) in the herbarium of the Johannes Gutenberg Univ. (Herb. MJG). Between 1964 and 1971, Stopp described some new taxa of *Ceropegia* L. (Apocynaceae) from Angola, including the endemic *C. damannii* Stopp (named for Damann) and *C. mendesii* Stopp (for Mendes, q.v.). **Angola:** November 1959–February 1960; Huambo (Água Clara, Canjangue, Chipipa, Huambo), Huíla, Namibe (Moçâmedes). **Herb.:** BM, COI, K, LISC (1 specimen), MJG (main). **Ref.:** Stopp, 1958, 1964, 1971; Vegter, 1986; Earnest & Earnest, 2006; Schunack & al., 2006; Meurer, 2007; Figueiredo & al., 2008, 2020; Polhill & Polhill, 2015. Figure 121.

T

Tanton, C.
(fl. 1997)
Angola: 1997, Luanda (Quiçama). **Herb.:** PRE. **Ref.:** Figueiredo & al., 2008.

Taruffi, Dino
(unknown–1929)
Bio.: d. 29 November 1929. Agronomist. Professor at the Univ. of Pisa, Italy, holding the chair of Rural Economy. In 1908, he was designated by the Italian *Istituto Agricolo Coloniale* to join a mission to Angola to assess the land resources with a view to determine the suitability of the country for settlement by Italians. The mission left Italy in August 1912 and returned in January 1913, having travelled more than 2000 km in Angola. In 1916, Taruffi published a report, which was translated into Portuguese in 1918, on the agricultural possibilities of the Benguela plateau (now the Bié plateau). Chiovenda (1917) cited collections made by Dino Taruffi in Angola. At the herbarium of the Natural History Museum, Univ. of Florence (Herb. FI), there are 42 specimens from Benguela (Chiara Nepi, pers. comm., October 2018). They were donated by “A. Taruffi” (note, not “D. Taruffi”) in November 1915. The two names likely refer to the same person as “Dino” is often used as an abbreviation of Alfredo. On the specimen labels (e.g., FI000819), the collections are recorded as having been made in 1914 (and received at Herb. FI in 1915). **Angola:** 1912–1913 (1914?). **Herb.:** FI (42 collections), FT. **Ref.:** Chiovenda, 1917; Taruffi, 1918; Anonymous, 1929; Arias, 1930; Bossard, 1993; Romeiras, 1999; Figueiredo & al., 2008.

Teixeira, Joaquim Martinho Lopes de Brito (1917–1969)

Bio.: b. Luanda, Angola, 11 November 1917; d. Lisbon, Portugal, 30 November 1969. Agronomist. After attending school in Luanda, he worked for a business owned by his godfather, who eventually sent him to Portugal to study at the *Instituto Superior de Agronomia* in Lisbon. From 1940 to 1945, he attended the *Instituto*, graduating as an agronomist. For the next four years, he acquired training at the *Estação Agronómica Nacional* in Oeiras near Lisbon. In 1949, he returned to Angola to join the *Serviços de Agricultura* where he assisted Gossweiler (q.v.), who was then the director of the *Gabinete de Botânica*. After Gossweiler's death in 1952, Teixeira was appointed director of the *Gabinete*. In 1958, he was invited by F.A. Mendonça (q.v.) to participate in an expedition to survey the east of the Cunene River. From 1958 to 1961, he was the head of the *Secção de Botânica e Ecologia* of the *Direcção de Agricultura e Florestas* and afterwards he was head of the *Divisão de Botânica e Ecologia* of the *Instituto de Investigação Agronómica de Angola* (IIAA) from 1961 to his early death at the age of 52. He reorganised and substantially expanded the herbarium of the IIAA (Herb. LUA) and implemented collecting activities in Angola. He was also in charge of the *Brigada de Cartografia da Vegetação*. His group training sessions with the staff at Herb. LUA, identifying the plants collected during their expeditions (plant family sorting) is still remembered by those who participated and learned from him, such as Cardoso de Matos (q.v.) and Pinto Basto (q.v.). He was a pleasant man, honest and unpretentious, and the first plant taxonomist born in Angola. He travelled to South Africa in 1963 to attend the Kirstenbosch Golden Jubilee in Cape Town during the apartheid era. This caused some trepidation among his staff as he was of African descent. Nevertheless, he was very well received and travelled in the country for a couple of weeks. During his career, Teixeira amassed a total of over 13,000 numbers that include many specimens collected by others (e.g., Cardoso de Matos [q.v.], Pinto Basto [q.v.]) during group expeditions. These collections are attributed to "Teixeira & al." on the labels. The main set is at Herb. LUA, with numerous duplicates in other herbaria. Typed copies of the collection records for numbers 1–13,239 are deposited at the Univ. of Lisbon and digitised at JSTOR Global Plants (JSTOR, 2021). He is commemorated in several plant names, such as *Euphorbia teixeirae* L.C.Leach (Euphorbiaceae). Among his c. 30 publications are the vegetation maps of Quiçama (Teixeira & al., 1967) and Bicuar (Teixeira, 1968), and the treatment of Hamamelidaceae for the *Conspectus florae Angolensis*, which was published posthumously (Teixeira, 1970). **Angola:** 1949–1969; Bié, Cabinda, Cuando Cubango, Cuanza Norte, Cuanza Sul, Huambo, Huíla, Luanda, Lunda, Malanje, Namibe, Uíge,

Fig. 122. Joaquim Brito Teixeira. Photographer unknown. From Centro de Botânica / Instituto de Investigação Científica Tropical (IICT), Universidade de Lisboa. Reproduced with permission.

Zaire: collected with numerous collectors. **Herb.:** BM, BR, COI, ELVE, IPA, K, LISC (6073 specimens), LISI, LMA, LUA, LUAI, MTJB, PO, PRE, SRGH, STR, TLA, WAG. **Ref.:** Mendonça, 1962a; Teixeira & al., 1967; Teixeira, 1968, 1970; Garcia, 1970; Vegter, 1988; Bossard, 1993; Liberato, 1994; Romeiras, 1999; Figueiredo & al., 2008; G. Cardoso de Matos and M.F. Pinto Basto, pers. comm., 2021. Figure 122.

Teixeira, Manuel Augusto de Pimentel
(1875–1950)
Bio.: b. Maçãs de D. Maria, Alvaiázere, Portugal, 24 August 1875; d. Moçâmedes, Namibe, Angola, 1950. Pharmacologist. Initially educated in Santarém, Portugal, he later graduated in pharmacology from the *Escola Médico-Cirúrgica* in Porto, Portugal. In 1902, he immigrated to the town of Moçâmedes in Namibe, Angola, where he operated a pharmacy until 1913. Note to be confused with J. Brito Teixeira (q.v.). **Angola:** Namibe. **Herb.:** BM, COI. **Ref.:** Bossard, 1993; Romeiras, 1999; Figueiredo & al., 2008.

Teusz, Julius Eduard
(1845–1912)
Bio.: b. Złotów, Poland, 8 January 1845; d. Lübben, Germany, 21 March 1912. Collector. He was working as an *Obergehilfe* (foreman) at the Botanical Garden and Botanical Museum Berlin-Dahlem, Germany (Herb. B), when he was

invited to join the expedition of Mechow (q.v.) to the Cuango River in Angola. On 19 September 1878, Teusz, Mechow, and a naval carpenter named Bugslag left Hamburg. They arrived at Luanda on 6 November 1878 and proceeded to Dondo, arriving there on 14 November. In January 1879, after leaving Bugslag in Dondo, Teusz and Mechow continued to Pungo Andongo where they were stationed until 21 June. They reached Malanje on 25 June and were based there preparing for the expedition for about a year, due to difficulties in securing the services of porters who were willing to venture into the interior. On 12 June 1880, the expedition finally departed with 110 porters. A month later, on 19 July 1880, they reached the Cuango River south of its confluence with the Cambo River. On 25 August, they started exploring the Cuango River and on 7 September landed on the right bank (D.R. Congo). Mechow travelled to the Musumb (capital), about three hours distant, and was received by the Mwat Yamv (in Portuguese *Muatiamvo*, *Muatiânvua*, the king), called Muene Puto Kassongo. After negotiating with the king, Mechow left Teusz and some other members of the expedition at the Musumb as a guarantee of his return. On 20 September 1880, Teusz saw off Mechow, Bugslag, and 19 porters as they proceeded by boat down the Cuango River. When they returned and rejoined Teusz, the expedition proceeded to the coast. They crossed the Cuango on 6 January 1881 and were back in Malanje on 20 February 1881. There they met Wissmann (q.v.), Pogge (q.v.), and Buchner (q.v.). Teusz returned to Luanda in April 1881. After this expedition, Teusz was contracted by the *Comité d'Études du Haut-Congo* to work as an agronomist along with Pechüel-Loesche (q.v.). In March 1882, Teusz arrived at Vivi, on the right bank of the Congo River, opposite Matadi (D.R. Congo). From there he continued to Kinshasa where, two years later, in 1884, he established the first coffee plantation. While travelling on the Congo River, Johnston (q.v.) met Teusz at Kinshasa. Later Johnston (1895) recalled "a surly German gardener attached to the expedition, who spent his spare time in collecting birds for certain museums". In Johnston's private notes, however, he referred to "that odious Teusz" (Teusz, 2018) who was in possession of a bird he had shot and (understandably) refused to sell or even allow it to be drawn, in fear that someone else might be the first to describe it. Although Teusz was born in Poland, he was often mistaken for a German and his name was (and still is) misspelled. In November 1884, he returned to Europe but was soon back in Africa and from 1885 he worked as director of the company *Kamerun-Land-und Plantagengesellschaft*, establishing a plantation near Limbe in Cameroon. He is commemorated in several names with the epithet "teucszii". Although Teusz was the natural history collector during Mechow's expedition, his name is generally not given on the labels. As a result, most material is databased as having been collected by

Mechow. From 1879 to 1881, Teusz collected 583 numbers in Angola and the D.R. Congo that were deposited at Herb. B. Collections from Pungo Andongo and Malanje, where Teusz was stationed until 12 June 1880, go up to around number 500; number 579 was collected in Malanje in February 1881, i.e., when the expedition returned. The numbering only partially follows a chronological order. Most specimens were likely destroyed at Herb. B, but duplicates exist in several herbaria. **Angola:** 6 November 1878–April 1881; Cuanza Norte, Luanda, Lunda Norte?, Malanje (Pungo Andongo, Malanje, Tembo-Aluma [Mangango], Calala Canginga [near Sunginge]). **Herb.:** B (main), BR, C, G, GH, GOET, JE, K, L, M, P, S, W, WU. **Ref.:** Mechow, 1882, 1884; Johnston, 1895; Urban, 1916; Gossweiler, 1939; Coosemans, 1951; Mendonça, 1962a; Vegter, 1988; Bossard, 1993; Liberato, 1994; Romeiras, 1999; Figueiredo & al., 2008, 2020; Teusz, 2018.

Thiébaud, Charles Émile
(1910–1995)

Bio.: b. Neuchâtel, Switzerland, 28 January 1910; d. Neuchâtel, Switzerland, 27 October 1995. Geologist. He graduated from the Univ. of Neuchâtel in 1931 and shortly afterwards, in 1932, joined Monard (q.v.) and Théodore Delachaux (1879–1949) on the second *Mission scientifique suisse en Angola* of the *Musée d'ethnographie de Neuchâtel*. The expedition took place from 1932 to 1933 and resulted in thousands of collections of ethnographic objects that were deposited at the *Musée*. Thiébaud made some plant collections and took photographs. Afterwards he co-authored a book, with Delachaux, on Angola. In 1937, after being awarded a doctorate from the Univ. of Neuchâtel, he joined the Shell Company as a geologist and in this capacity developed his career in countries such as Egypt, Venezuela, Borneo, and Iraq, until his retirement in 1967. **Angola:** 1932–1933; collected also with Monard (q.v.). **Herb.:** M, Z+ZT (main). **Ref.:** Delachaux & Thiébaud, 1934; Dozy, 1996; Anonymous, 2021a.

Thompson, Louis Clifford ("Cliff", "Potty")
(1920–1997)

Bio.: b. Haenertsburg, Limpopo, South Africa, 29 February 1920; d. 19 November 1997. Farmer and amateur naturalist. He was only 19 years old when World War II broke out and he went to Zimbabwe to train as a pilot in the Royal Air Force. He became a flight sergeant and served in Europe. On 29 July 1942, he was piloting a Bristol Beaufort bomber from a base in Malta when he was shot down over the Mediterranean and reported as missing. With the navigator and two gunners, he survived in an emergency dinghy for a few days until they were captured by the Italian navy and became prisoners of war. After two failed attempts to escape the prison camp, he succeeded on the third attempt, crossing the line to the allied front near Monte Cassino in October 1943. Back in South Africa after the war, he spent a brief period beekeeping in the Cape and taking courses on pruning and tree grafting. He then returned to the family farm at Wegraakbos, Magoebaskloof, where he settled, marrying in 1953 and raising his family there. He was an active member of the Mountain Club of South Africa and explored mountains and deserts of southern Africa. When travelling, he collected and sent specimens to his sister, the nurserywoman Sheila C. ("Box") Thompson (1917–1998). According to herbarium databases, he collected in Angola at least in 1965 and 1994. The 1965 collections were likely made during a family trip to Angola, as his mother Edith A. Eastwood Thompson (1895–1991) recalled travelling with her son's family to the country. She did not specify the date but mentioned she had borrowed a plant press from her neighbour Lynette Davidson (1916–1996), who was in charge of the Moss Herbarium. Davidson was curator of that herbarium from 1965 to 1982. Edith put the plants "in an apple box, together with a pile of *Farmer's Weekly* for blotting paper and [...] sat on it throughout the journey" (Wongtschowski, 2003). **Angola:** 1965, 1994; Huíla, Namibe. **Herb.:** CM, J, PRE. **Ref.:** Gunn & Codd, 1981; Codd & Gunn, 1985; Vegter, 1988; Figueiredo & al., 2008; Figueiredo & Smith, 2011.

Thorold, Charles Aubrey
(1906–1998)

Bio.: b. Wiltshire, England, 9 November 1906; d. London, England, 2 March 1998. Agronomist and pathologist. He was educated at Marlborough College, Wiltshire, England. In 1928, he graduated with a B.A. from Trinity College, Oxford, obtaining an M.A. in 1934, and much later, in 1975, a D.Sc. From 1930 to 1937, he was a mycologist in Kenya. Afterwards, in 1939, he worked in Trinidad as a plant pathologist. He returned to Africa in 1947 to assume a position in Ghana and in 1949 moved to Nigeria where he worked on diseases of the cocoa tree, *Theobroma cacao* L. (Malvaceae) until 1956. He collected from 1932 to 1956 in several countries in tropical West Africa (Cameroon, Ghana, and Nigeria) and in the island of São Tomé in the Gulf of Guinea, and published several papers on the diseases of, and epiphytes on, cocoa trees. Later, after retirement, he published a book on the diseases of cocoa trees. He returned to England to be the officer in charge of Woburn Experimental Station, Bedfordshire, from 1957 to 1968 when he retired. **Angola:** No information recorded. **Herb.:** BM, BR, EA, IMI, K, M, P, UPS. **Ref.:** Thorold, 1975; Stafleu & Cowan, 1986: 295–296; Vegter, 1988; Romeiras, 1999; Figueiredo & al., 2008; Polhill & Polhill, 2015.

Fig. 123. Jules Timperman. Photographer unknown. From Leteinturier & Malaisse (2001). Reproduced with the permission of the Botanic Garden Meise, Belgium.

Timperman, Jules
(1930–2017)
Bio.: b. Boitsfort, Brussels, Belgium, 7 March 1930; d. Woluwe-Saint-Lambert, Brussels, Belgium, 15 September 2017. Technician. From 1952 to 1995, he held the position of technician at the *Laboratoire de Botanique systématique et de Phytogéographie* at the Univ. Libre de Bruxelles (ULB) in Brussels. He attended several botany courses at ULB and accompanied Duvigneaud (q.v.) on an expedition to the D.R. Congo in 1956. He published on *Crotalaria* L. (Fabaceae), authoring c. 10 species names in that genus. **Angola:** 1956; Moxico (Luau); a few collections with Duvigneaud (q.v.). **Herb.:** BR, BRLU. **Ref.:** Duvigneaud & Timperman, 1959; Leteinturier & Malaisse, 2001; Figueiredo & al., 2008. Figure 123.

Tisserant, Charles Michel
(1886–1962)
Bio.: b. Châtel-sur-Moselle, Nancy, France, 14 October 1886; d. Paris, France, 27 September 1962. Missionary with the Congregation of the Holy Spirit (Spiritans). In 1904, he started his novitiate with the Congregation of the Holy Spirit (Spiritans) in Chevilly, Loiret, France, which he interrupted to do military service from 1905 to 1906. Afterwards he returned to Chevilly

and continued his studies in linguistics and botany with Charles Sacleux (1856–1943), with whom he would remain in contact. Tisserant was ordained a priest on 28 October 1910. In 1911, he embarked for Africa, being attached to the *Préfecture apostolique* of the French colony Ubangi-Chari (now Central African Republic, C.A.R.) where over the next decades he worked in several missions. In June 1942, being prevented to return to Europe as a result of World War II, he was sent to the *Missão do Huambo* in Angola, to recover from illness. He remained there for five months. After the war ended, he was appointed head of the botany section of the *Station Centrale* at Boukoko (C.A.R.). He collected over 10,000 specimens that became the foundation of his catalogue of the flora of Ubangi-Chari, one of his many publications. After more than 40 years of missionary work, in 1954 he returned to the Spiritan's headquarters in Paris. He continued studying his collections until his death in 1962. His notebooks and extensive correspondence on scientific matters with Sacleux and others are deposited at the Spiritan's Archive. He received several decorations and is commemorated in numerous species names and in the genus *Tisserantia* Humbert (Asteraceae). **Angola:** 1942; Huambo. **Herb.:** A, BM, BR, CN, COI, G, H, HBG, K, LISC (2 specimens), MO, P, PC, PRE, WAG. **Ref.:** Tisserant, 1950; Vegter, 1988; Bossard, 1993; Liberato, 1994; Romeiras, 1999; Figueiredo & al., 2008; Boulvert, 2011b; Calhoun, 2018; Anonymous, s.d.(h).

Torre, Albano Rocha da – see **Rocha da Torre**

Torre, António Rocha da
(1904–1995)
Bio.: b. Meadela, Viana do Castelo, Portugal, 11 June 1904; d. Santo Tirso, Porto, Portugal, 20 January 1995. Botanist and pharmacologist. He graduated in natural sciences at the Univ. of Coimbra, Portugal, and afterwards graduated in pharmaceutical sciences at the same university. Having been appointed colonial pharmacist, in May 1933 he embarked for Mozambique where he arrived at Lumbo (Nampula, Mozambique) in July 1933 to work at Vila Cabral (now Lichinga, Niassa). With the encouragement of Carrisso (q.v.), he started collecting plants in his spare time. In 1934, he was commissioned to do botanical surveys of the region up to the Rovuma River. From 1935 to 1937, he directed the state pharmacy at Nampula and afterwards, from 1938 to 1939, he was transferred to Inhambane. After an extended leave in Portugal, he returned to Mozambique and directed the state pharmacy at Maputo until March 1940 when he was appointed for four years to undertake surveys for the phytogeographical map of Mozambique. From then on he dedicated

himself to botany full time. In 1945, he returned to Lisbon and in 1946 he was appointed research assistant at the *Junta das Missões Geográficas e de Investigações Coloniais*. In 1947, he participated in the third expedition of the *Missão Botânica de Moçambique*. In this expedition there were three brigades, led respectively by José Gonçalves Garcia (1904–1971), Barbosa (q.v.), and Torre. Torre was assisted by Cavaco (q.v.). In 1955, Torre joined the first expedition of the *Missão Botânica de Angola e Moçambique* to southern Angola (*Campanhas de Angola 1955–1956*, q.v.), which he led from 13 December 1955 to 2 March 1956. In 1964, he replaced F.A. Mendonça (q.v.) as leader of the *Missão Botânica de Angola e Moçambique*. During the 1960s and early 1970s, Torre was in charge of several expeditions in Mozambique. His last expedition took place in 1973 and consisted of a survey of the area that would be flooded for the Cahora-Bassa Dam. He retired in 1974, having collected c. 19,000 numbers; his collections from Angola range from numbers 8200 to 8863. The transcripts of his collection records are at the Univ. of Lisbon and digitised at JSTOR Global Plants (JSTOR, 2021). He published several papers, produced Flora accounts for the *Conspectus florae Angolensis*, and published over 100 new plant names. He is commemorated in about 28 names, for example in *Aloe torrei* I.Verd. & Christian (Asphodelaceae). **Angola:** 1955–1956; Cunene,

Fig. 124. António Rocha da Torre. Photographer unknown. From Centro de Botânica/ Instituto de Investigação Científica Tropical (IICT), Universidade de Lisboa. Reproduced with permission.

Huíla, Namibe. **Herb.:** B, BM, BR, C, COI, EA, FHO, FI, G, K, LD, LISC (574 specimens), LMA, LMU, LUA, LUAI, LUBA, M, MO, P, PO, SRGH, WAG. **Ref.:** Torre, 1940; Mendonça, 1962a, 1962b; Exell & Hayes, 1967; Gomes e Sousa, 1971a, 1971b; Vegter, 1988; Bossard, 1993; Liberato, 1994; Gonçalves, 1996; Romeiras, 1999; Figueiredo & al., 2008, 2018a; Saraiva & al., 2012. Figure 124.

Torrinha, Camacho
(fl. 1965)
Note: Likely António Carlos Camacho Torrinha, who graduated from the Agricultural School of Tchivinguiro (Huíla, Angola) in 1964–1965. **Angola:** 1965; Benguela (Ganda). **Herb.:** LISC (1 specimen). **Ref.:** Figueiredo & al., 2008.

Trovão, José Ferreira
(1925–2016)
Bio.: b. Póvoa do Varzim, Portugal, 1925; d. Póvoa do Varzim, Portugal, 28 March 2016. Veterinarian. He graduated from the Univ. of Lisbon in 1947. In 1951, he accepted an appointment with the veterinary services in Angola and remained in the territory until 1968. Following an illness, he retired from government service and returned to his hometown. He later undertook a political career with the local administration and social organisations and became a well-known local figure. **Angola:** 1954; Cuando Cubango. **Herb.:** COI, LISC (1 specimen), LUA, PRE. **Ref.:** Matias, 2002; Figueiredo & al., 2008.

Tucker
(unknown)
Note: At least one collection is cited in the *Conspectus florae Angolensis* as having been collected by Tucker and deposited at the Gray Herbarium, Harvard Univ., U.S.A. (Herb. GH). It is cited as "Dondi, *Tucker 28*" under *Biophytum jessenii* R.Knuth (Oxalidaceae) (Exell & Mendonça, 1937–1951: 265). The collection cited is undated. This locality refers to Dondi, near Cachiungo, Huambo, where a mission was established in 1914 and operated by the United Church of Canada and the American Congregational Church. After its foundation in 1914, the mission was directed by the missionary John Taylor Tucker (1883–1958), who lived on site with his family. Later, in the late 1920s, it was run by M. Childs's (q.v.) husband. Tucker's son, Theodore L. Tucker, and his family, continued living at Dondi until 1956. It is not known which member of the family collected plant material. **Angola:** Huambo (Dondi). **Herb.:** GH. **Ref.:** Tucker, 1927; Reed, 2010.

[Turner, Billie Lee]
(1925–2020)
Note: American botanist. In 1958, he was director of the herbarium of the Univ. of Texas (Herb. TEX) in Austin, subsequently renamed the Plant Resources Center and now called Billie L. Turner Plant Resources Center. From 1955 to 1956, he collected with Shantz (q.v.) while on an expedition to resurvey the study sites of Shantz's Smithsonian expedition (1919–1920) and was co-author of the resulting report. Much later, he published a personal account of the trip where Turner (2016) dealt mostly with their altercations and some private matters. The map accompanying the account shows an itinerary reaching Namibe in Angola. This map is not correct as their itinerary led to Lüderitz, Namibia, not to Namibe, Angola, as depicted. Likewise, on the east coast, the itinerary led to Maputo, not Beira, Mozambique, as depicted. Turner (2016: 130) mentioned that Shantz had been in Namibe in the previous expedition and had taken photographs there. However, no photographs taken in Angola exist in Shantz's collection and his notes do not indicate that he was there during the 1919–1920 trip. Turner (2016) stated that, overall, during that expedition he collected 314 numbers in three sets that were deposited at Herb. TEX (main), Royal Botanic Gardens, Kew (Herb. K), and Univ. of Arizona (Herb. ARIZ). However, the main set appears to be the one at Herb. K (G. Yatskievych, pers. comm., January 2019). Turner did not collect in Angola.
Ref.: Shantz & Turner, 1958; Gunn & Codd, 1980; Turner, 2016; Adams & al., 2020; G. Yatskievych, pers. comm., December 2018.

Valles, Edgar Francisco da Purificação
(1912–1993)
Bio.: b. Goa, India, 10 June 1912; d. Portugal, 23 December 1993. Agronomist. He was born in Goa, which was then an overseas territory of Portugal, now India. He graduated from the Pune College of Agriculture, Maharashtra, and after working for a while in India, in the 1940s he obtained the equivalence to agronomist and forester at the *Instituto Superior de Agronomia* in Lisbon. He went to Angola from 1943 to 1945 to assume a position with the *Serviços de Agricultura* in Cuanza Norte. Afterwards he was transferred to Goa but returned to Angola with his wife in 1949 to settle in Cabinda. From 1952 to 1953, he was in Bié, then in Benguela for two years, and finally in Luanda where he held the position of director of agriculture from 1961 to his retirement. He

Fig. 125. Edgar Valles. Photographer unknown. Private collection of Edgar Valles filius. Reproduced with permission.

published several articles, especially in *Gazeta Agrícola*. **Angola:** 1952–1953; Cabinda, Cuanza Norte. **Herb.:** COI, LISC (42 specimens), LUA, PRE, US. **Ref.:** Vegter, 1988; Bossard, 1993; Romeiras, 1999; Figueiredo & al., 2008; Figueiredo, 2014; E. Valles filius, pers. comm., November 2018. Figure 125.

Van der Waal, C.
(fl. 1996)
Angola: 1996; Luanda (Quiçama). **Herb.:** PRE. **Ref.:** Figueiredo & al., 2008.

Vanderyst, Hyacinthe Julien Robert
(1860–1934)
Bio.: b. Tongeren, Belgium, 12 September 1860; d. Kisantu, D.R. Congo, 14 November 1934. Agronomist, missionary, and secular priest. In 1885, he graduated from the Catholic Univ. of Leuven, Belgium. In 1891, he joined the Congo missions as a voluntary secular assistant. He was stationed at Banana (D.R. Congo) for a while but had to return to Belgium because of illness. He became an assistant inspector and later inspector, at the Ministry of Agriculture, a position that he left in 1902 to study theology in Rome, Italy, to become a priest. After four years of study, he was ordained at Leuven,

Belgium, and became a Scheut Missionary, then joining the Jesuit missionaries of the Kwango Mission as an auxiliary priest. The Kwango Mission was based at Kisantu and had been founded in 1893 by Belgian Jesuits. Vanderyst arrived in April 1906 and spent the following periods there: 26 April 1906–19 March 1909; 22 November 1909–17 July 1912; 2 April 1913–26 September 1919; 15 August 1920–9 July 1923; 15 October 1924–14 April 1928; 29 March 1930–8 April 1931; and from 12 January 1932 to his death on 14 November 1934. He published extensively on various subjects and collected profusely, amassing c. 10,000 numbers. **Angola:** July 1923, January 1932; Benguela (1923), Huambo (1923), Luanda (1923), Benguela (Lobito, 1932); his collections in Angola were recorded by Bamps (1973) as numbers 13055–13190 and 13351–13359 in 1924, and numbers 28013–28132 in January 1932; however, the specimens from Angola databased and presently available online at Herb. BR consist of numbers 13039–13378 in July 1923 and numbers 28013–28134 in January 1932; a specimen numbered 34329 may indicate a later visit, as the previous number 34329 was collected in September 1932 (in Yokolo, D.R. Congo). **Herb.:** B, BM, BR (main), C, K, MO, P, UPS. **Ref.:** De Wildeman, 1935; Van de Casteele, 1952; Bamps, 1973; Stafleu & Cowan, 1986: 662–663; Vegter, 1988; Bossard, 1993; Romeiras, 1999; Figueiredo & al., 2008.

[Vasse, Guillaume]
(1868–1930)
Note: French collector, naturalist, and hunter who worked for the *Muséum national d'Histoire naturelle* in Paris, and for the botanical gardens of Le Havre and Marseille, both in France. He undertook an expedition to Mozambique with his wife in 1904, arriving in June of that year and remaining until November 1906. During the expedition he made several collections. The plant collections are at Herb. P. One of them (P02968890) dating from 1905 has been mistakenly databased at Herb. P as a collection of "G. Vane" in Angola. Vasse did not collect in Angola. **Ref.:** Vasse, 1909; Gomes e Sousa, 1942; Alice Gomes e Sousa, 1949; Exell, 1960; Tinley, 1977.

[Vatke, Georg Carl Wilhelm]
(1849–1889)
Note: Botanist who studied at the Univ. of Berlin and was an assistant at the botanic garden of Berlin from 1876 to 1879 and later a private scholar. He collected in central and eastern Europe from 1868 to 1876 and had a private herbarium that was acquired by the Haussknecht Herbarium, Jena (Herb. JE). Vatke co-founded the *Berliner Botanischer Tauschverein* and distributed also material from Africa collected by others. This material included the collections

of Mechow/Teusz (q.v.) from Angola, Johann Maria Hildebrandt (1846–1881) from Madagascar, and Hoepfner (q.v) from Namibia (J. Müller, curator at Herb. JE, pers. comm., 2018). Widespread information (e.g., JSTOR, 2021) indicates that Vatke collected in Angola and Madagascar. The reference for this seems to be Jackson's (1901) list of collectors with material held at Herb. K in 1899. In this list, Vatke appears as: "Vatke, Wilhelm, (per). Angola, Madagascar". The term "per" indicates that the material was received through Vatke, not collected by him. As far as is known, Vatke never visited Angola (nor Madagascar). **Ref.:** Hoffmann, 1889; Figueiredo & al., 2020; J. Müller, pers. comm., November 2018.

Vesey-Fitzgerald, Leslie Desmond Edward Foster (1910–1974)

Bio.: b. Dunleer, Ireland, 7 June 1910; d. Nairobi, Kenya, 3/4 May 1974. Entomologist. In 1930, he graduated from the Agricultural College, Wye, England. From 1932 to 1946, he was with the Colonial Service, initially in Trinidad working on the biological control of sugar cane pests, and afterwards stationed in the Seychelles and visiting Madagascar and coastal East Africa. In 1939, as World War II was about to start, he joined the Rubber Research Institute in Malaya. He served with the Federal Malay States Volunteers and afterwards, from 1942 to 1947, worked with the Middle East Anti-Locust Unit in Saudi Arabia and Oman. He returned to Africa for two years as a game warden in Kenya and subsequently was with the Red Locust Control Unit in Zambia where he remained for 15 years. In 1964, he moved to Tanzania to assume a position as ecologist with the National Parks. With vast field experience, he became an expert on the grasslands of Central and East Africa and made numerous plant collections. He is commemorated in *Aloe veseyi* Reynolds (Asphodelaceae). Vesey-Fitzgerald was in Angola in June 1963 and collected near Luanda; only one specimen has been located, at Herb. US. **Angola:** 1963; Luanda. **Herb.:** B, BM, BR, EGR, FHO, K, L, M, NU, P, SRGH, S. U, US, UZL, WAG. **Ref.:** Lanjouw & Stafleu, 1957 (sub "Fitzgerald, D.V."); Vegter, 1988; Polhill & Polhill, 2015.

Vicente, José dos Santos (fl. 1960)

Angola: 1960. **Herb.:** COI. **Ref.:** Figueiredo & al., 2008.

Vilão, Manuel (fl. 1955)

Angola: 1955; Cabinda. **Herb.:** COI, LISC (1 specimen). **Ref.:** Figueiredo & al., 2008.

Villain, Félix
(1879–1937)
Bio.: b. Villedieu-les-Poêles, Manche, France, 13 September 1879; d. Huíla, Angola. 18 June 1937. Missionary with the Congregation of the Holy Spirit (Spiritans). After studying at the seminary in Mortain, France, he took orders with the Congregation of the Holy Spirit (Spiritans) in 1899. He was ordained as a priest in 1903, at Chevilly, France. After a short stay in Portugal, where he learned Portuguese, he embarked for Angola to join the *Missões Portuguesas do Espírito Santo da Huíla* in 1904. During his stay in Angola, he was stationed in several places: Huíla (Huíla) until 1906, Quihita (Huíla) from 1906 to 1907, Humbe (Cunene) from 1907 to 1924, Quihita from 1924 to 1928, Huíla from 1929 to 1932, and Sendi (Huíla) from 1932 to 1937. He died in Huíla in 1937. He was especially interested in the Nyaneka-Nkhumbi people, studying their society and learning the Nyaneka language. He published an article on the peoples and missions of Cunene in 1929 and left a typescript on the Nyaneka-Nkhumbi society. He was also interested in plants and collected with Bonnefoux (q.v.), who was his superior at Humbe. **Angola:** 1904–1937; Cunene. **Herb.:** BM, P. **Ref.:** Villain, 1929; Gossweiler, 1939; Lanjouw & Stafleu, 1954 (sub "Bonnefoux & Villain"); Bossard, 1993; Romeiras, 1999; Figueiredo & al., 2008; Anonymous, s.d.(i).

[Vilmorin, Henry Lévêque de]
(1843–1899)
Note: French horticulturalist who worked on plant selection, especially on the improvement of wheat by hybridisation. In *Conspectus florae Angolensis*, at least two specimens (of *Adenocarpus* DC. and *Droogmansia* De Wild., both Fabaceae) are cited as collected by Vilmorin in Namibe, Angola, and deposited at the herbarium of the *Muséum national d'Histoire naturelle*, Paris (Herb. P). Two specimens by Vilmorin can be examined online, both also from Namibe (MNHN-P-P019239 and MNHN-P-P03482473). Both were in Vilmorin's herbarium, but were collected by Berthelot (q.v.). Vilmorin is also listed in JSTOR Global Plants (JSTOR, 2021) as having collected in South Africa. There is indeed a collection dated 1895 from the Cape under his name at Herb. P. It is probable that this was also acquired from another collector as there is no indication in the literature that Vilmorin was ever in southern Africa. **Ref.:** Gunn & Codd, 1981.

[Viraz, B.]
Note: Listed in Figueiredo & al. (2008) based on a record in the Herb. PRE database of a specimen collected in 1946. It likely refers to a Gossweiler (q.v.) specimen.

Wagemans, Jean
(1914–1981)
Bio.: b. 1914; d. 1981. Forester (*ingénieur forestier*) who was director of the Mayumbe forest station at Luki, in Mayombe, D.R. Congo in the 1950s. The station was created in 1948. He made some collections in Cabinda, at least in 1953, which are deposited at the herbarium of Botanic Garden Meise, Belgium (Herb. BR). **Angola:** September 1953; Cabinda. **Herb.:** BR, C, K, P. **Ref.:** Lebrun, 1969; Vegter, 1988 (d. "1982"); Romeiras, 1999; Figueiredo & al., 2008.

Wallenstein, Franz Paul
(fl. 1949–1959)
Bio.: Agronomist with *Serviços de Agricultura* in Angola. He published a report on the *Estação Agrícola da Humpata* in 1956. In 1959, he was the head of the *Serviços de Agricultura* at Malange. He took Reynolds (q.v.) to the locality where *Aloe paedogona* A.Berger (Asphodelaceae) occurs. **Angola:** 1949–1950; Cunene, Huíla. **Herb.:** LUA. **Ref.:** Wallenstein, 1956; Reynolds, 1960; Bossard, 1993; Romeiras, 1999; Figueiredo & al., 2008.

Ward, Cecil James ("Roddy")
(1926–2015)
Bio.: b. Durban, KwaZulu-Natal, South Africa, 5 September 1926; d. Durban, KwaZulu-Natal, South Africa, 5 July 2015. Ecologist. Father of J.D. Ward (q.v.) and Mark Charles Ward. Roddy Ward spent his formative years on the beaches, and at the rivers and estuaries of the Isipingo-Mlazi-Mbokodweni area of the KwaZulu-Natal south coast. He attended Michaelhouse School in the KwaZulu-Natal Midlands from 1939 to 1943. In 1944, he enrolled at the Natal Univ. College, Pietermaritzburg, a forerunner of the Univ. of Natal in Pietermaritzburg, now the Univ. of KwaZulu-Natal. In the same year, he departed for service in World War II in North Africa and Italy with the Natal Carbineers. In 1946, he returned to South Africa from the war and demobilisation service to resume tertiary studies at the Natal Univ. College in Pietermaritzburg. He completed a B.Sc. (Hons) degree in 1950, and from 1951 to mid-1953 lectured in botany and zoology in the School of Pharmacy, Natal Technical College, Durban, now the Durban Univ. of Technology, followed by an appointment as the first ecologist for the Natal Parks, Game and Fish Preservation Board (Natal Parks Board) from mid-1953 to mid-1963,

based out of the Hluhluwe Game Reserve. From mid-1963 to end-1987, he was a lecturer, later senior lecturer, in ecology and systematics at the Univ. of Durban-Westville (UDW), where he was curator of the herbarium of the botany department and occasionally acted as head of the department (UDW is now integrated with the Univ. of KwaZulu-Natal). The herbarium of UDW was subsequently named the C.J. Ward Herbarium. After retirement, from 1988 to his death in mid-2015, he worked as an independent consulting plant ecologist, primarily for impact assessments and providing base-line data for monitoring changes, as well as advice to provincial, governmental, and private sector bodies or local authorities for management of natural resources or for optimum development for residential, industrial, and recreational purposes and/or associated infrastructure. In 1972, he graduated with an M.Sc. degree from the then Univ. of Natal in Pietermaritzburg with a thesis entitled *The plant ecology of the Isipingo Beach Area, Natal, South Africa*, for which he was awarded the Captain Scott Memorial Medal by the South African Biological Society. His plant collections numbered over 17,000 specimens, meticulously preserved and labelled, in both joint and individual collections, several of which were with his sons, John D. Ward (q.v.) and the late Mark Ward. His collections spanned KwaZulu-Natal and several other parts of South Africa, Swaziland (Eswatini), Mozambique (notably the Gorongosa area in 1971 and 1972), Angola (notably the southern region), and Namibia (notably

Fig. 126. Cecil James Ward in 2012. Photographer unknown. Private collection of John Ward. Reproduced with permission.

the central and northern Namib). These collections have been well-distributed amongst a number of herbaria. From the 1940s through to 2015, he took over 200,000 photographs, all of which were dated and labelled, providing long-term monitoring and ecological records. **Angola:** December 1974 to January 1975; Namibe; collected with J.D. Ward (q.v.). **Herb.:** K, MO, NH, NU, PRE. **Ref.:** Gunn & Codd, 1981: 371; Vegter, 1988; Figueiredo & al., 2008; Glen & Germishuizen, 2010: 452; Carnie, 2012, 2015: 4; Dutton, 2017. Figure 126.

Ward, John Douglas
(1956–)

Bio.: b. Durban, KwaZulu-Natal, South Africa, 13 January 1956. Geologist. He spent his formative years in the Hluhluwe Game Reserve, where his father, C.J. Ward (q.v.) was the regional ecologist for the then Natal Parks Board, and Ubizane Game Ranch outside Hluhluwe, under the late Norman Deane. John D. Ward attended Glenwood High School (1968–1972), and later the Univ. of Natal, now the Univ. of KwaZulu-Natal in Pietermaritzburg, and graduated with B.Sc. (1975) and B.Sc. (Hons) degrees (1976) in botany, followed by a Ph.D. in geology in 1984. During the long university holidays, he worked under Huntley (q.v.) in the Quissama and Iona National Parks in Angola (1973/74 and 1974/75), followed by a stint at the Desert Ecological Research Unit, Gobabeb, Namibia (1975/76) under Chris Bornman and Seely (q.v.). From 1979 to 1982,

Fig. 127. John Ward in 2010. Photographer unknown. Private collection of John Ward. Reproduced with permission.

he was a research assistant on the Kuiseb Project of the Co-operative Scientific Programme of South Africa's Council for Scientific and Industrial Research (CSIR) in the central Namib desert, which led to his Ph.D. in 1984. His joint plant collections were undertaken in KwaZulu-Natal as well as in central and southern Angola and Namibia, some jointly with his late father, C.J. Ward and his late brother, Mark Charles Ward. John D. Ward has practised mostly as a geologist, having worked for the Geological Survey of Namibia (1984–1989), De Beers Africa Exploration (1989–1991; 2000–2006), CDM/Namdeb (1992–1999), Gem Diamonds (2006–2009), Namakwa Diamonds (2010–2011), and since 2012 has been an independent consultant and co-business partner on several ventures in Africa. **Angola:** 1973–1974 and 1974–1975; central and southern Angola. **Herb.:** K, NH, NU, PRE, US. **Ref.:** Figueiredo & al., 2008; John D. Ward, pers. comm., 19 September 2019. Figure 127.

Wawra, Heinrich
(1831–1887)

Bio.: b. Brünn, Austria (now Brno, Czech Republic), 2 February 1831; d. Baden, near Vienna, Austria, 24 May 1887. Naval surgeon and botanist. After graduating in medicine at the Univ. of Vienna in 1855, he joined the Austrian navy as a surgeon in 1856. During his 22 years of service with the navy, he undertook many voyages during which he collected plants. On 31 April 1857, he left from Trieste (now Italy) on the corvette *Carolina*, visiting Madeira, then Brazil and Argentina, and from there across the Atlantic to Cape Town, South Africa, where he arrived on 11 December 1857 and sojourned for a month. He left on 12 January 1858 to navigate north along the west coast of Africa, calling at Benguela (21–28 January), Luanda (31 January), Ascension (23 February), and Santiago, Cape Verde Islands (21–26 March). The ship was back in Trieste on 16 May 1858. Afterwards he joined two expeditions with Emperor Maximilian of Mexico, to Brazil on the war steamer *Elisabeth* (1859–1860) and to Mexico on the frigate *Novara* (1864). He also sailed around the world on the frigate *Donau* (1868–1871) and later in a circumnavigation from 1872 to 1873. After retiring from the navy in 1878, he was a private tutor and travelled again to Brazil in 1879. He was awarded the honorary title of *Ritter* (Baronet) von Fernsee in 1873. He is commemorated in *Fernseea* Baker (Bromeliaceae), *Neowawraea* Rock (Phyllanthaceae), and many species names. Wawra's one-week stay on the Benguela coast included a visit to nearby Catumbela where he also collected. Although he had intended to collect also in Luanda, an outbreak of fever among the ship's crew required his attention and he could not engage in much collecting. Nevertheless, while at Luanda he met Welwitsch (q.v.) with whom he had an agreeable meeting, appreciating

Fig. 128. Heinrich Wawra in 1867. By Eduard Kaiser. From Wikimedia Commons. Public domain.

Welwitsch's knowledge and generosity in giving him plants for his herbarium. Wawra thought his Angolan collections were the most interesting of those he had amassed on the voyage on the *Carolina*, and so, with Johann Peyritsch (1835–1889), he published the work *Sertum Benguelense* in 1860. In this publication, he included an account of the trip and the descriptions of twenty new taxa, based on his collections. Wawra's collection numbers in Angola seem to range from 210 to 342, according to this account. **Angola:** January 1857; Benguela (Catumbela), Luanda. **Herb.:** B, BP, BRNM, G, H, K, LE, M, NY, P, PC, PRE, TUR, US, W (main). **Ref.:** Wawra & Peyritsch, 1860; Gossweiler, 1939; Gunn & Codd, 1981; Stafleu & Cowan, 1988: 111–113; Vegter, 1988; Bossard, 1993; Romeiras, 1999; Figueiredo & al., 2008. Figure 128.

Wellman, Frederick Creighton (c. 1870–1960)

Bio.: b. Fredonia, Kansas, U.S.A., 3 January 1870 [or 1871]; d. Chapel Hill, North Carolina, U.S.A., 3 September 1960. Medical doctor, writer, and artist. Also known as Cyril Kay-Scott and Richard Irving Carson, pseudonyms under which he published plays, novels, short stories, and poems. He graduated from the Kansas City Medical Hospital as a medical doctor. With his first wife, he went to Angola in 1896 as a medical missionary. In addition to his medical duties, he made collections of plants and insects. The couple divorced in 1911 and he soon remarried. In 1912, he worked for a while in Honduras, but in the same year he returned to the U.S.A. to hold the chair of tropical medicine at Tulane Univ., New Orleans. In 1913, after two marriages and four children, he eloped with the still underage Elsie Dunn (1893–1963), who would also later become a poet, playwright, and novelist in her own right. To escape the authorities, they changed their names to Cyril Kay-Scott and Evelyn Scott and moved to London. Still in fear of being recognised, he then made arrangements to collect entomological specimens in Latin America for the British Museum. They travelled to Brazil, but he was however unable to carry out his collection of specimens and ended up working as a bookkeeper at a Singer sewing machine store. He was promoted to auditor and then superintendent, and they moved to Natal (Rio Grande do Norte), Brazil. After a son was born in 1914, the family moved to Cercadinho (Tocantins) to run a sheep

Fig. 129. Frederick Wellman. Photographer unknown. Lola Ridge papers, Sophia Smith Collection, SSC-MS-00131, Smith College Special Collections, Northampton, Massachusetts, U.S.A. Reproduced with permission.

ranch until 1917, then to Vila Nova (Rio Grande do Sul) where he worked for the International Ore Corporation. In 1919, they returned to New York where they lived for a few years. After a stay in Bermuda, c. 1923 they travelled throughout Europe and while in Paris, he studied art. In 1928, he returned to the U.S.A. and got divorced from Elsie. He became an art teacher, setting up an art school in Santa Fe, New Mexico, and marrying for the fourth time. He left the art school in 1931 to become director of the Denver Art Museum and later the Dean at the College of Fine Arts at the Univ. of Denver. He retired in 1934 and afterwards worked for a while with one of his sons on a Works Progress Administration project. He died at the age of 90, having published several papers in the medical field and many literary works. He had a full life that was described in an autobiography fittingly entitled *Life is too short* (Kay-Scott, 1943). There is conflicting information regarding dates when documents and biographical accounts are compared. His papers are at the Howard-Tilton Memorial Library in Tulane Univ., New Orleans. His plant collections from Angola may amount to over 1800 numbers, considering the numbers stated in the labels, and include several types. A list of material received at the herbarium of the Royal Botanic Gardens, Kew (Herb. K) in 1906, ranging from number 1765 to 1827, bears a note saying the collections were made at 15°05′E 12°44′S at an elevation of 1360 m. Wellman is commemorated in several names, for example the Angolan endemic *Cissus wellmanii* Gilg & M.Brandt (Vitaceae). **Angola:** 1906–1908. **Herb.:** B, K, P, US (grass type fragments). **Ref.:** Gossweiler, 1939; Kay-Scott, 1943; Vegter, 1988; Bossard, 1993; Romeiras, 1999; Davis, 2007; Figueiredo & al., 2008; Anonymous, s.d.(j). Figure 129.

Welwitsch, Friedrich Martin Joseph (1806–1872)

Bio.: b. Maria Saal, Austria, 5 February 1806; d. London, England, 20 October 1872. Botanist and collector extraordinaire. He attended the Univ. of Vienna with the intention of graduating in law, but eventually moved to the medical faculty and started to develop his interest in botany. After a period working on cholera for the government, and travelling with a nobleman as a tutor, he returned to Vienna and finished his studies graduating in medicine in 1836. He spent another period as tutor and, in 1839, he was appointed by the German scientific society *Unio Itineraria* to explore and collect plants in the Azores and Cape Verde Islands. With that aim, he travelled to Lisbon, but eventually he remained in Portugal and did not travel to the islands. He resided in Portugal from 1839 to 1853. In 1847, he undertook a brief residency in the Algarve, but for the rest of the time he was based in Lisbon, settling in the valley of Alcântara in 1848. During his stay he was in charge of several gardens, including the Royal Garden and Museum at Ajuda. He undertook numerous excursions to neighbouring areas and also further into the interior of Portugal to collect plants that he sent to the *Unio Itineraria*. The plant collections of that period that he kept formed the basis of his Lusitanian herbarium. Part of this herbarium (428 specimens) is currently housed in the *Museu Nacional de História Natural e da Ciência* (Herb. LISU) in Lisbon. In 1850, after having lived in Portugal for 11 years, the Portuguese Government appointed him as naturalist to undertake an expedition to Angola. The preparations took a long time, and it was only on 8 August 1853 that he left Lisbon to initiate an exploration of Angola that lasted for seven years. He arrived in Luanda on 30 September 1853, about seven weeks after having departed from Portugal. He immediately started collecting while exploring the coastal area. A year later, on 10 September 1854, he left Luanda and travelled along the Bengo River (Zenza) to Golungo Alto (Cuanza Norte). He made his base at Sange (the town of Golungo Alto) and explored the Golungo Alto region for two years, until 11 October 1856. He then left to explore Pungo Andongo (Malanje), where he arrived on 18 October 1856. He explored this area and its surroundings for eight months until he returned to Golungo Alto on 11 June 1857. However, at this time he was unable to proceed with his explorations because his men deserted him or refused to continue. He therefore left for Luanda on 27 August 1857 after having explored the Angolan interior for three years. Once back in Luanda in September 1857, Welwitsch fell ill and remained based in the city until June 1859. The only significant trip he undertook during that 20-month period was to Libongo (Bengo) in September 1858. In June 1859, after having recovered, he travelled from Luanda to Moçâmedes (Namibe) by sea. On the way there, he stopped at Benguela where he made some collections. For the next few months, he explored the area north and south of Moçâmedes, as far south as Baía dos Tigres. In October he travelled to Huíla. At Lopolo, he was detained for two months while the settlement was under siege. He returned to Moçâmedes in June 1860, and later embarked for Luanda and from there for Lisbon in September 1860. On 20 October 1863, he arrived in London from Lisbon, with 42 packages of plant and animal specimens with a volume of c. 9 m^3. It was estimated that these collections included over 5000 plant species and 3000 insect species. For the next nine years, until his death in 1872 (also on 20 October), Welwitsch worked on his plant collections. He identified and named numerous species. Many names published by him during his lifetime were not recorded in indexes as having been authored by him, an error that remains to be corrected. Throughout Welwitsch's lifetime, his plant and other collections from Angola remained in his possession. His plant collections amounted to c. 10,000 numbers with numerous duplicates. After his death, the ownership of his collections had to be determined through litigation, as the Portuguese government contested Welwitsch's will. Eventually, in 1875, the court ruled that the collection was to be divided into separate sets. The first set and duplicates were to be sent to Herb. LISU in Lisbon with the exception of the second-best set that was to be kept at the herbarium of the present-day Natural History Museum (Herb. BM). When the collections arrived in Lisbon, Herb. LISU kept the first set and one to three duplicates, and from 1877 to 1883 distributed over 22,000 duplicates to several herbaria. Welwitsch is commemorated in arguably his most noteworthy discovery, the plant that was to be known as *Welwitschia* Hook.f. In Welwitsch's words, it was on the plateau north of Cabo Negro (Namibe) that, in September 1859, he came across a "dwarf tree […] particularly remarkable" which he called *Tumboa* Welw., but that eventually was named after Welwitsch by J.D. Hooker, as *Welwitschia*. He is also commemorated in numerous species names. **Angola:** 30 September 1853–September 1860;

Fig. 130. Friedrich Welwitsch. Photographer unknown. From Centro de Botânica/Instituto de Investigação Científica Tropical (IICT), Universidade de Lisboa. Reproduced with permission.

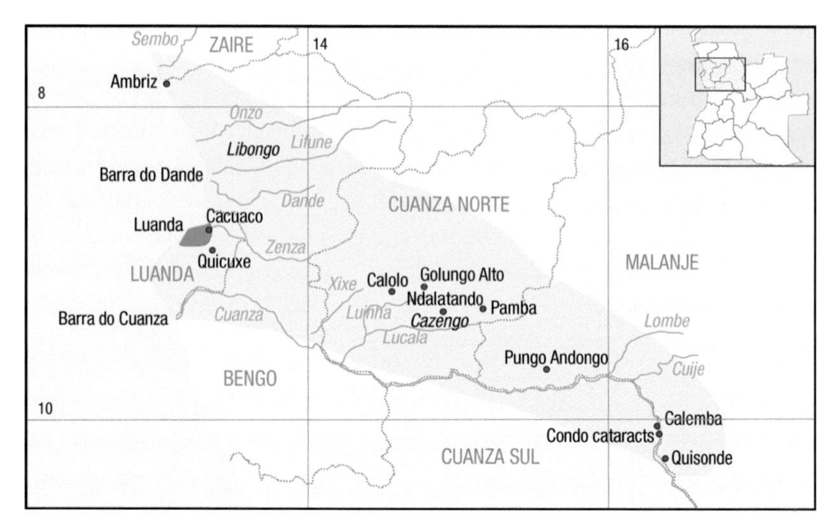

Fig. 131. The area where Friedrich Welwitsch may have made collections in northern Angola.

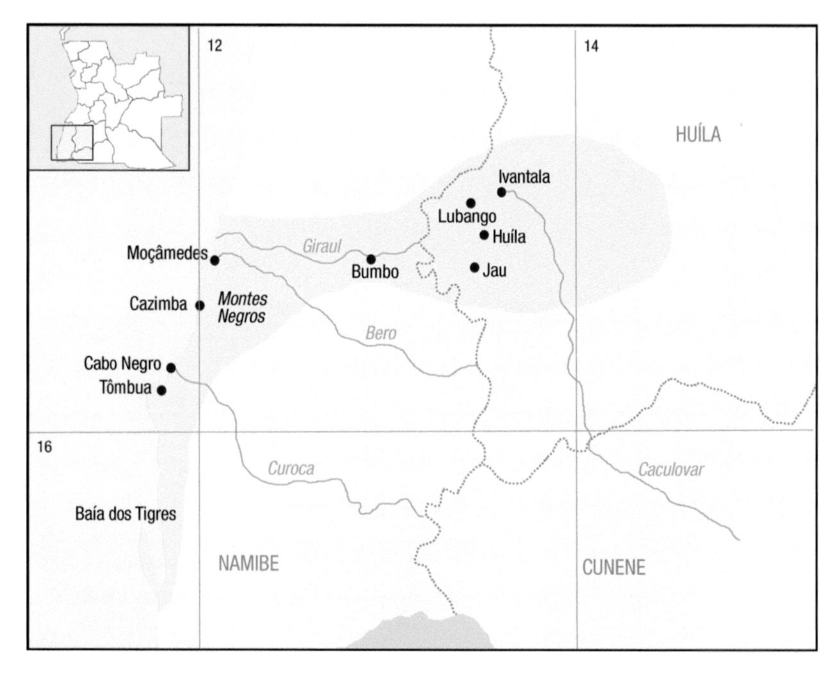

Fig. 132. The area where Friedrich Welwitsch may have made collections in southern Angola.

Bengo, Cuanza Norte, Cuanza Sul, Huíla, Luanda, Malanje, Namibe, Zaire. **Herb.:** B, BAS, BM, BR, C, CGE, CN, COI, DBN, E, F, FI, G, G-DC, GOET, H, HAL, K, KIEL, LE, LISU (main, c. 15,250 specimens of vascular plants; A.I. Correia, pers. comm., 2019), LY, M, MANCH, MEL, MO, MPU, NU, OXF, P, PC, PRE, REG, RG, STE, STU, TUR, US, W, WRSL, Z+ZT. **Ref.:** Hiern, 1896; Gossweiler, 1939; Romariz, 1952; Mendonça, 1962a; Swinscow, 1972; Dolezal, 1974; Stafleu & Cowan, 1988: 174–178; Vegter, 1988; Bossard, 1993; Liberato, 1994; Romeiras, 1999; Albuquerque, 2008: 2, 3; Albuquerque & al., 2009; Figueiredo & al., 2018b; Figueiredo & Smith, 2020b, 2020c. Figures 130, 131, and 132.

Chronology and itinerary of Welwitsch's travels in Africa

8 August 1853 – Embarked in Lisbon, Portugal.

12–14 August 1853 – Madeira Island.

20–24 August 1853 – Cape Verde Islands.

29 August–6 September 1853 – Freetown, Sierra Leone.

15–22 September 1853 – Príncipe Island.

23 September 1853 – São Tomé Island.

30 September 1853 – Arrived at Luanda; for a year explored the coast from the mouth of the Quizembo (Sembo) River to the mouth of the Cuanza River.

10 September 1854 – Left Luanda along Bengo River (Zenza) for Golungo Alto region where he set base at Sange (Golungo Alto) for two years and explored surroundings as far as Cazengo region.

11 October 1856 – Left Golungo Alto towards Pungo Andongo, passing Ambaca (Pamba).

18 October 1856 – Arrived at Presídio de Pungo Andongo where he was based for eight months.

November and December 1856 – Explored Pedras Negras.

January and February 1857 – Explored Pungo Andongo region.

March 1857 – Expedition to the Cuanza River, including Calemba Islands, forests from Quisonde to Condo, near the cataracts of Cuanza River, which was the furthest east he reached; returned to Pungo Andongo, passing salt lakes near Cuije River and forests on right bank of Cuanza River.

7 June 1857 – Returned to Golungo Alto via Ambaca (Pamba) and to his base at Sange (Golungo Alto).

11 June 1857 – Arrived at Golungo Alto.

27 August 1857 – Left Sange (Golungo Alto) to return to Luanda.

7 September 1857 – Arrived in Luanda; ill for five weeks.

September 1858 – Excursion to Libongo.
June 1859 – Left Luanda by sea; stopped at Benguela, where he explored vicinity and then continued by sea to Little Fish Bay (Moçâmedes).
June to August 1859 – Explored surroundings of Little Fish Bay (Moçâmedes), including banks of Bero River, mouth of Giraul River, Montes Negros.
August 1859 – Pinda (Tômbua), Baía dos Tigres.
September 1859 – Excursions along the coast as far as Cabo Negro.
September 1859 – Between Porto de Pinda (Tômbua) and Curoca River.
10 October 1859 – Left Moçâmedes to Huíla via Bumbo.
End of October 1859 – Reached Huíla plateau; explored the region for c. seven months and was under siege at Lopolo for two months.
End of May 1860 – Left Lopolo.
Beginning of June 1860 – Back at Moçâmedes and afterwards by sea to Luanda.
September 1860 – Collected along Bengo River and afterwards embarked for Lisbon.
December 1860 – São Tomé Island.
January 1861 – Arrived in Lisbon.

West, Oliver
(1910–unknown)
Bio.: b. Grahamstown, Eastern Cape, South Africa, 10 August 1910. Ecologist. He studied at Rhodes Univ., Grahamstown, graduating with a B.Sc.; later graduated from the Univ. of the Witwatersrand with a B.Sc. (Hons) (1936), M.Sc. (1937), and D.Sc., the last for a vegetation study of Weenen County, KwaZulu-Natal, which was published as volume 23 of the *Memoirs of the Botanical Survey of South Africa*, an early serial work of the South African National Biodiversity Institute. After completing his Master's degree, he was appointed to the Division of Plant Industry in South Africa, working at the Estcourt and Tabamhlope Research stations in KwaZulu-Natal. During World War II, he was stationed in Kenya, Somalia, and Ethiopia from 1940 to 1943. In April 1945, he was appointed to the then Rhodesian department of agriculture and placed in charge of the Matopos Pasture Research Station. From 1951, he was stationed at the Marandellas Research Station and in 1956 promoted to Chief Pasture Research Officer, a position he held until his retirement on 10 August 1972. He published significant works on the role of fire in pasture management (West, 1965, 1971). Two years after retiring he published a field guide to the aloes of what was then Rhodesia (West, 1974). The foreword to this book was written by Wild (q.v.). Eighteen years later,

West's book, the first on the aloes of the country, and also the first specialist book that dealt with a specific genus of succulent plants and that was written, printed, and published in Zimbabwe, was updated by Michael J. Kimberley (1934–2020) and republished as *Aloes of Zimbabwe*. Kimberley (1992: 2–3) credits West as having played an important role in the eventual promulgation of Zimbabwean environmental legislation aimed at the protection and conservation of the flora of the country. **Angola:** Ca. 1955. One specimen is databased at PRE as originating from southern Angola. The specimen was collected on the banks of the Cunene River, Cafu, and living material was cultivated at the Grasslands Research Station. It is uncertain whether the specimen was collected by West or only cultivated by him. It is further uncertain whether the Grassland Research Station referred to is the one in South Africa or the Marandellas Research Station in Zimbabwe, where West was stationed at the time. **Herb.:** BM, BR, Cedara Agricultural College, DAO, K, M, MRSH, NH, P, PRE (main), SRGH (main set of Zimbabwean specimens), US (1 specimen). **Ref.:** West, 1965, 1971, 1974; Gunn & Codd, 1981: 375; Vegter, 1988; Kimberley, 1992.

Wild, Hiram
(1917–1982)
Bio.: b. Sheffield, England, 15 March 1917; d. [Johannesburg, South Africa?] 28 April 1982. Botanist. He graduated with a B.Sc. in 1939 and a Ph.D. in 1945 from the Imperial College of Science and Technology in London. In 1945, he was appointed as a government botanist in then Rhodesia (now Zambia and Zimbabwe) and for the next 20 years he increased the holdings of the Government Herbarium, at Harare, Zimbabwe (Herb. SRGH), from 11,000 to 160,000 specimens. In 1955, Wild, A.W. Exell (q.v.), and F.A. Mendonça (q.v.) visited the five territories that would be covered by the *Flora Zambesiaca* project that they then initiated. In 1956, Wild travelled to the Royal Botanic Gardens, Kew, and remained there for three years preparing the first two volumes of the *Flora*. He became a Professor of botany at the Univ. College (now Univ. of Zimbabwe) in Harare in 1965. He founded the journal *Kirkia* and published 113 publications that were listed by Efrst (1983). These include the vegetation map for the *Flora Zambesiaca* region that he co-authored with Barbosa (q.v.). In 1980, he retired to England. Wild (1953, later updated by Wild & al., 1972) published one of the first books produced on the vernacular plant names of a south-tropical African country. The first edition of this work (Wild, 1953) appeared in the same year that Gossweiler (q.v.) produced a similar compilation for the plants of Angola. Wild (1953) and Wild & al. (1972) were later updated by Kwembeya & Takawira ([199-?]). Wild is

Fig. 133. Hiram Wild (right) in 1955 with Arthur Exell (left) and Francisco Mendonça (middle). Photographer unknown. South African National Biodiversity Institute. Reproduced with permission.

commemorated in numerous species names, such as *Aloe wildii* (Reynolds) Reynolds (Asphodelaceae). **Angola:** 1964; Benguela, Cuanza Sul, Huambo, Huíla; collected with Barbosa & (R.M.) Santos (q.v.). **Herb.:** BM, BOL, EA, COI, DAO, FHO, K, LISC (27 specimens), LMA, M, MO, NY, P, PRE, SRGH (main), UPS. **Ref.:** Wild, 1953; Exell & Hayes, 1967; Wild & al., 1972; Efrst, 1983; Vegter, 1988; Figueiredo & al., 2008; Polhill & Polhill, 2015. Figure 133.

Williams, Ion James Muirhead (1912–2001)

Bio.: b. Kenilworth, Cape Town, South Africa, 29 June 1912; d. Hermanus, Western Cape, January 2001. Engineer and botanist. He completed his secondary schooling at St Andrew's College, Grahamstown, Eastern Cape, and thereafter studied at the Univ. of Cape Town, from 1930 to 1934, graduating with a B.Sc. in civil engineering. During his engineering career he was involved in several projects in the Cape Town dockyard and at the city's power station, as well as in bridge-building, and during World War II he worked on fortifications and observation posts. Two years after the war ended he moved to Hermanus, a coastal town in the Western Cape, with his wife, Sheila, and two children. While living in Hermanus he was a driving force behind the upgrading of several existing or the establishment of new natural and cultural history, facilities that are still in operation, including the old Hermanus harbour, the Fernkloof Nature Reserve, and the construction of Rotary Way and the Cliff Path, and the Vogelgat Nature Reserve. The sheep farm *Vogelgat*, which Williams purchased in 1969, was eventually completely cleared of exotic invasive trees and became a model private nature reserve. In addition, the properties adjacent to *Vogelgat* also have been cleared of alien vegetation, so preserving the fynbos of the area in prime condition. In 1962, Williams initiated his academic botanical studies and ten years later, in 1972, graduated with a doctorate for a dissertation on the genus *Leucadendron* R.Br. (Proteaceae) (Williams, 1972). It was an unprecedented move by the Univ. of Cape Town to award a Ph.D. in botany that was backed by an engineering degree awarded four decades earlier (Bean, 2002: 4). Williams also became an expert on the southern African Rutaceae, including *Agathosma* Willd., known as *boegoe* in Afrikaans. Ion Williams received the Freemanship of Hermanus in 1997, 50 years after he moved to the town. Sheila Williams, the wife of Ion, was also an enthusiastic environmentalist and played a leading role in preparing specimens for deposition in the S.L. Williams Herbarium in the Fernkloof Nature Reserve. **Angola:** 1971; Huíla. **Herb.:** BM, BOL, K, M, NBG, PRE, STE. **Ref.:** Williams, 1972; Gunn & Codd, 1981; Vegter, 1988; Bean, 2002; Du Toit, [2004]; Glen & Germishuizen, 2010; Schoeman, 2017: 144–148.

Williams, John George Lyal
(1913–1997)

Bio.: b. Cardiff, Wales, 4 April 1913; d. Leicester, England, 28 December 1997. Naturalist and ornithologist. After leaving school, he worked for a while at a shipping company and then trained as a taxidermist at the National Museum of Wales. During World War II, he served with the Royal Air Force (RAF), in North Africa, the Middle East, Turkey, and the Balkans. During that time, he met his future wife, a medical doctor also with the RAF. They married in Cairo in 1945 and settled in Kenya, where she practised, and he became curator of birds at the Coryndon Museum (now the National Museum) in Nairobi. He held that position for twenty years, undertaking many expeditions and discovering new species, and became known as an authority on sunbirds. As a taxidermist he was known for the high quality of his specimens. Concerned with conservation, he was influential in the creation of the Lake Nakuru National Park. Early on he also saw the need for field guides and in 1963 published the extremely popular *A field guide to the birds of East and Central Africa*. More field guides followed, on butterflies, national parks, and orchids, for Africa, but also for Europe and North America. In 1978, he returned to England with his wife, and they settled in Oakham. Williams collected in Angola; on 21 August 1957 he collected the type of *Monadenium angolense* P.R.O.Bally (Euphorbiaceae) at Morro Moco in Huambo. **Angola:** 1957; Huambo (Morro Moco). **Herb.:** B, BR, EA (main), K, P. **Ref.:** Bally, 1961; Williams (J.G.), 1963; Turner, 1998; Romeiras, 1999; Williams (A.), 1999; Figueiredo & al., 2008; Polhill & Polhill, 2015.

Williams, Louis Otho
(1908–1991)

Bio.: b. Jackson, Wyoming, U.S.A., 16 December 1908; d. Rogers, Arkansas, U.S.A., 1991. Botanist. In 1928, he entered the Univ. of Wyoming, graduating with a Master's degree in 1933 and then continued his studies at Washington Univ. in St. Louis, Missouri, where he was awarded a doctorate. He then joined the Oakes Ames Orchid Herbarium at Harvard Univ., Cambridge, Massachusetts. During World War II he contributed to the war effort by working in Brazil from 1942 to 1945 on a rubber procurement programme. Afterwards he accepted an appointment in Honduras where he remained for eleven years. He returned to the U.S.A. and took up employment at the Plant Industry Station at Beltsville, Maryland. In 1958, he undertook a collecting expedition to the (present-day) D.R. Congo, Rwanda, Angola, and Namibia. When embarking on this trip, Williams left the U.S.A. with the agronomist Norris W. Gilbert on 20 January 1958. After visiting the D.R. Congo, they flew to Luanda and from there to Lubango in southern Angola. They travelled by car from Lubango via Caconda to Huambo where they remained for a few days while collecting. They returned to Luanda by air and then left the country. They returned to the U.S.A. on 22 April (Gilbert) and 5 May (Williams), respectively. In 1960, Williams joined the Field Museum of Natural History in Chicago as Curator of Central American Botany. He was later appointed as Chief Curator at the Field Museum and revived the department of botany with collecting activities and the development and conclusion of the *Flora of Guatemala* project. He retired in 1972 and moved to the Ozark Mountains in Arkansas. During his lifetime he collected 43,000 specimens that are distributed among numerous herbaria. Only two collections from Angola have been located at the U.S. National Herbarium (Herb. US). There are no collections in the main National Arboretum Herbarium (Herb. NA) or the National Seed Herbarium (Herb. BARC). He also collected for the U.S. National Plant Germplasm System, two of these collections continue to be maintained there. **Angola:** April 1958; Luanda, Huambo, Huíla. **Herb.:** A, AMES, ARIZ, BM, BR, C, DAO, DS, EAP, ENCB, F (main), G, GB, GH, K, LLC, MEXU, MICH,

MO, NY, P, PH, R, RM, S, SEL, TEX, TRT, UC, US, UTC, W, WIS. **Ref.:** Williams & Gilbert, 1958; Burger, 1991.

Wissmann, Hermann von
(1853–1905)
Bio.: b. Frankfurt an der Oder, Germany, 4 September 1853; d. Weißenbach, near Liezen, Austria, 16 June 1905. Military officer and geographer. After studying in the school for cadets in Berlin, he joined the fusiliers and attended the military academy at Anklam, Germany. He received his commission as an officer in 1874 and became a lieutenant in the infantry at Mecklenburg, Germany. He was arrested for duelling and while in prison met Buchner (q.v.), who had been arrested for the same reason. In 1879, Wissmann was stationed at Rostock, where he met Pogge (q.v.), who enthralled him for African exploration and recommended him to the *Afrikanische Gesellschaft in Deutschland.* As a consequence, Wissmann joined Pogge's second expedition to Africa as a geographer for which he prepared by studying at the *Seemannsschule* (naval school) and at the Univ. of Rostock. The aims of the expedition were to reach the Lunda Kingdom, establish a research station, and then to proceed north. Wissmann and Pogge left Hamburg on 19 November 1880 and disembarked at Luanda on 7 January 1881. Two weeks later, on 25 January, they were at Malanje. There they met Mechow (q.v.), who was on his way back from Cuango, and Buchner, who was returning from Lunda. After this meeting they decided to head northeast to the country of the Luluwa (now the D.R. Congo). They left Malanje at the end of May or at the beginning of June 1881 and reached Mona Quimbundo on 20 July. On 1 August, they proceeded in a northerly direction along the Luele and Chicapa rivers with 69 porters. They reached the Cassai River on 2 October 1881. Crossing it, they entered what is now territory of the D.R. Congo. On 23 October 1881, they parted ways for a while but reunited shortly afterwards. By mid-April 1882, they were at the Lualaba River near Nyangwe (D.R. Congo) and from there they took different routes. Wissmann went east on 1 June 1882 and reached the east coast of Africa on 14 November 1882. He returned to Germany in 1883. In 1889, he published the narrative of this first expedition. His second expedition was a venture of Leopold II of Belgium and the *Association internationale Africaine.* On 13 November 1883, Wissmann left Germany with the meteorologist and photographer Franz Müller (who died during the expedition); Müller's brother, the zoologist, botanist and (later) photographer Hans Müller; the medical doctor and anthropologist Ludwig Wolf; the geographer Curt von François; as well as a carpenter and two armourers. The objective of the allegedly scientific expedition was in fact the subjugation of territories along the Cassai

River to Leopold II. The expedition arrived at Luanda on 17 January 1884 and from there took the regular route to Malanje. There they met Pogge, who was returning at great cost to the coast, almost unrecognisable due to illness. They also met Marques (q.v.) and Henrique Dias de Carvalho (1843–1909), who were preparing their expedition. Wissmann left Malanje on 16 July with 320 porters, heading east. He crossed the Cuango River on 17 August, and by 18 October 1884 he had reached the Cassai River, and from there continued to what is now the D.R. Congo. The return journey was by boat, which had been carried disassembled, and 28 canoes. On 5 June, they navigated down the Cassai River and on 9 July down the Congo River, ending in Leopoldville (now Kinshasa, D.R. Congo) on 17 July 1885. The narrative was published in 1891. From 1886 to 1887, Wissmann undertook another expedition, crossing the continent from Kanaga (D.R. Congo) to near Quelimane (Mozambique). From 1888 to 1893, he visited German East Africa (now Tanzania) several times. In 1889, he was appointed Imperial Commissioner with the task of subduing Arab uprisings in the region. His success in recovering territory for Germany resulted in several decorations, an honorary doctorate from the Univ. of Rostock and a nobility title in 1890, after which he added "von" to his surname. He was also involved in initiatives to deter slave traffic in 1892, clearing a region of slave-hunters. He was governor of German East Africa

Fig. 134. Hermann Wissmann in 1898. Photographer J.C. Schaarwächter. From Wikimedia Commons. Public domain.

from August 1895 to December 1896. In 1897, he became president of the *Deutsche Gesellschaft für Geographie*. He returned to Africa in 1898, for a hunting safari in German South West Africa, now Namibia, and from 1899, he settled at Weißenbach, near Liezen in Austria, where he died after a hunting accident at 51 years of age. Wissmann has been listed in the literature as having collected in Angola during the period 1880–1885. If any collections were made during his expeditions, they would have been sent to the Botanic Garden and Botanical Museum Berlin-Dahlem (Herb. B). However, none is mentioned in the account of the collections received at Herb. B during the period 1878–1891 (Urban, 1892). No specimens have been traced in other herbaria. On the first expedition, collections were made by Pogge. On Wissmann's second expedition (1883–1885), scientific endeavour was authorised only if it was compatible with the main objective of subjugating territories. Wissmann & al. (1891) listed insect collections, but not plant collections. **Angola:** 1880–1885; apparently, he did not collect. **Herb.:** B. **Ref.:** Armand, 1884; Wissmann, 1889; Wissmann & al., 1891; E.G.R., 1905; Schnee, 1920: 721; Vegter, 1988; Bossard, 1993; Romeiras, 1999; Figueiredo & al., 2008, 2020; Heintze, 2010; Anonymous, 2021b. Figure 134.

Woodward, Edson F.
(fl. 1949–1950)
Bio.: A qualified pharmacist and chief pharmacognosist with the private company S.B. Penick Co. He participated in the "Upjohn-Penick Expedition for Botanical Exploration" to Africa from 1949 to 1950. While on this expedition he travelled to Angola and made some collections with Brass (q.v.). **Angola:** March–April 1950; with Brass (q.v.). **Herb.:** NY. **Ref.:** Anonymous, 1955: 158.

Wrede
(unknown–1842)
Bio.: d. Benguela, Angola, January 1842. Botanist. In 1841, as a young German botanist from Hanover, he joined a commercial expedition to Angola. He went to Angola on the brig *Camões*, one of several vessels in an expedition to the Angolan coast led by José Ribeira dos Santos (1798–1842). Santos was the Portuguese Consul-general at Altona (now a district of Hamburg), Germany, and also a partner in the Hamburg-based commercial company Santos & Monteiro. Six vessels full of merchandise were fitted out for the expedition and sailed from Altona in July 1841. In addition to Wrede and Santos, the expeditionary team consisted of a medical doctor named Georg Tams, a German linguist from Hamburg named Grossbendner, and a Portuguese linguist-secretary in charge of facilitating transactions. The voyage was undertaken in

luxury, including entertainment provided by a band of six musicians, an Italian chef, and a library. They arrived at the coast of Benguela in October 1841. The trip proved fatal for most of them: Santos died in Benguela in January 1842. By then Wrede, Grossbendner, and the secretary had already died. An account of the expedition was written by the surviving doctor, Tams, and became a well-known book, especially for the description of the slave traffic, which Tams especially opposed, and the treatment of slaves in the country. It was translated into English and Portuguese. The expedition was purportedly destined to trade merchandise for local products, but there were suspicions that it was a cover for participating in the illegal slave trade. After concerns were raised in the British Parliament, a statement from the Hamburg authorities to the British Foreign Secretary, Lord Aberdeen, dated November 1841, declared that "it clearly appears that again Mr dos Santos's Altonian expedition has been converted into slave-ships". Recent literature indicates that the intention of Santos was to trade in *urzela* (*Roccella tinctoria* DC., a lichen used for dyeing). Nevertheless, his dealings with notorious slave traffickers and his long stays at slave trafficking ports raise suspicion and the matter remains unresolved. During the expedition they came across one of the warships of the fleet of H.M.S. *Waterwitch*, where Curror (q.v.) was then in service. The warship was giving chase to a slave ship. To escape, the captain of the latter wrecked it at the Ambriz coast of Angola. The slaves could not be saved, and the captain escaped from the British by hiding in the brig *Camões*, unbeknownst to Santos, according to Tams. During the expedition Tams collected artefacts that were deposited at the Museum of Ethnology of Leipzig in Germany. Bossard (1993) listed Wrede as a collector in Angola specifying as collecting localities: Ambriz, Benguela, Luanda, Namibe, and Sumbe. However, Wrede is not cited as a collector in the literature. If plant collections were indeed made, they might have been sent to Leipzig, where the artefacts collected by Tams were deposited. The herbarium of the Univ. of Leipzig (Herb. LZ), including all its specimens, literature and documents, was completely destroyed by bombing in 1943 during World War II. Had they been preserved there, none of Wrede's botanical collections would be extant. **Angola:** No specimens located. **Herb.:** LZ? **Ref.:** Tams, 1845; Bossard, 1993; Romeiras, 1999; Figueiredo & al., 2008, 2020; Heintze, 2010; Wissenbach, 2011.

Wulfhorst, August
(1861–1936)
Bio.: b. Gütersloh, Germany, 12 March 1861; d. Gütersloh, Germany, 28 September 1936. Missionary. In 1890, after becoming a missionary of the

Rheinische Missionsgesellschaft, he travelled to Namibia (then German South West Africa) to establish a Rhenish mission. The following year, in 1891, Wulfhorst and another Rhenish missionary, Friedrich Meisenholl (1864–1938), established a mission station at Ondjiva in Cunene, Angola. Afterwards another mission was established at Omupanda, south of Ondjiva, and Wulfhorst was stationed there from 1899 to 1907 with a break of two years from 1900 to 1902, when he was in Germany. In 1915, during World War I, the German Omupanda mission had to be closed. Wulfhorst moved to Omaruru and later to Karibib where he was stationed from 1919 to 1927. Later he was sent to Swakopmund, and in 1931 he returned to Germany where he died in his hometown. During his stay in Africa, Wulfhorst collected ethnographical objects and plant specimens. The latter were sent to Schinz (q.v.) at the Univ. of Zürich, Switzerland. In 1898, Schinz received a total of 172 numbers from Wulfhorst. The database of the Zürich herbaria (Herb. Z+ZT) lists 135 specimens, most of which are imaged. Collecting details of 55 of these sheets were listed by Figueiredo & al. (2013). The highest collection number encountered was 191. Numbering is not chronological, and the date given on the label is often the date on which the material was received at Zürich. Wulfhorst also took photographs, a few of which were published by Tönjes (1911). According to Vilhunen (1995), these photographs did not survive the destruction of

Fig. 135. August Wulfhorst. Photographer unknown. © Archiv- und Museumsstiftung der VEM, Wuppertal. Reproduced with permission.

the archives of the *Rheinische Missionsgesellschaft* during World War II. However, the Archives and Museum Foundation of the *Vereinte Evangelische Mission* (*VEM*), where the *Rheinische Missionsgesellschaft* archives are kept, list in its holdings “Bilder aus d. Missionsarbeit in Ovamboland, c. 1910” as part of Wulfhorst’s bequest to the society. His extensive correspondence, reports, and diaries, as well as his first wife’s travelogue, are also extant at the *VEM* archives. Wulfhorst married Thusnelda Härlin in 1892 and after her death in 1922, he wedded her sister, Johanna. **Angola:** 1894–1898; Cunene (Omupanda). **Herb.:** BM, K, Z+ZT (main). **Ref.:** Tönjes, 1911; Olpp, 1937; Gunn & Codd, 1981; Vegter, 1988; Vilhunen, 1995; Dierks, 2003–2004; Figueiredo & al., 2013, 2020. Figure 135.

Young, Marion Emma Blenkiron Norwood (1903–2000)

Bio.: b. Benwell, Newcastle-on-Tyne, Northumberland, England, 22 June 1903; d. Johannesburg, South Africa, 24 October 2000. Née Blenkiron. Botanist. She was one of eleven sisters, one of whom was the artist (painter) Edith Hilder (b. 11 November 1904, married Rowland Hilder in 1929, d. 5 August 1992 in London). Marion Emma (later known as “Marion Norwood Young”) studied at Leeds Univ. and afterwards immigrated to South Africa. She became a teacher in Johannesburg and later a lecturer in botany at the Univ. of the Witwatersrand (1927–1928). She then met R.G.N. Young (q.v.), who was a postgraduate student, and they married in 1928, but later divorced. After the divorce Marion was independently wealthy. During World War II, she served as a nurse and afterwards worked in an office. During the war she also studied fossil pollen (William Stucke, pers. comm., 16 July 2019). Later, she was the Librarian at the British Consulate in Johannesburg, South Africa. She used to relate that she refused to retire unless she was given a pension. Even after she was made a Member of the British Empire (M.B.E.) in 1970, she carried on working at the consulate until she was 75. Thereafter she moved to the house called Culverdown, near Sudbury, Suffolk, England, which belonged to a family member. When the family moved to South Africa in the early 1980s, she joined them. She took part in numerous expeditions to, at the time, little-explored areas, including South Africa’s Transkei region (eastern parts of the Eastern Cape Province) and Angola, and it has been recorded that in the 1930s she twice “went up the Zambezi in a dug-out canoe in search of a rare water-plant". She also collected specimens from the Far East as well as from all over southern Africa (Fox & Norwood Young, 1982: inside back cover dust jacket text). In her 80s, she spent a lot of time in Darnis, Lot, France, with her daughter whose family had been running *L'Ancienne Auberge de Darnis* since 1986. A room on the first floor of the *Auberge* is known as "Granny's Room" in honour of Norwood Young. In 1982, with Francis William Fox, she co-authored the well-known book *Food from the veld*, which was published in South Africa. The first print of the book (Fox & Norwood Young, 1982) had indices for English, Afrikaans, and scientific names. The 1988 revised reprint had indices in numerous languages, including non-South African ones (Fox & Norwood Young, 1988: 377–422). These indices were prepared, based on Norwood Young's dataset, by William Stucke. From July to November 1932, M.E. Norwood Young accompanied her then husband on an expedition sponsored by the Natural History Museum, London, to Zambia, the D.R. Congo, and Angola. The collections are under the name of her husband. While in South Africa she also made collections with A.G. Grant. Her collections appear under the names M.E. Blenkiron, M.E. Young, and M.E.N. Young. She is commemorated in *Cissus marionae* Exell & Mendonça (Vitaceae). **Angola:** July–November 1932; Lunda Norte, Lunda Sul, Malanje, Moxico; collections under R.G.N. Young (q.v.). **Herb.:** BM (main), J, NH, PRE. **Ref.:** Anonymous, 1970; Gunn & Codd, 1981; Fox & Norwood Young, 1982, 1988; Vegter, 1988; William Stucke, pers. comm., 16 July 2019. Figure 136.

Fig. 136. Marion Young. Photographer unknown. Private collection of Ken Stucke. Reproduced with permission.

Young, Ralph George Norwood (1904–1979)

Bio.: b. Florence, Italy, 26 June 1904; d. Florence, Italy, 4 July 1979. Botanist. He was born into a family of means who owned Villa Schifanoia, now the Robert Schuman Centre for Advanced Studies, in Florence. The property was bought at the beginning of the 20th century by a wealthy Australian, John Norwood Young, who made it a hub for Florence's Anglophone community. Ralph Young studied briefly at the Univ. of Lausanne, Switzerland, from 1921 to 1922, and then at Cambridge Univ. in England, from 1922 to 1925, graduating with a B.A. He worked for a while at the Royal Botanic Gardens, Kew, but immigrated to South Africa in 1926. There he studied at the Univ. of the Witwatersrand and graduated with a B.Sc. in 1927 and an M.Sc. in 1929. In 1928, he worked at the Transvaal Museum and at the National Herbarium in Harare, Zimbabwe. In that year, he married Marion Emma Blenkiron (q.v., sub M.E.N. Young). After a brief sojourn farming in the Transvaal (1930–1931), the couple undertook an expedition to Angola, Zambia, and the D.R. Congo under the auspices of the British Museum. Being of mildly independent income and therefore not requiring permanent employment, he took several temporary or short-term positions, working again at the Transvaal Museum

Fig. 137. Ralph Young. Photographer unknown. South African National Biodiversity Institute. Reproduced with permission.

as acting botanist (1933–1934) and at the Univ. of Pretoria as temporary officer (1934–1945), teaching briefly in Johannesburg in 1956, and then taking teaching positions in Geneva, Switzerland from 1956 to 1965. In 1965, he returned to South Africa and until 1974 he worked in various posts as a teacher or clerk. He died when on holidays in Italy visiting his birthplace. His collections consist of c. 13,400 numbers of which 10,000 are from the Zambia-D.R. Congo-Angola expedition. Although collected with his then wife, Marion Young (q.v.), the collections are under his name only. During this expedition, he collected first in Zambia and the D.R. Congo and then proceeded to Angola. There he collected at Luau (Moxico), then continued northwards to Dala and Saurimo (Lunda Sul), Dundo (Lunda Norte), south again to Saurimo and then west to Caculo and Xassengue (Lunda Sul). Finally, crossing the Cuango River, he collected in Malanje. The specimens collected in Angola date from July to November 1932 and the numbers range from at least 287 to 1418, based on examined data. He published a short account of the expedition in 1933. The main set of this collection is at the herbarium of the Natural History Museum in London (Herb. BM), while the main set of the South African material is at the National Herbarium in Pretoria (Herb. PRE). He also collected bulbs and seeds that he sent to Kew. He is commemorated in many species' names, the types of which he collected in Angola. **Angola:** July–November 1932; Lunda Norte, Lunda Sul, Malanje, Moxico. **Herb.:** A, BM (main), BR, COI, J, K, LISC (146 specimens), MO, NH, NY [incl. DPU], P, PRE (main), S, SRGH, US. **Ref.:** Young, 1933; Anonymous, 1934, 2016; Gossweiler, 1939; Exell, 1960; Codd, 1980; Gunn & Codd, 1981; Vegter, 1988; Bossard, 1993; Liberato, 1994; Romeiras, 1999; Figueiredo & al., 2008; William Stucke, pers. comm., 5 October 2021. Figure 137.

ACKNOWLEDGEMENTS

We thank the following individuals who provided information or images (the names are arranged alphabetically according to first name, given the different ways in which especially Portuguese surnames are often cited): Alessia Guggisberg (Eidgenössische Technische Hochschule Zürich), Ana Isabel Correia (Univ. of Lisbon), António Coutinho (Univ. of Coimbra), Arlindo Cardoso (Univ. of Coimbra), Beppi Hart (Department of Agriculture, Land Reform and Rural Development), Braam van Wyk (Univ. of Pretoria), Branca Moriés (Univ. of Lisbon), Carla Vieira (Univ. Nova, Lisbon), Catarina Madruga (Museum für Naturkunde, Berlin), Ceri Humphries (Archives, Natural History Museum, London), César Garcia (Univ. of Lisbon), Chiara Nepi (Univ. of Florence), Chris Willis (SANBI), Christian Froese (Archiv- und Museumsstiftung der Vereinten Evangelischen Mission), Colin Walker (Open Univ., Milton Keynes), Cristiana Vieira (Univ. of Porto), Dale Kruse (Texas A&M Univ.), Daleen Maree (SANBI), Daniel Spalink (Texas A&M Univ.), David Goyder (Royal Botanic Gardens, Kew), David John (Natural History Museum, London), David Luna de Carvalho, Edgar Valles, Elmar Robbrecht (Botanic Garden Meise), Ernie Kekana (Department of Agriculture, Land Reform and Rural Development, South Africa), Fernanda Lages (ISCE, Huíla), Filipe Covelo (Univ. of Coimbra), Francisco Maiato Gonçalves (ISCE, Huíla), Frederico Tátá Regala, George Yatskievych (Univ. of Texas), Gerry Moore (United States Department of Agriculture), Gilberto Cardoso de Matos, Graziela Valente, Hans Beeckman (Royal Museum for Central-Africa, Tervuren), Harlan Svoboda (U.S. National Arboretum), Ilse Tuebben, Jacqueline Curro (Mid-Atlantic Germanic Society), Jason Hinshaw (United States Department of Agriculture), Joaquim Santos (Univ. of Coimbra), Jochen Müller (Univ. of Jena), Johannes Wahl (ETH Zürich, ETH-Bibliothek), John Wiersema (Smithsonian Institution), Jorge Paiva (Univ. of Coimbra), José António Quinto Barcelos, Julia Besten (Archiv- und Museumsstiftung der Vereinten Evangelischen Mission), Katherine Harrington (Royal Botanic Gardens, Kew), Ken Stucke, Kitty Grandvaux Barbosa, Kyle Simpson (Texas A&M Univ.), Laurence J. Dorr (Smithsonian Institution), Leonel Pereira (Univ. of Coimbra), Luis Catarino (Univ. of

Lisbon), Luis Ceríaco (Univ. of Porto), Luis Gaivoto, Madalena Ponte, Maria Fernanda Pinto Basto, Marike Trytsman (Department of Agriculture, Land Reform and Rural Development, South Africa), Mark Carine (Natural History Museum, London), Mark Garland (United States Department of Agriculture), Nadine Kamlah (Universitätsarchiv Rostock), Nicholas Turland (Botanischer Garten und Botanisches Museum, Berlin), Nuno Porto (Univ. of British Columbia), Pat Herendeen (Chicago Botanic Garden), Paulo Silveira (Univ. of Aveiro), Peter van Welzen (Naturalis Biodiversity Centre, Leiden), Reto Nyffeler (Univ. of Zürich), Rui Correia (filius), Sandra Turck (SANBI), Stefan Dressler (Forschungsinstitut Senckenberg), Steven Dessein (Botanic Garden Meise), Susana Matos (Univ. of Lisbon), Tony Dold (Selmar Schonland Herbarium, Albany Museum), Victor Silva (Missionários do Espírito Santo, Lisbon).

The following institutions are thanked for granting permission to use images:

Archiv- und Museumsstiftung der Vereinten Evangelischen Mission, Germany; Arquivo Histórico dos Museus da Universidade de Lisboa and the Portuguese Research Infrastructure of Scientific Collections, Portugal; Botanischer Garten und Botanisches Museum, Berlin, Freie Universität Berlin, Germany; ETH Zürich, ETH-Bibliothek, Switzerland; Mid-Atlantic Germanic Society, U.S.A.; Missionários do Espírito Santo, Portugal; Selmar Schonland Herbarium, Albany Museum, Grahamstown, South Africa; South African National Biodiversity Institute, South Africa; Universidade de Lisboa, Portugal; and Universitätsarchiv Rostock, Germany.

LITERATURE CITED

Abreu, J.A., Sampaio Martins, E. & Catarino, L. 2014. New species of *Maerua* (Capparaceae) from Angola. *Blumea* 59: 19–25. https://doi.org/10.3767/000651914X681964

Adams, R.P., Averett, A., Ayers, T., Barrie, F., Blackwell, M., Blackwell, W.H., Bierner, M.W., Clary, K., Delprete, P.G., Elisens, W., Irwin, D., Lavin, M., Northington, D., Olmstead, R., Powell, M., Plettman Rankin, S., Raven, P.H., Robinson, H., Scott, R., Strother, J.L., Stuessy, T.F., Tomb, S., Turner, B.L., II, Turner, M.W., Walker, J. & Yatskievych, G. 2020. Obituary and tribute to Billie L. Turner: Botanist, teacher, mentor, philosopher, friend. *Phytologia* 102: 88–105.

Aguiar, F.[Q.]B. 1984. *Bibliografia pedológica e agro-ecológica de Angola.* Huambo: Instituto de Investigação Agronómica.

Aguiar, F.Q.B. 2021. Fernando Queiroz de Barros Aguiar. Curriculum on Linkedin.com. https://pt.linkedin.com/in/fernando-queiroz-de-barros-aguiar-5400b572 (accessed 7 December 2021).

Albuquerque, S. 2008. Friedrich Welwitsch. P. 3 in: Figueiredo, E. & Smith, G.F., *Plants of Angola / Plantas de Angola.* Strelitzia 22. Pretoria: South African National Biodiversity Institute.

Albuquerque, S., Brummitt, R.K. & Figueiredo, E. 2009. Typification of names based on the Angolan collections of Friedrich Welwitsch. *Taxon* 58: 641–646. https://doi.org/10.1002/tax.582028

Almaça, C. 2005. *Albert Monard e o Museu Bocage.* Publicações Avulsas, Museu Bocage, ser. 2, 9. Lisbon: Museu Bocage.

Almeida, P.F. 1956. Espécies vegetais de valor económico nos países da África ocidental: Aspectos da sua exploração e melhoramento. Pp. 59–68 in: Conselho Científico para a África ao Sul do Sara (org.), *6ª Conferência internacional dos africanistas ocidentais*, São Tomé, 1956, vol. 3, *Comunicações botânica e biologia vegetal.* [S.l.]: Comissão de Cooperação Técnica na África ao Sul do Sara.

Almeida, P.F. 1961. Fixação de dunas na Baía dos Tigres. *Agron. Angol.* 13: 79–90.

Andrade, A.A.B. 1985. *O naturalista José de Anchieta.* Estudos de História e Cartografia Antiga, Memórias, 24. Lisbon: Instituto de Investigação Científica Tropical.

Anonymous. 1835. *The Navy list, corrected to the 20th March, 1835.* London: John Murray. https://archive.org/details/navylist15admigoog (accessed 7 December 2021).

Anonymous. 1852. *Relação e indice alphabetico dos estudantes matriculados na Universidade de Coimbra e no Lyceu no anno lectivo de 1852 para 1853.* Coimbra: Imprensa da Universidade. https://am.uc.pt/item/50389

Anonymous. 1862. *Quarenta e cinco dias em Angola.* Porto: Typographia de Sebastião José Pereira. https://archive.org/details/quarentaecincodi00port

Anonymous. 1878. Joachim John Monteiro. *Nature* 17: 425–426. https://doi.org/10.1038/017425d0

Anonymous. 1889. *Die Forschungsreise S.M.S. "Gazelle" in den Jahren 1874 bis 1876 unter Kommando des Kapitän zur See Freiherrn von Schleinitz*, vol. 1. Berlin: Ernst Siegfried Mittler. https://doi.org/10.18452/31

Anonymous. 1905. *Farmacêuticos estabelecidos no continente, ilhas e colónias.* Handwritten manuscript available from Centro de Documentação Farmacêutica (CDF-OF) at http://www.cdf.pt/archeevo/details?id=1001127 (accessed 7 December 2021).

Anonymous. 1912. Accession of tropical African plants from 1907–1912. *Bull. Misc. Inform. Kew* 1912: 316–320. https://doi.org/10.2307/4104546

Anonymous. 1913. William Harvey Brown. *Ann. Iowa* 11: 234. https://doi.org/10.17077/0003-4827.3865

Anonymous. 1916. Current note: Conducted by the associate editor. *J. Econ. Entomol.* 9: 513–516. https://doi.org/10.1093/jee/9.5.513

Anonymous. 1917. Vice-almirante Hermenegildo Capelo. *Ilustração Portuguesa*, ser. 2, 587, 21 May 1917: 414.

Anonymous. 1929. Prof. Comm. Dino Taruffi. *Bull. Reale Soc. Tosc. Ortic.*, ser. 4, 14(9/12): 39.

Anonymous. 1933. Prof. José Joaquim de Almeida. *O Século*, 5–6 May 1933.

Anonymous. 1934. Review of the work of the Royal Botanic Gardens, Kew, during 1933. *Bull. Misc. Inform. Kew* 1933 (Appendix): 1–56. https://www.jstor.org/stable/4113436

Anonymous. 1937. *Anuário do Império Colonial Português*, 3rd ed. Lisbon: Empresa do Anuário Comercial.

Anonymous. 1950. Exploradores e naturalistas da flora de Moçambique. *Moçambique: Documentário Trimestral* 61: 97–112. http://memoria-africa.ua.pt/Library/MDT.aspx

Anonymous. 1955. (C.C.D. 1698) Chemical Specialties Co., Inc. *v.* United States. Pp. 155–163 in: *United States Customs Court Reports*, vol. 34. Washington, D.C.: U.S. Government Printing Office.

Anonymous. 1963. *Essências florestais do Maiombe português—Angola*, vol. 1. Lisbon: Jardim e Museu Agrícola do Ultramar.

Anonymous. 1970. M.B.E. [Members of the Order of the British Empire]. P. 6384 in: Supplement to *The London Gazette*, 13th June 1970. https://www.thegazette.co.uk/London/issue/45117/supplement/6384/data.pdf (accessed 7 December 2021).

Anonymous. 1981. *Testemunhos de José Cristóvão Henriques (Engenheiro-Silvicultor).* Lisbon: Junta de Investigações Científicas do Ultramar.

Anonymous. 1985. *Kirstenbosch Botanic Garden.* Cape Town, Claremont: National Botanic Gardens.

Anonymous. 1996–. Doutores Honoris Causa pela Universidade do Porto. António Barros Machado. Universidade do Porto. https://sigarra.up.pt/up/pt/web_base.gera_pagina?p_pagina=doutores%20honoris%20causa%20pela%20u.porto%20-%20ant%C3%B3nio%20barros%20machado

Anonymous. 1997. Liz Matos. *SABONET News* 2(2): 46–48.

Anonymous. 1998. Ninety years of fruitful living. *Record East*, 2 October 1998: 15.

Anonymous. 2000. Esperança da Costa. *SABONET News* 5(3): 136–137.

Anonymous. 2001. Teresa Martins. *SABONET News* 6(2): 70–71.

Anonymous. 2006. Fonds A2203 – Papers of Roux family. Historical Papers Research Archive, The Library, University of the Witwatersrand, Johannesburg, South Africa. http://historicalpapers-atom.wits.ac.za/papers-of-roux-family (accessed 7 December 2021).

Anonymous. 2007. O P. Ernesto Lecomte, Prefeito Apostólico da Cimbebásia. *Missão Espiritana* 12(12): 107–124. https://dsc.duq.edu/missao-espiritana/vol12/iss12/11 (accessed 7 December 2021).
Anonymous. 2015. Nicolas Hallé: Une vie cousue de multiples passions. *Ouest France* [online-version]. https://www.ouest-france.fr/normandie/cherbourg-octeville-50100/nicolas-halle-une-vie-cousue-de-multiples-passions-3644750 (accessed 7 December 2021).
Anonymous. 2016. *A short history of Villa Schifanoia.* Florence: Robert Schuman Centre for Advanced Studies. https://www.eui.eu/Documents/ServicesAdmin/Logistics/EUI-Campus/schifanoia-fly-historyweb-2.pdf (accessed 7 December 2021).
Anonymous. 2018a. Jacob Azancot de Menezes: Percursos em prol de Angola e da Ciência. *Jornal Tornado* [online-version], 15 December 2018. https://www.jornaltornado.pt/jacob-azancot-de-menezes/ (accessed 7 December 2021).
Anonymous. 2018b. Edward "Eddie" Roux. In: South African History Online. https://www.sahistory.org.za/people/edward-eddie-roux (accessed 7 December 2021).
Anonymous. 2019. Mary Seely. In: Namibiana Buchdepot. https://www.namibiana.de/namibia-information/who-is-who/autoren/infos-zur-person/mary-seely.html (accessed 7 December 2021).
Anonymous. 2021a. Charles Emile Thiébaud (1910–1995). In: Musée d'ethnographie de Neuchâtel. https://www.men.ch/fr/histoires/portraits/charles-emile-thiebaud/charles-emile-thiebaud-detail/ (accessed 7 December 2021).
Anonymous. 2021b. Wißmann, Hermann von. Index entry in: *Deutsche Biographie*. Munich: Bayerische Staatsbibliothek. https://www.deutsche-biographie.de/pnd118769537.html (accessed 7 December 2021).
Anonymous. s.d.(a). Le Père Marius Bonnefoux. In: Missionnaires Spiritains: Figures. http://spiritains.forums.free.fr/defunts/bonnefouxm.htm (accessed 7 December 2021).
Anonymous. s.d.(b). António Branquinho d'Oliveira. In: Mortágua Município. http://www.cm-mortagua.pt/modules.php?name=Sections&sop=viewarticle&artid=59 (accessed 7 December 2021).
Anonymous. s.d.(c). Le Père Charles Duparquet. In: Missionnaires Spiritains: Figures. http://spiritains.forums.free.fr/defunts/duparquetc.htm (accessed 7 December 2021).
Anonymous. s.d.(d). Le Père Charles Estermann. In: Missionnaires Spiritains: Figures. http://spiritains.forums.free.fr/defunts/estermannc.htm (accessed 7 December 2021).
Anonymous. s.d.(e). Professor Carlos Eugénio de Melo Geraldes (1878–1962). https://www.isa.ulisboa.pt/files/id/carlos-melo-geraldes/Carlos_Eugenio_de_Melo_Geraldes.pdf (accessed 7 December 2021).
Anonymous. s.d.(f). Photograph of 'Messers Hatton and Cookson's House at Cabenda, the Resting Place of Mr Stanley's Expedition'. Ref.: COPY 1/39/216. The National Archives, Kew, Richmond, U.K. https://discovery.nationalarchives.gov.uk/details/r/C16363902 (accessed 7 December 2021).
Anonymous. s.d.(g). Miss Diana Powell-Cotton. Biography on the website of The British Museum. https://www.britishmuseum.org/collection/term/BIOG134214 (accessed 7 December 2021).
Anonymous. s.d.(h). Le Père Charles Tisserant. In: Missionnaires Spiritains: Figures. http://spiritains.forums.free.fr/defunts/tisserantc.htm (accessed 7 December 2021).
Anonymous. s.d.(i). Le Père Félix Villain. In: Missionnaires Spiritains: Figures. http://spiritains.forums.free.fr/defunts/villainf.htm (accessed 7 December 2021).

Anonymous. s.d.(j). Kay-Scott, C., b. 1879. In: Archives and Special Collections at Tulane University. [New Orleans: Tulane University Libraries]. https://archives.tulane.edu/agents/people/1198 (accessed 7 December 2021).

Archer, R.H. 1997. Leslie Charles Leach (1909–1996). *Bothalia* 27: 91–96. https://doi.org/10.4102/abc.v27i1.663

Archives West. 2007. Gladwyn Murray Childs papers, 1882–1972. http://archiveswest.orbiscascade.org/ark:/80444/xv95681 (accessed 7 December 2021).

Arias, G. 1930. Commemorazione del prof. Dino Taruffi. *Atti Reale Accad. Georgof. Firenze* 27: vii–xlv.

Armand, P. 1884. Nouvelles des voyageurs. *Bull. Soc. Géogr. Marseille* 8: 54–88.

Australian National Herbarium. 2021. Brass, Leonard John (1900–1971). https://www.anbg.gov.au/biography/brass-leonard.html (accessed 7 December 2021).

Azevedo, J.M. 2014. *A colonização do sudoeste angolano do deserto de Namibe ao planalto da Huíla 1849–1900.* Ph.D. Dissertation. University of Salamanca, Salamanca, Spain. https://doi.org/10.14201/gredos.125978

Baker, J.G. 1903 ("1904"). Loganiaceae. Pp. 503–544 in: Thiselton-Dyer, W.J. (ed.), *Flora of Tropical Africa*, vol. 4(1). London: Lovell Reeve & Co. https://doi.org/10.5962/bhl.title.42

Balarin, M.G., Brink, E. & Glen, H. 1999. Itinerary and specimen list of M.A. Pocock's botanical collecting expedition in Zambia and Angola in 1925. *Bothalia* 29: 169–201. https://doi.org/10.4102/abc.v29i1.587

Bally, P.R.O. 1961. *The genus Monadenium.* Berne: Benteli Publishers.

Bamps, P. 1973. Collections botaniques en Angola déposées dans l'herbier de Bruxelles. *Garcia de Orta, Sér. Bot.* 1: 43–44.

Bamps, P. 1975a. Plantes nouvelles ou rares de l'Angola. *Garcia de Orta, Sér. Bot.* 2: 71–76.

Bamps, P. 1975b. *Rhigozum angolense*, Bignoniacée nouvelle de l'Angola. *Bull. Jard. Bot. Natl. Belg.* 45: 149–153. https://doi.org/10.2307/3667593

Bandeira, S. & Leal, F.C. 1864. Angola. Mapa coordenado pelo Visconde de Sá da Bandeira, Tenente General, Ministro da Guerra e por Fernando da Costa Leal, Tenente Coronel, Governador de Mossâmedes, 2nd ed. Map c. 1 : 2600000. Lisbon: [s.n.]. https://purl.pt/4497/3/

Barbosa, L.A.G. 1970. *Carta fitogeográfica de Angola.* Luanda: Instituto de Investigação Científica de Angola.

Barros Machado, A. 1954. Révision systématique des Glossines du groupe palpalis (Diptera). *Publicações Culturais da Companhia de Diamantes de Angola* 22: 1–189.

Barros Machado, A. & Machado, B.B. 1945. Inventário das cavernas calcárias de Portugal. *O Instituto* 105: 198–245.

Bassett Family Association. 2006–2021. BassettBranches.org: Home of the Bassett Family Association. http://www.bassettbranches.org/ (accessed 7 December 2021).

Bastian, A. 1859. *Afrikanische Reisen: Ein Besuch in San Salvador der Hauptstadt des Königreichs Congo; Ein Beitrag zur Mythologie und Psychologie.* Bremen: Heinrich Strack. https://books.google.at/books?id=DTEUAAAAIAAJ

Baum, H. 1903. Reisebericht. Pp. 1–153 in: Warburg, O. (ed.), *Kunene-Sambesi-Expedition.* Berlin: Kolonial-Wirtschaftliches Komitee. https://doi.org/10.5962/bhl.title.37083

Bean, A. 2002. Ion James Muirhead Williams: 1912–2001. *Veld Fl. (1975+)* 88(1): 4–5.

Beard, J.S. 1958. The *Protea* species of the summer rainfall region of South Africa. *Bothalia* 7: 41–65, pl. 1. https://doi.org/10.4102/abc.v7i1.1647

Beard, J.S. [1993]. *The proteas of tropical Africa.* [S.l.]: Kangaroo Press.

Belck, W. 1901. Dr. Karl Hoepfner. *Z. Elektrochemie* 7: 415–417. https://doi.org/10.1002/bbpc.19010072807
Bellorini, C. 2016. *The world of plants in Renaissance Tuscany: Medicine and botany.* London: Routledge. https://doi.org/10.4324/9781315551395
Beolens, B., Watkins, M. & Grayson, M. 2014. *The eponym dictionary of birds.* London: Bloomsbury.
Bernaschina, P. & Ramires, A. 2007. *Missão botânica transnatural: Angola 1927–1937.* [S.l.]: Artez.
Bertrand, H. 1966. Larves de coléoptères aquatiques de l'Angola (Insecta, Coleoptera). *Museu do Dundo: Subsídios para o Estudo da Biologia da Lunda* 72: 135–161.
Beukes, P. 1996. *Smuts the botanist: The Cape Flora and the grasses of Africa.* Cape Town: Human & Rousseau.
Bevilacqua, J.R.S. 2016. *De caçadores a caça: Sobas, Diamang e o Museu do Dundo.* Ph.D. Dissertation. Universidade de São Paulo, São Paulo, Brazil. https://doi.org/10.11606/T.8.2016.tde-25082016-132727
Birmingham, D. 2015. *A short history of modern Angola.* Johannesburg & Cape Town: Jonathan Ball.
Bletchley Park. 2021a. Roll of Honour – Mr Arthur Wallis Exell. https://bletchleypark.org.uk/roll-of-honour/2941/ (accessed 7 December 2021).
Bletchley Park. 2021b. Roll of Honour – Mrs Mildred Alice Exell. https://bletchleypark.org.uk/roll-of-honour/2942/ (accessed 7 December 2021).
Bocage, J.V.B. 1897. José d'Anchieta. *J. Sci. Math. Phys. Nat.*, ser. 2, 5: 126–132.
Bonnefoux, B.M. 1941 ("1940"). *Dicionário olunyaneka-portuguès.* Huíla: Missão da Huíla.
Bornman, H. & Hardy, D. 1971. *Aloes of the South African veld.* Johannesburg: Voortrekkerpers.
Boss, G. 1927. Beiträge zur Zytologie der Ustilagineen. *Planta* 3: 597–627. https://doi.org/10.1007/BF01916502
Bossard, E. 1993. Collecteurs botaniques en Angola. *Mem. Soc. Brot.* 29: 85–104.
Boulvert, Y. 2011a. René Gustave Letouzey (1918–1989): Spécialiste incontesté de la botanique des forêts du Cameroun. Pp. 479–481 in: Serre, J. (ed.), *Hommes et destins*, vol. 11, *Afrique noire.* Paris: L'Harmattan.
Boulvert, Y. 2011b. Charles Michel Tisserant (Révérend Père) (1886–1962). Pp. 729–733 in: Serre, J. (ed.), *Hommes et destins*, vol. 11, *Afrique noire.* Paris: L'Harmattan.
Bournaud, M. 1980. Disparition du Docteur Henri P.I. Bertrand (1892–1978). *Trichoptera Newslett.* 7: 5. https://www.zobodat.at/pdf/TRI_07_0005.pdf (accessed 7 December 2021).
Bradley, J.C. 1919. An entomological cross-section of the United States. *Sci. Monthly* 8: 356–377, 403–420, 514–526.
Brandão, J.M. 2008. Missão Geológica de Angola: Contextos e emergência. *Memórias e Notícias*, n.s., 3: 285–292.
Brandstetter, A.-M. & Hierholzer, V. (eds.) 2017. *Nicht nur Raubkunst! Sensible Dinge in Museen und universitären Sammlungen.* Göttingen: V&R unipress. https://doi.org/10.14220/9783737008082
Brásio, A. 1940. *A missão e seminário da Huíla.* Colecção Pelo Império 64. Lisbon: Agência Geral das Colónias.
Braun, J. 1995. *Eine deutsche Karriere: Die Biographie des Ethnologen Hermann Baumann (1902–1972).* Munich: Akademischer Verlag.
Brown, W.H. 1899. *On the South African frontier: The adventures and observations of an American in Mashonaland and Matabeleland.* New York: Charles Scribner's Sons. https://doi.org/10.5479/sil.788163.39088018050716

Buch, L. von. 1826. Biographical memoir of the late Christian Smith, M.D. naturalist to the Congo Expedition. *Edinburgh New Philos. J.* 1: 209–216.
Buchan, U. 2013. *A green and pleasant land: How England's gardeners fought the Second World War.* New York: Random House.
Buchner, M. 1881. *Recepção e conferencia do Exmo Dr. Max Buchner explorador allemão na Sessão d'Assembleia da Sociedade em 1 de Setembro de 1881.* Loanda: Typographia do Mercantil.
Burger, A. 1955. A produção de óleos atereos em Angola. *Bol. Assoc. Industr. Angola* 7(25): 41.
Burger, W. 1991. Louis Otho Williams (1908–1991). *Taxon* 40: 355–356. https://doi.org/10.1002/j.1996-8175.1991.tb01161.x
Büttner, R. 1890a. *Reisen im Kongolande: Ausgeführt im Auftrage der Afrikanischen Gesellschaft in Deutschland*, 2nd ed. Leipzig: I.C. Hinrichs'sche Buchhandlung. https://books.google.at/books?id=3Vc5AQAAIAAJ
Büttner, R. 1890b. Neue Arten von Guinea, dem Kongo und dem Quango I. *Verh. Bot. Vereins Prov. Brandenburg* 31: 64–96.
Büttner, R. 1891. Neue Arten von Guinea, dem Kongo und dem Quango II. *Verh. Bot. Vereins Prov. Brandenburg* 32: 34–54.
Caldas, F.B. 1984. In memoriam: Arnaldo Deodato da Fonseca Rozeira, 29-4-1912 a 8-3-1984. *Anais Fac. Ci. Univ. Porto* 65: 5–10.
Calhoun, D. 2018. Colonial collectors: Missionaries' botanical and linguistic prospecting in French colonial Africa. *Canad. J. African Stud. / Rev. Canad. Études Africaines* 52: 205–228. https://doi.org/10.1080/00083968.2018.1483834
Cambefort, Y. 2008. Renaud Paulian (1913–2003): Un naturaliste extraordinaire. *Zoosystema* 30: 749–756.
Capelo, H. & Ivens, R. 1881. *De Benguella às terras de Iácca*, 2 vols. Lisbon: Imprensa Nacional. https://doi.org/10.5962/bhl.title.53713
Capelo, H. & Ivens, R. 1882. *From Benguela to the territory of Yacca*, trans. A. Elwes. London: Sampson Low, Marston, Searle, & Rivington. https://books.google.at/books?id=VjMQAAAAYAAJ
Capelo, H. & Ivens, R. 1886. *De Angola à contra-costa*, 2 vols. Lisbon: Imprensa Nacional. https://doi.org/10.5962/bhl.title.60528
Carnie, T. 2012. 'Plant man' honoured for a life in ecology. *The Mercury*, 4 April 2012.
Carnie, T. 2015. Renowned Durban botanist dies. *The Mercury*, 8 July 2015: 4.
Carpenter, F. 1982. Joseph Charles Bequaert. *Psyche* 89: 1–2. https://doi.org/10.1155/1982/71204
Carrisso, L.W. 1928. *O problema colonial perante a nação: Conferência proferida na sala dos Capelos da Universidade de Coimbra em 2 de Março de 1928.* Coimbra: Imprensa da Universidade. https://digitalis-dsp.uc.pt/jspui/handle/10316.2/11466
Carrisso, L.W. 1930. A missão botânica da Universidade de Coimbra à colónia de Angola, em 1927. *Bol. Soc. Brot.* 6: 309–312.
Carrisso, L.W. 1932. Colecção de fotografias diapositivas de Angola (I). *Revista Fac. Ci. Univ. Coimbra* 2: 74–99.
Carrisso, L.W. 1937. Contribuições para o conhecimento da flora de África. *Bol. Soc. Brot.*, sér. 2, 12: 5.
[Carvalho, H.A.D. (text) & Aguiar, M.S.A. (photographs)]. [1887]. Album da expedição ao Muatianvua. [287 gelatin prints with handwritten text on 95 lvs.]. https://purl.pt/23726 (accessed 7 December 2021).
Castanheira Diniz, A. 1973. *Características mesológicas de Angola: Descrição e correlação dos aspectos fisiográficos; Dos solos e da vegetação das zonas agrícolas angolanas.* Nova Lisboa: Missão de inquéritos agrícolas de Angola.

Castanheira Diniz, A. & Cardoso de Matos, G. 1986–1994. Carta de zonagem agro-ecológica e da vegetação de Cabo Verde. I–X. *Garcia de Orta, Sér. Bot.* 8: 39–82; 9: 35–70; 10: 19–72; 11: 9–29; 12: 69–120.

Castelo, C. 2011. Hélder Lains e Silva. In: Roque, R. (org.), *History and anthropology of "Portuguese Timor", 1850–1975: An online dictionary of biographies.* http://www.historyanthropologytimor.org/wp-content/uploads/2012/01/SILVA_Helder_Lains_e_C_Castelo_2011.pdf (accessed 7 December 2021).

Cavaco, A. 1959. *Contribution à l'étude de la flore de la Lunda d'après les récoltes de Gossweiler (1946–1948).* Publicações Culturais do Museu do Dundo 42. Dundo: Museu do Dundo.

Ceríaco, L.M.P., Marques, M.P., André, I., Afonso, E., Blackburn, D.C. & Bauer, A.M. 2020. Illustrated type catalogue of the "lost" herpetological collections of Museu do Dundo, Angola. *Bull. Mus. Comp. Zool.* 162(7): 379–440. https://doi.org/10.3099/0027-4100-162.7.379

Chaudhri, M.N., Vegter, I.H. & Wal, C.M. de. 1972. *Index herbariorum*, part 2(3), *Collectors I–L.* Regnum Vegetabile 86. Utrecht: International Bureau for Plant Taxonomy.

Child, G. (engraver). 1745. Bay of Kabinda / Map of the Mouth of the River Kongo or Zayre, with the Adjacent Coast. http://www.atlasofmutualheritage.nl/en/Map-bay-Kabinda-mouth-Congo-river.7901 (accessed 7 December 2021).

Chiovenda, E. 1917. Piante dei dintorni di Bailundo (Benguella) m. 1500–1700 s.m., raccolte dal Prof. Dino Taruffi nel 1914. *Bull. Soc. Bot. Ital.* 1917: 28–31.

Chiovenda, E. 1924. Piante nuove dell'Angola raccolte dal Dott. N. Mazzocchi-Alemanni. *Bull. Soc. Bot. Ital.* 1924: 38–46.

Chiovenda, E. 1925. La collezione di piante fatta dal Comm. Nallo Mazzocchi-Alemanni nell'Angola nel 1923. *Agric. Colon.* 18: 378–391.

Coates Palgrave, M., with contributions from Bingham, M., Grosvenor, R., Hyde, M., Kimberly, M., Loveridge, J., Simon, B., Timberlake, J. & Williamson, G. 2009. Robert Baily Drummond (1924–2008). *Bothalia* 39: 117–119. https://doi.org/10.4102/abc.v39i1.239

Codd, L.E.W. 1951. *Trees and shrubs of the Kruger National Park.* Botanical Survey Memoir 26. [Pretoria]: Division of Botany and Plant Pathology, Department of Agriculture.

Codd, L.E.[W.]. 1980. R.G.N. Young. *Forum Bot.* 18(1): 1.

Codd, L.E.[W.] & Gunn, M. 1985. Additional biographical notes on plant collectors in southern Africa. *Bothalia* 15: 631–654. https://doi.org/10.4102/abc.v15i3/4.1832

Comando Geral da Armada. 2021. Livro Mestre – Classe Marinha: K [1909–1940]. PT/BCM-AH/30A/01/001-007/02487. https://arquivohistorico.marinha.pt/viewer?id=2369&FileID=6232

Compton, R.H. 1965. *Kirstenbosch: Garden for a nation; Being the story of the first 50 years of the National Botanic Gardens of South Africa 1913–1963.* Cape Town: Tafelberg-Uitgewers.

Coosemans, M. 1951. Teusz (Édouard). Col. 903–904 in: Institut Royal Colonial Belge (ed.), *Biographie coloniale belge*, vol. 2. Brussels: Institut Royal Colonial Belge.

Costa, E., Martins, T. & Monteiro, F. 2004. *A checklist of Angolan grasses—Checklist das Poaceae de Angola.* Southern African Botanical Diversity Network Report 28. Pretoria: SABONET.

Costa, M.S. 2004. Homenagem. *Floresta & Amb.* 64: 5.

Couceiro, H.P. 1948. *Angola (Dois anos de Governo Junho de 1907–Junho de 1909): História e comentários.* Edição comemorativa do terceiro centenário da restauração de Angola. Lisbon: Edições F. Gama. https://www.fd.unl.pt/Anexos/Investigacao/1741.pdf (accessed 7 December 2021).

Crawford Cabral, J. 1967. Mamíferos da reserva do Luando. *Bol. Inst. Invest. Ci. Angola* 4(2): 33–44.

Crawford Cabral, J.C.M. 1970. Alguns aspectos da ecologia da Palanca Real (*Hippotragus niger variani* Thomas). *Bol. Inst. Invest. Ci. Angola* 7(1): 9–42.

Crawford-Cabral, J.C.M. 1998. *The Angolan rodents of the superfamily Muroidea: An account of their distribution.* Lisbon: Instituto Nacional de Investigação Científica.

Crawford Cabral, J.C.M. & Mesquitela, L.M. 1989. *Indice toponímico de colheitas zoológicas em Angola.* Lisbon: Instituto de Investigação Científica Tropical.

Crawford-Cabral, J.C.M. & Veríssimo, L.N. 2005. *The ungulate fauna of Angola.* Estudos, Ensaios e Documentos 163. Lisbon: Instituto de Investigação Científica Tropical.

Crouch, N.R., Smith, G.F., Klopper, R.R., Figueiredo, E., McMurtry, D. & Burns, S. 2015. Winter-flowering maculate aloes from the Lowveld of southeastern Africa: Notes on *Aloe monteiroae* Baker (Asphodelaceae: Alooideae), the earliest name for *Aloe parvibracteata* Schönland. *Bradleya* 33: 147–155. https://doi.org/10.25223/brad.n33.2015.a20

Cruz, A.M. 1967. O povo Ovakwambundu. *Bol. Inst. Invest. Ci. Angola* 4(2): 67–88.

Curror, A.B. 1843. Letter to William Jackson Hooker. Ref.: KADC7434. Directors' Correspondence 58/27 Library and Archives at Royal Botanic Gardens, Kew. https://plants.jstor.org/stable/10.5555/al.ap.visual.kadc7434 (accessed 7 December 2021) [sub "Burrows"].

Curror, A.B. 1844. Letter to William Jackson Hooker. Ref.: KADC7431. Directors' Correspondence 58/24 Library and Archives at Royal Botanic Gardens, Kew. https://plants.jstor.org/stable/10.5555/al.ap.visual.kadc7431 (accessed 7 December 2021) [sub "Burrows"].

Curtis, C.P., Jr. & Curtis, R.C. 1925. *Hunting in Africa East and West.* Boston: Houghton Mifflin.

D'Aguilar, J. 2013. Henri Bertrand entomologiste jusqu'au bout. *Insectes* 170: 31–32.

D'Orey, J.D.S. 1982. Marantaceae colhidas por John Gossweiler em Angola existentes em LISJC. *Garcia de Orta, Sér. Bot.* 5: 47–57.

Daire, M.-Y., López-Romero, E. & Le Gall, C. 2013. Théodore Monod (1902–2000) et l'archéologie bretonne: Note sur un épisode méconnu de la vie du «fou du désert». *Rev. Archeol. Ouest* 30: 289–301. https://doi.org/10.4000/rao.2193

Danckelman, A. von. 1884. Ein Besuch in den portugiesischen Kolonien Südwestafrikas (Sommer 1883). *Deutsche Geogr. Blätt.* 7: 31–62, t. 2.

Dandy, J.E. 1958. *The Sloane herbarium: An annotated list of the Horti sicci composing it; with Biographical details of the principal contributors.* London: The British Museum.

Darby, M. 2019. Balfour-Browne, John ('Jack') William Alexander Francis. In: Biographical dictionary of British coleopterists. http://www.coleoptera.org.uk/biographical-dictionary

Darbyshire, I., Kordofani, M., Farag, I., Candiga, R. & Pickering, H. (eds. & comps.). 2015. Notes on some of the principal collectors in the Sudan Region and their historical setting. Pp. 13–26 in: *The plants of South Sudan: An annotated checklist.* Richmond: Kew Publishing, Royal Botanic Gardens, Kew.

Daskalos, S. 2000. *Um testemunho para a história de Angola (do Huambo ao Huambo).* Lisbon: Vega.

David, E.M.T.S.S. 2013. *Artes plásticas angolanas dos últimos 10 anos: Duas gerações de artistas e um contexto de paz.* Masters thesis. Universidade do Porto, Porto, Portugal. http://biblioteca.fba.up.pt/docs/Elsa_David/ElsaDavid_tese.pdf (accessed 7 December 2021).

Davis, L. 2007. Frederick Creighton Wellman. In: Find a grave: Memorials. https://www.findagrave.com/memorial/19977082/frederick-creighton-wellman (accessed 7 December 2021).

De Wildeman, E. 1935. Le R.P. Hyacinthe Vanderyst (1860–1934). *Bull. Séances Inst. Roy. Colon. Belge* 6: 28–38.

De Winter, B. & Germishuizen, G. 2000. Leslie Edward Wostall Codd (1908–1999). *Bothalia* 30: 111–115. https://doi.org/10.4102/abc.v30i1.547

Decelle, J. 1982. Assemblée mensuelle du 1 décembre 1982. (décès de J. Ghesquière). *Bull. Ann. Soc. Roy. Belge Entomol.* 118: 212.

Delachaux, T. & Thiébaud, C.E. 1934. *Pays et peuples d'Angola.* Paris: Édition Victor Attinger.

Desmond, K. 2017. *Planet savers: 301 extraordinary environmentalists.* Abingdon-on-Thames: Routledge/Taylor & Francis.

Desmond, R. 1994. *Dictionary of British and Irish botanists and horticulturists.* London: Taylor & Francis; The Natural History Museum.

Desmond, R. 1999. *Sir Joseph Dalton Hooker: Traveller and plant collector.* Woodbridge: Antique Collectors' Club with the Royal Botanic Gardens, Kew.

Dias, G.S. 1937. O Padre Bonnefoux. *Bol. Geral Colon.* 148: 93.

Dias, G.S. 1939. *José de Anchieta.* Colecção Pelo Império 38. Lisbon: Agência Geral das Colónias.

Dierks, K. 2003–2004. Wulfhorst, August. In: Biographies of Namibian personalities. https://www.klausdierks.com/Biographies/Biographies_W.htm (accessed 7 December 2021).

Diniz, M.A. 1993. *Conspectus florae Angolensis: Bignoniaceae.* Lisbon: Instituto de Investigação Científica Tropical.

Dixon, K.W. 2006. Celebration of a life in botany: Dr John Stanley Beard – On the occasion of his 90th birthday. *J. Roy. Soc. Western Australia* 89: 93–99.

Dold, T. & Cocks, M. 2012. *Voices from the forest: Celebrating nature and culture in Xhosaland.* Auckland Park, South Africa: Jacana Media.

Dold, T. & Kelly, J. (comps. & eds.). 2018. *Bushmen, botany and baking bread: Mary Pocock's record of a journey with Dorothea Bleek across Angola in 1925.* Grahamstown: NISC.

Dolezal, H. 1974. *Friedrich Welwitsch: Vida e obra.* Lisbon: Junta de Investigações Científicas do Ultramar.

Domico, T. 2018. *The great cactus war: True story of the greatest plant invasion in human history.* Friday Harbour, Washington: Green Flash Books.

Dorr, L.J. 1997. *Plant collectors in Madagascar and the Comoro Islands.* Richmond: Royal Botanic Gardens, Kew.

Dorr, L.J. & Nicolson, D.H. 2008. *Taxonomic literature: A selective guide to botanical publications and collections with dates, commentaries and types*, 2nd ed., suppl. 7, *F–Frer.* Ruggell: Gantner. https://doi.org/10.5962/bhl.title.48631

Dorr, L.[J.], Stauffer, F.W. & Rodríguez, L. 2017. Albert Mocquerys in Venezuela (1893–1894): A commercial collector of plants, birds, and insects. *Harvard Pap. Bot.* 22: 17–26. https://doi.org/10.3100/hpib.v22iss1.2017.n5

Dozy, J.J. 1996. Charles Émile Thiébaud (1910–1995). *Bull. Angew. Geol.* 1(2): 183–184.

Dravers, P. 2014. George Lawson obituary. *The Guardian*, 23 September 2014. https://www.theguardian.com/theguardian/2014/sep/23/george-lawson-obituary (accessed 7 December 2021).

Du Toit, S.J. [2004]. Dr Ion Williams: Hermanus' saviour. In: hermanus.co.za. https://www.hermanus.co.za/info/history/dr-ion-williams (accessed 7 December 2021).

Duarte, A.J. 1964. *Elementos de entomologia agrícola para Angola.* Luanda: Instituto de Investigação Científica de Angola.

Duarte, A.J. 1966. *Pragas importantes do algodoeiro em Angola.* Luanda: Junta Provincial de Povoamento de Angola.

Duparquet, C. 1953. *Viagens na Cimbebásia.* Luanda: Museu de Angola.

Dutton, P. 2017. *Spirit of the wilderness.* Pinetown, South Africa: 30 Degrees South Publishers.

Duvigneaud, P.A. 1956. Novidades da flora de Angola — V. *Bol. Soc. Brot.*, sér. 2, 29: 85–86.

Duvigneaud, P.[A.] & Timperman, J. 1959. Études sur la végétation du Katanga et de ses sols métallifères: Communication no. 3; Études sur le genre *Crotalaria. Bull. Soc. Roy. Bot. Belgique* 91: 135–176. https://www.jstor.org/stable/20792301

E.G.R. 1905. Hermann von Wissmann. *Geogr. J. (London)* 26: 227–230. https://www.jstor.org/stable/1776225

Earnest, C. & Earnest, R. 2006. In remembrance Klaus Stopp and Donald Shelley. *Der Kurier (Mid-Atlantic Germanic Society)* 24(3): [9]. https://magsgen.com/upload/files/Der_Kurier_issues/2006-09-24-3.pdf

Efrst, W.H.O. 1983. In memoriam Hiram Wild. *Vegetatio* 51: 125–128. https://doi.org/10.1007/BF00129431

Ellenberger, V. 1938. *A century of mission work in Basutoland (1833–1933).* Morija: Sesuto Book Depot.

Ellenberger, V. 1953. *La fin tragique des bushmen.* Paris: Amiot-Dumont.

Esdorn, I. 1972. Ilse Baronin von Nolde (1889–1970). *Willdenowia* 6: 415–418. https://www.jstor.org/stable/3995564

Estermann, C. 1941. Contribuição dos missionários do Espírito Santo para a exploração científica do sul de Angola. *Bol. Geral Colon.* 196: 3–15.

Estermann, C. 1956–1961. *Etnografia do sudoeste de Angola*, 3 vols. Lisbon: Junta de Investigações do Ultramar.

Exell, A.W. 1938a. Missão Botânica do Dr Carrisso a Angola. *Bol. Geral Colon.* 153: 3–24.

Exell, A.W. 1938b. Dr. Carrisso's botanical mission to Angola. *J. Bot.* 76: 121–134.

Exell, A.W. 1939. Collections from Angola in the Sloane Herbarium. *J. Bot.* 77: 146–147.

Exell, A.W. 1944. *Catalogue of the vascular plants of S. Tomé (with Príncipe and Annobon).* London: British Museum (Natural History).

Exell, A.W. 1952. John Gossweiler. *Taxon* 1: 93–94. https://doi.org/10.1002/j.1996-8175.1952.tb01409.x

Exell, A.W. 1960. History of botanical collecting in the Flora Zambesiaca area. Pp. 23–34 in: Exell, A.W. & Wild, H. (eds.), *Flora Zambesiaca*, vol. 1(1). London: Crown Agents for Overseas Governments and Administrations.

Exell, A.W. 1962. Pre-Linnean collections in the Sloane Herbarium from Africa south of the Sahara. *Compt. Rend. Réun. Plén. Assoc. Pour Étude Taxon. Fl. Afrique Trop.* 4: 47–49.

Exell, A.W. 1973. Angiosperms of the islands of the Gulf of Guinea (Fernando Po, Príncipe, S. Tomé, and Annobón). *Bull. Brit. Mus. (Nat. Hist.), Bot.* 4(8): 327–411.

Exell, A.W. 1984. In memory of Francisco de Ascenção Mendonça. *Garcia de Orta, Sér. Bot.* 6: 1–6.

Exell, A.W. & Hayes, G.A. 1967. A list of botanical collectors in the Flora Zambesiaca area. *Kirkia* 6: 85–104.
Exell, A.W. & Mendonça, F.A. 1937–1951. *Conspectus florae Angolensis*, vol. 1. Lisbon: Junta de Investigações Coloniais.
Exell, A.W. & Mendonça, F.A. 1956. Supplement to the introduction (1937). Pp. ix–xi in: Exell, A.W. & Mendonça, F.A., *Conspectus florae Angolensis*, vol. 2(2). Lisbon: Junta de Investigações do Ultramar.
Exell, M.A. 1937a. Leguminosae from Mozambique collected by Gomes e Sousa. *Bol. Soc. Brot.*, sér. 2, 12: 6–16.
Exell, M.A. 1937b. Diary of the 1937 expedition to Angola. Unpublished typescript. Copy at the University of Lisbon.
Exell, M.A. 1939. Contribuições para o conhecimento da flora de África: II. Two new species of *Copaifera* from Angola. *Bol. Soc. Brot.*, sér. 2, 13: 322–325.
Fernandes, A. 1939. Notícia sôbre a vida e a obra do Prof. Luiz Wittnich Carrisso. *Bol. Soc. Brot.*, sér. 2, 13: iii–lxxii.
Fernandes, A. 1954. John Gossweiler (1873–1952). *Vegetatio* 4: 334–335. https://doi.org/10.1007/BF00301801
Fernandes, A. 1959. Artur Augusto Taborda de Morais (1900–1959). *Anuário Soc. Brot.* 25: 11–19.
Fernandes, A. 1984. Lembrando funcionários do Museu, Laboratório e Jardim Botânico. *Anuário Soc. Brot.* 50: 9–35.
Fernandes, A. 1993. *A universidade de Coimbra e o estudo da flora e da vegetação dos países africanos de língua oficial portuguesa*. Coimbra: Departamento de Botânica, Faculdade de Ciências e Tecnologia da Universidade de Coimbra.
Ferrão, J.E.M. 1993. A evolução do ensino agrícola colonial. *Anais Inst. Super. Agron.* 43: 35–73.
Ferreira, F.A. 2015. Investimentos privados de brasileiros na África Portuguesa: O caso da Western Africa Malachite Copper Mines Company. In: *XI Congresso Brasileiro de História Econômica e 12ª Conferência Internacional de História de Empresas, Vitória/2015.* https://www.abphe.org.br/arquivos/2015_frederico_antonio_ferreira_investimentos-privados-de-brasileiros-na-africa-portuguesa-o-caso-da-western-africa-malachite-copper-mines-company.pdf (accessed 7 December 2021).
Ferreira, R. 2015. The conquest of Ambriz: Colonial expansion and imperial competition in Central Africa. *Mulemba* 5: 221–242. https://doi.org/10.4000/mulemba.439
Ficalho, C. & Hiern, W.P. 1881. On Central-African plants collected by Major Serpa Pinto. *Trans. Linn. Soc. London, Bot.*, ser. 2, 2: 11–36. https://doi.org/10.1111/j.1095-8339.1881.tb00002.x
Figueira, R. & Lages, F. 2019. Museum and herbarium collections for biodiversity research in Angola. Pp. 513–542 in: Huntley, B.J., Russo, V., Lages, F. & Ferrand, N. (eds.), *Biodiversity of Angola: Science and conservation; A modern synthesis.* Berlin: Springer Nature. https://doi.org/10.1007/978-3-030-03083-4_19
Figueiredo, E. & Smith, G.F. 2008. *Plants of Angola / Plantas de Angola*. Strelitzia 22. Pretoria: South African National Biodiversity Institute.
Figueiredo, E. & Smith, G.F. 2010. The colonial legacy in African plant taxonomy. *S. African J. Sci.* 106(3/4): 5–7. https://doi.org/10.4102/sajs.v106i3/4.161
Figueiredo, E. & Smith, G.F. 2011. Who's in a name: Eponymy of the name *Aloe thompsoniae* Groenew. (Asphodelaceae), with notes on naming species after people. *Bradleya* 29: 121–124. https://doi.org/10.25223/brad.n29.2011.a14
Figueiredo, E. & Smith, G.F. 2012. *Common names of Angolan plants*, 1st ed. Pretoria: Inhlaba Books.

Figueiredo, E. & Smith, G.F. 2017. *Common names of Angolan plants*, 2nd ed. Pretoria: Protea Book House.

Figueiredo, E. & Smith, G.F. 2019. Typification of the name *Hoodia currorii* (Apocynaceae). *Phytotaxa* 423: 297–300. https://doi.org/10.11646/phytotaxa.423.5.5

Figueiredo, E. & Smith, G.F. 2020a. Andrew Beveridge Curror (1811–1844): Collecting natural history specimens while preventing the slave trade along the west coast of Africa. *Phytotaxa* 436: 141–156. https://doi.org/10.11646/phytotaxa.436.2.4

Figueiredo, E. & Smith, G.F. 2020b. Friedrich Welwitsch and his overlooked contributions to the *Flora of Tropical Africa*, with a discussion of the names of *Kalanchoe* (Crassulaceae subfam. Kalanchooideae) that should be ascribed to Welwitsch alone. *Phytotaxa* 458: 83–100. https://doi.org/10.11646/phytotaxa.458.1.5

Figueiredo, E. & Smith, G.F. 2020c. Friedrich Welwitsch and his contributions to the exploration and study of the flora of São Tomé and Príncipe, Gulf of Guinea, with typification of three names described from his collections. *Phytotaxa* 459: 227–234. https://doi.org/10.11646/phytotaxa.459.3.4

Figueiredo, E. & Smith, G.F. 2021a. An overview of plant collecting in Angola from 1690 to 2000. *Phytotaxa* 523: 32–54. https://doi.org/10.11646/phytotaxa.523.1.2

Figueiredo, E. & Smith, G.F. 2021b. Joaquim José da Silva (c. 1755–1810): His life, natural history collecting activities, and involvement in the so-called first scientific expedition in the interior of Angola. *Candollea* 76: 125–138. https://doi.org/10.15553/c2021v761a13

Figueiredo, E., Matos, S., Cardoso, J.F. & Sampaio Martins, E. 2008. List of collectors / Lista de colectores. Pp. 4–11 in: Figueiredo, E. & Smith, G.F., *Plants of Angola / Plantas de Angola*. Strelitzia 22. Pretoria: South African National Biodiversity Institute.

Figueiredo, E., Soares, M., Seibert, G., Smith, G.F. & Faden, R.B. 2009. The botany of the Cunene-Zambezi Expedition with notes on Hugo Baum (1867–1950). *Bothalia* 39: 185–211. https://doi.org/10.4102/abc.v39i2.244

Figueiredo, E., Smith, G.F. & Nyffeler, R. 2013. August Wulfhorst (1861–1936) and his overlooked contributions on the flora of Angola. *Candollea* 68: 123–131. https://doi.org/10.15553/c2013v681a17

Figueiredo, E., Smith, G.F. & Amaral, P.B. 2017. Notes on António de Figueiredo Gomes e Sousa, a nearforgotten collector of succulent plants in Mozambique. *Bradleya* 35: 186–194. https://doi.org/10.25223/brad.n35.2017.a21

Figueiredo, E., Silva, V., Coutinho, A. & Smith, G.F. 2018a. Twentieth century vascular plant taxonomy in Portugal. *Willdenowia* 48: 303–330. https://doi.org/10.3372/wi.48.48209

Figueiredo, E., Silva, V. & Smith, G.F. 2018b. Friedrich Welwitsch and the horticulture of succulents in Portugal in the 19th century. *Bradleya* 36: 200–211. https://doi.org/10.25223/brad.n36.2018.a15

Figueiredo, E., Smith, G.F. & Ceríaco, L.M.P. 2019a. The vascular plant collections of Francisco Newton (1864–1909) in Angola. *Phytotaxa* 413: 207–224. https://doi.org/10.11646/phytotaxa.413.3.2

Figueiredo, E., Smith, G.F. & Silva, V. 2019b. Flávio Ferreira Pinto de Resende (1907–1967): A Portuguese scientist and hero-botanist, and his forays into systematics. *Taxon* 68: 420–423. https://doi.org/10.1002/tax.12053

Figueiredo, E., Williams, D. & Smith, G.F. 2019c. The identity of John Rattray, diatomist and collector on the *Buccaneer* expedition (1885–1886) to West Africa. *Phytotaxa* 408: 296–300. https://doi.org/10.11646/phytotaxa.408.4.7

Figueiredo, E., Smith, G.F. & Dressler, S. 2020. The botanical exploration of Angola by Germans during the 19th and 20th centuries, with biographical sketches and notes on collections and herbaria. *Blumea* 65: 126–161. https://doi.org/10.3767/blumea.2020.65.02.06

Figueiredo, L. 2014. *Sita Valles: Revolucionária, comunista até à morte.* Lisbon: Alêtheia Editores.

Fischer, W. 1972. Hoepfner, Carl. P. 348 in: *Neue Deutsche Biographie*, vol. 9 [online version]. https://www.deutsche-biographie.de/pnd116928581.html#ndbcontent (accessed 7 December 2021).

Fonte, B. 1998. *Dicionário dos mais ilustres transmontanos e alto durienses,* vol. 1. Guimarães: Ed. Cidade Berço.

Fontinha, M. 1983. *Desenhos na areia dos quiocos do nordeste de Angola.* Estudos, Ensaios e Documentos 143. Lisbon: Instituto de Investigação Científica Tropical.

Fontinha, M. & Videira, A. 1963. *Cabaças gravadas da Lunda.* Subsídios para a História, Arqueologia e Etnografia dos Povos da Lunda 57. Dundo: Museu do Dundo.

Fox, F.D. 1912. Some notes on big game in Portuguese Angola, S.W. Africa. *African Affairs* 11: 430–437. https://doi.org/10.1093/oxfordjournals.afraf.a099528

Fox, F.W. & Norwood Young, M.E. 1982. *Food from the veld: Edible wild plants of southern Africa botanically identified and described.* Johannesburg: Delta Books.

Fox, F.W. & Norwood Young, M.E. 1988. *Food from the veld: Edible wild plants of southern Africa botanically identified and described*, 2nd reprint. Johannesburg: Delta Books.

Fox, M. 1912. Mildred Fox to Royal Botanic Gardens, Kew, 16 October 1912. Directors' Correspondence 185/420, ID KADC1656. Library and Archives at Royal Botanic Gardens, Kew. https://plants.jstor.org/stable/10.5555/al.ap.visual.kadc1656

Frade, E.C. 1965. Plantações de eucaliptos da companhia do caminho de ferro de Benguela: Secagem de lenha da *Eucalyptus rostrata* Schlechtendal. *Agron. Angol.* 22: 193–202.

Frade, F. & Sieiro, D.M. 1960. Palanca preta gigante de Angola. *Garcia de Orta* 8: 21–38.

Frahm, J.-P. & Eggers, J. 2001. *Lexikon deutschsprachiger Bryologen.* Norderstedt: Books on Demand.

Frank, J. 2007. Remembering Gordon D. Gibson: Prominent anthropologist made his mark in many areas. *San Diego-Tribune*, 2 October 2007. https://www.sandiegouniontribune.com/sdut-remembering-gordon-d-gibson-prominent-2007oct02-story.html (accessed 7 December 2021).

Freitas, A.S.B. 1908a. *Indice para a obra Plantas uteis da Africa Portugueza pelo Conde de Ficalho.* Lisbon: Sociedade de Geographia de Lisboa. https://digitalis-dsp.uc.pt/html/10316.2/9775/P5.html (accessed 7 December 2021).

Freitas, A.S.B. 1908b. Augusto S.B. de Freitas to Júlio Henriques, 7 March 1908. Fundo JH – Pasta A-Ba (70) – BARdFRE (ASI)-1. Arquivo de Botânica, Universidade de Coimbra. https://digitalis-dsp.uc.pt/html/10316.2/10620/item1_index.html (accessed 7 December 2021).

Freitas, H. 2017. Foreword to the 2nd edition. Pp. 7–9 in: Figueiredo, E. & Smith, G.F., *Common names of Angolan plants*, 2nd ed. Pretoria: Protea Book House.

Froidevaux, H. 1912. La Mission Rohan-Chabot dans l'Angola. *Géographie* 26: 359–361.

Gagnepain, F. (Rédacteur Principal). 1944. Humbert (Henri). P. 43 in: Humbert, H. (Éditeur Scientifique), *Flore générale de l'Indo-Chine,* tome préliminaire, *Introduction – Tables générales.* Paris: Masson. https://bibliotheques.mnhn.fr/medias/doc/EXPLOITATION/IFD/FLINC_S000_1944_T000_N000/

Garcia, J.G. 1970. Eng. Agr. Joaquim Martinho Lopes de Brito Teixeira. *Bol. Soc. Brot.*, sér. 2, 44: vii–xvi.

Gardiner, B.G. 1989. [Note of the editor]. P. 30 in: Furley, D.D., Notes on the correspondence of E. M. Holmes (1843–1930). *Linnean* 5(3): 23–30.

Gereau, R., Taylor, C.M., Croat, T.B., Hoch, P.C. & Miller, J.S. 2021. Walter Lewis (1930–2020). *Pl. Sci. Bull.* 67: 124–127. https://botany.org/userdata/IssueArchive/issues/originalfile/PSB_2021_67_2.pdf (accessed 7 December 2021).

Gilbert, G. 1951. Hommage à Pierre-Edouard Luja. *Bull. Soc. Roy. Bot. Belgique* 83: 263–265.

Gimingham, C. 2005. Noel Marshall Pritchard. *J. Bryol.* 27: 171–172. https://doi.org/10.1179/jbr.2005.27.2.171

Gito, E. 2016. Empresários invadem terras do Cunene com a ajuda da polícia: "Até cemitérios estão a destruir". *O Novo Jornal*, 7 June 2016. http://novojornal.co.ao/sociedade/interior/empresarios-invadem-terras-do-cunene-com-a-ajuda-da-policia-ate-cemiterios-estao-a-destruir-31196.html (accessed 7 December 2021).

Glen, H.F. 1998. David Spencer Hardy (1931–1998). *Bothalia* 28: 239–247. https://doi.org/10.4102/abc.v28i2.644

Glen, H.F. & Germishuizen, G. (comps.). 2010. *Botanical exploration of southern Africa*, 2nd ed. Strelitzia 26. Pretoria: South African National Biodiversity Institute.

Golubic, S. & Wilmotte, A. 2014. The phycologist Pierre Compère: His contribution to cyanobacterial studies. *Pl. Ecol. Evol.* 147: 307–310. https://doi.org/10.5091/plecevo.2014.1044

Gomes e Sousa, Alice 1949. Exploradores e naturalistas da fauna de Moçambique. *Moçambique: Documentário Trimestral* 57: 69–72. http://memoria-africa.ua.pt/Library/MDT.aspx

Gomes e Sousa, A.F. 1930. Subsídios para o conhecimento da Guiné Portuguesa. *Mem. Soc. Brot.* 1: 1–94, tt. 1–44, 1 map.

Gomes e Sousa, A.F. 1939. Exploradores e naturalistas da flora de Moçambique. *Moçambique: Documentário Trimestral* 20: 33–69. http://memoria-africa.ua.pt/Library/MDT.aspx

Gomes e Sousa, A. F. 1942. Exploradores e naturalistas da flora de Moçambique. *Moçambique: Documentário Trimestral* 30: 53–67. http://memoria-africa.ua.pt/Library/MDT.aspx

Gomes e Sousa, A.F. 1943. Exploradores e naturalistas da flora de Moçambique. *Moçambique: Documentário Trimestral* 36: 39–44. http://memoria-africa.ua.pt/Library/MDT.aspx

Gomes e Sousa, A.F. 1949. Exploradores e naturalistas da flora de Moçambique. *Moçambique: Documentário Trimestral* 57: 59–67. http://memoria-africa.ua.pt/Library/MDT.aspx

Gomes e Sousa, A.F. 1966–1967. Dendrologia de Moçambique. Estudo Geral, 2 vols. *Mem. Inst. Invest. Agron. Moçambique* 1: 1–822.

Gomes e Sousa, A.F. 1971a. Exploradores e naturalistas da flora de Moçambique. *Bol. Soc. Estud. Colón. Moçambique* 40(166/167): 99–103.

Gomes e Sousa, A.F. 1971b. Exploradores e naturalistas da flora de Moçambique. *Moçambique: Documentário Trimestral* 105: 97–112. http://memoria-africa.ua.pt/Library/MDT.aspx

Gonçalves, A.E. 1996. António Rocha da Torre (1904–1995). *Garcia de Orta, Sér. Bot.* 13: 4–5.

Gonçalves, M.M. 1985. In memoriam H. Lains e Silva (1921–1984). *Garcia de Orta, Sér. Estudos Agron.* 12(1/2): VII–XIV.

Gossweiler, J. 1939. Elementos para a história da exploração botânica de Angola. *Bol. Soc. Brot.* 13: 283–305.

Gossweiler, J. 1948. Flora exótica de Angola: Nomes vulgares e origem das plantas cultivadas ou sub-espontâneas. *Agron. Angol.* 1: 121–198.

Gossweiler, J. 1949. Flora exótica de Angola: Nomes vulgares e origem das plantas cultivadas ou sub-espontâneas. *Agron. Angol.* 2: 173–255.

Gossweiler, J. 1950. Flora exótica de Angola: Nomes vulgares e origem das plantas cultivadas ou sub-espontâneas. *Agron. Angol.* 3: 143–167.

Gossweiler, J. 1953. Nomes indígenas das plantas de Angola. *Agron. Angol.* 7: 1–587.

Gossweiler, J. & Mendonça, F.A. 1939. *Carta fitogeográfica de Angola.* Lisbon: Ministério das Colónias.

Gouveia, A.C.P. 1966. *Vegetação do Cunene (entre Matala e Vila Roçadas): Contribuição para o seu estudo.* Lisbon: Instituto Superior de Agronomia.

Goyder, D.J. & Gonçalves, F.M.P. 2019. The flora of Angola: Collectors, richness and endemism. Pp. 79–96 in: Huntley, B.J., Russo, V., Lages, F. & Ferrand, N. (eds.), *Biodiversity of Angola: Science and conservation; A modern synthesis.* Berlin: Springer Nature. https://doi.org/10.1007/978-3-030-03083-4_5

Grace, O.M., Klopper, R.R., Figueiredo, E. & Smith, G.F. 2011. *The aloe names book.* Strelitzia 28. Pretoria: South African National Biodiversity Institute; Richmond: Royal Botanic Gardens, Kew.

Gruvel, A. 1911. Mission Gruvel sur la Côte occidentale d'Afrique (1909–1910). *Ann. Inst. Océanogr.* 3(4): 1–56.

Gu, C., Peng, C.-I. & Turland, N.J. 2007. Begoniaceae. In: Wu, Z.Y., Raven, P.H. & Hong, D.Y. (eds.), *Flora of China*, vol. 13 *(Clusiaceae through Araliaceae).* Beijing: Science Press; St. Louis: Missouri Botanical Garden Press. http://www.efloras.org/florataxon.aspx?flora_id=2&taxon_id=242412320

Gunn, M. & Codd, L.E.[W.]. 1981. *Botanical exploration of southern Africa.* Cape Town: A.A. Balkema.

Güssfeldt, P., Falkenstein, J. & Pechuël-Loesche, E. 1888. *Die Loango-Expedition: Ausgesandt von der Deutschen Gesellschaft zur Erforschung Aequatorial-Africas 1873–1876.* Leipzig: E. Baldamus. https://doi.org/10.5962/bhl.title.58908

Hamet, R. 1913. Sur un *Kalanchoe* nouveau de l'Herbier de Stockholm. *Ark. Bot.* 13(11): 1–5, t. 1.

Hamy, E.-T. 1908. La mission de Geoffroy Saint-Hilaire en Espagne et en Portugal (1808): Histoire et documents. *Nouv. Arch. Mus. Hist. Nat.*, sér. 4, 10: 1–66.

Hardy, D.[S.] [text] & Fabian, A. [paintings]. 1992. *Succulents of the Transvaal.* Halfway House: Southern Book Publishers.

Harries, J. 2007. *Butterflies & barbarians: Swiss missionaries and systems of knowledge in South-East Africa.* Oxford: James Currey.

Hedgpeth, J.W. 1946. The voyage of the *Challenger*. *Sci. Monthly* 63: 194–202.

Heim, R. 1968. Notice nécrologique sur Henri Humbert (1887–1967), membre de la section de botanique. *Compt. Rend. Hebd. Séances Acad. Sci., Sér. D* 266: 36–39.

Heintze, B. 1999a. *Ethnographische Aneignungen: Deutsche Forschungsreisende in Angola.* Frankfurt am Main: Lembeck.

Heintze, B. (ed.). 1999b. *Max Buchners Reise nach Zentralafrika 1878–1882: Briefe, Berichte, Studien.* Köln: Köppe.

Heintze, B. 1999c. Die Konstruktion des angolanischen "Eingeborenen" durch die Fotografie. *Fotogeschichte* 19(71): 3–14.

Heintze, B. 2007. *Deutsche Forschungsreisende in Angola*, 2nd ed. Frankfurt am Main: Lembeck.

Heintze, B. 2010. *Exploradores Alemães em Angola (1611–1954): Apropriações etnográficas entre comércio de escravos, colonialismo e ciência.* Translated by R. Coelho-Brandes & M. Santos. Frankfurt: Universitätsbibliothek Johann Christian Senckenberg.

Heintze, B. (ed.). 2011. *Eduard Pechuël-Loesche: Tagebücher von der Loango-küste (Zentralafrika) (24.2.1875–5.5.1876); sowie Stichworte zu den Tage-buchaufzeichnungen vom 10.7. bis 19.8.1874.* Frankfurt am Main: Universitätsbibliothek Johann Christian Senckenberg. http://nbn-resolving.de/urn/resolver.pl?urn:nbn:de:hebis:30:3-229806 (accessed 7 December 2021).

Heintze, B. 2018. *Ein preussischer Major im Herzen Afrikas: Alexander v. Mechows Expeditionstagebuch (1880–1881) und sein Projekt einer ersten deutschen Kolonie.* Berlin: Reimer.

Hellmich, W. 1957. Herpetologische Ergebnisse einer Forschungsreise in Angola. *Veröff. Zool. Staatssamml. München* 5: 1–92.

Henriques, J. 1885. Contribuição para o estudo da flora d'algumas possessões portuguezas: Plantas colhidas por F. Newton na África occidental. *Bol. Soc. Brot.* 3: 129–140.

Henriques, J. 1887. Contribuições para o estudo da flora da costa occidental d'Africa. *Bol. Soc. Brot.* 5: 220–221.

Henriques, J.A. 1899. Subsídios para o conhecimento da flora d'Africa Occidental: Catálogo das plantas colhidas por Agostinho Sizenando Marques, subchefe da expedição portuguesa às terras de Muata-Iamvo. *Bol. Soc. Brot.* 16: 35–76.

Henriques, J.C. 1968. Acerca da regeneração natural da floresta densa húmida (Maiombe, Angola): Um caso de inventariação e sua análise. *Garcia de Orta* 16: 467–492.

Hering, E.M. 1959. In memoriam Karl H.E. Jordan * 7.XII.1861 † 12.I.1959. *Mitt. Deutsch. Entomol. Ges.* 18: 23–25.

Hertel, H. & Schreiber, A. 1988. Die Botanische Staatssammlung München 1813–1988. *Mitt. Bot. Staatssamml. München* 26: 81–512.

Hess, H. 1952. Über einige neue *Strophanthus*-Arten und -Bastarde aus Angola (Afrika). *Ber. Schweiz. Bot. Ges.* 62: 79–103.

Hess, H. 1953. Über die Familien der Podostemonaceae und Hydrostachyaceae in Angola. *Ber. Schweiz. Bot. Ges.* 63: 360–383.

Hess, H. 1955. Zur Kenntnis der Eriocaulaceae von Angola und dem unteren Belgischen Kongo. *Ber. Schweiz. Bot. Ges.* 65: 115–204.

Heuertz, M. 1953. In memoriam Edouard Luja. *Bull. Inform. Grand-Duché Luxembourg* 9(10/11): 143–145.

Heuertz, M. 1955. Nécrologie Edouard Luja (1875–1953). *Bull. Soc. Naturalistes Luxemb.* 59: 8–11.

Hiern, W.P. 1896. *Catalogue of the African plants collected by Dr Friedrich Welwitsch in 1853–61*, vol. 1(1). London: Printed by order of the Trustees [of the British Museum]. https://doi.org/10.5962/bhl.title.10876

Hoffmann, O. 1889. W. Vatke. *Ber. Deutsch. Bot. Ges.* 7: 21–24.

Holmgren, P.K., Holmgren, N.H. & Barnett, L.C. 1990. *Index herbariorum*, part 1, *The herbaria of the world*, 8th ed. Regnum Vegetabile 120. New York: New York Botanical Garden for the International Association for Plant Taxonomy.

Hooker, W.J. 1844. *Scytanthus currori*. Hook. *Hooker's Icon. Pl.* 7: tt. 605–606.

Hulot, [E.] [Baron]. 1912. La Mission Rohan-Chabot dans l'Angola. *Géographie* 26: 203–207.

Huntley, B.J. 2012. *Kirstenbosch: The most beautiful garden in Africa.* Cape Town: Struik Nature.

Huntley, B.J. 2017. *Wildlife at war in Angola.* Pretoria: Protea Book House.

Huntley, B.J. & Matos, E.M. 1994. Botanical diversity and its conservation in Angola. Pp. 53–74 in: Huntley, B.J. (ed.), *The botanical diversity of southern Africa*. Strelitzia 1. Pretoria: National Botanical Institute of South Africa.

Huntley, B.J., Russo, V., Lages, F. & Ferrand, N. (eds.). 2019. *Biodiversity of Angola: Science and conservation; A modern synthesis*. Berlin: Springer Nature. https://doi.org/10.1007/978-3-030-03083-4

Instituto de Investigação Científica Tropical. 2003. *Plano de actividades 2004*. Lisbon: Ministério da Ciência e do Ensino Superior.

Jackson, B.D. 1901. A list of the collectors whose plants are in the herbarium of the Royal Botanic Gardens, Kew, to 31st December, 1899. *Bull. Misc. Inform. Kew* 1901: 1–80. https://doi.org/10.2307/4113200

Jacobs, M. 1984. *Herman Johannes Lam (1892–1977): The life and work of a Dutch botanist*. Amsterdam: Rodopi.

Jacobs, N.J. 2015. Marriage, science, and secret intelligence in the life of Rudyerd Boulton (1901–1983): An American in Africa. *Kronos* 41: 287–313.

Jacot Guillarmod, A. 1978. Mary Agard Pocock (1886–1977). *Phycologia* 17: 440–445. https://doi.org/10.2216/i0031-8884-17-4-440.1

Jaeger, F. 1937. Jessens Forschungen in Angola. *Geogr. Z.* 43: 104–106.

Jammart, R.M. 2021. Dacrémont, Alfred. In: Mijn doodsprenten - 100.000. https://www.jammart.be/honderdduizend/honderdduizend-keys.htm (accessed 7 December 2021).

Jacquat, M.S. 2008. Albert Monard. In: Dictionnaire historique de la Suisse (DHS). Version of 16 September 2008, translated from French. https://hls-dhs-dss.ch/fr/articles/048755/2008-09-16/ (accessed 7 December 2021).

Jessen, O. 1936. *Reisen und Forschungen in Angola*. Berlin: Reimer.

John, D.M. 2014. George W. Lawson. *Phycologist* 87: 32–34.

John, D.M. 2017. Career of a globe trotting phycologist. *Phycologist* 92: 42–45.

John, D.M., Lawson G.W. & Ameka, G.L. 2003. *The marine macroalgae of the tropical West Africa sub-region*. Stuttgart: E. Schweizerbart'sche Verlagsbuchhandlung.

Johnson, K.R. 2003. *Karl Jordan: A life in systematics*. Ph.D. Dissertation. Oregon State University, Corvallis, Oregon, U.S.A. https://ir.library.oregonstate.edu/concern/graduate_thesis_or_dissertations/1j92gb61j (accessed 7 December 2021).

Johnson, K.R. 2012. *Ordering life: Karl Jordan and the naturalist tradition*. Baltimore: John Hopkins University Press.

Johnston, H.H. 1895. *The river Congo from its mouth to Bolobo; with A general description of the natural history and anthropology of its western basin*, 4th ed. London: Sampson Low, Marston. https://doi.org/10.5962/bhl.title.58200

Johnston, H.H. 1923. *The story of my life*. Indianapolis: Bobbs-Merrill.

Johnston, I.M. 1924. New plants of Portuguese west Africa collected by Mrs. Richard C. Curtis. *Contr. Gray Herb.* 73: 31–40.

Jones, D. 2006. Okavango Pete: An appreciation of Peter Smith. *Botswana Notes Rec.* 33: 147–148.

Jönsson, P. 2017. Lars Erik Kers, 1931–2017. *Lustgården* 97: 105.

Jordan, K. 1936. Dr. Karl Jordan's expedition to South-West Africa and Angola: Narrative. *Novit. Zool.* 40: 17–62.

JSTOR. 2021. Global Plants. New York: Ithaka. Database at https://plants.jstor.org

Jurion, F. 1977. Becquet (Augustin-Jean-Marie). Col. 19–26 in: Academie Royale des Sciences d'Outre-Mer (ed.), *Biographie Belge d'Outre-Mer*, vol. 7-B. [Brussels]: Academie Royale des Sciences d'Outre-Mer.

Kay-Scott, C. 1943. *Life is too short: An autobiography*. New York: Lippincott.

Kean, S. 2019. Science's debt to the slave trade: Historians confront the tainted origins of key plant and animal collections. *Science* 364: 16–20. https://doi.org/10.1126/science.364.6435.16

Keil, K. 1957. Danckelman, Alexander Freiherr von. P. 502 in: *Neue Deutsche Biographie*, vol. 3 [online version]. https://www.deutsche-biographie.de/pnd135729564.html#ndbcontent (accessed 7 December 2021).

Keraudren, M. & Aymonin, G.G. 1968. Le Professeur Henri Humbert (1887–1967). *Vegetatio* 16: 220–223. https://doi.org/10.1007/BF00261364

Kerr, W.E. 1984. Virgilio de Portugal Brito Araújo (1919–1983). *Acta Amazon.* 14: 327–328. https://doi.org/10.1590/1809-43921984142327 (accessed 7 December 2021).

Kiepert, R. (ed.). 1882. Major von Mechow's Kuango-Reise. Map, 1 : 3 000 000. Berlin: Gesellschaft für Erdkunde. http://bibliotheque-numerique.chambery.fr/idurl/1/24529 (accessed 7 December 2021).

Kimberley, M.J. 1971. Gilbert Westacott Reynolds, Hon. D.Sc. (Cape), F.L.S. *Excelsa* 1: 3–6.

Kimberley, M.J. 1988. Leslie Charles Leach: Euphorbarum botanicus insignis. *Euphorbia J.* 5: 4–5.

Kimberley, M.J. 1992. *Aloes of Zimbabwe*, 2nd ed. Bundu Series. Harare: Longman Zimbabwe.

Kingsbury, N. 2009. *Hybrid: The history and science of plant breeding.* Chicago and London: The University of Chicago Press. https://doi.org/10.7208/chicago/9780226437057.001.0001

Klaassen, E.S. & Craven, P. 2003. *Checklist of grasses in Namibia.* Southern African Botanical Diversity Network Report 20. Pretoria & Windhoek: SABONET.

Koch, C. 1862. G.W. Ackermann. *Wochenschr. Gärtnerei Pflanzenk.* 5: 233.

Kotze, T.J. 1965. Black-thorn (*Acacia detinens*) proves its worth. *Farming S. Afr.* 41(8): 13, 15, 17, 19, 21.

Kruger, F. 2013. Richard James Poynton (23 June 1925 – 11 April 2013). *South. Forests* 75: 2, iii–iv. https://doi.org/10.2989/20702620.2013.804335

Kwembeya, E.G. & Takawira, R. [199-?]. *A checklist of Zimbabwean vernacular plant names.* Harare: National Herbarium and Botanic Garden.

Lachenaud, O. & Fabri, R. 2020. In memoriam: Paul Bamps (1932–2019). *Pl. Ecol. Evol.* 153: 177–180. https://doi.org/10.5091/plecevo.2020.1689

Lains e Silva, H. 1956a. Capinzais secundários de Angola. *Garcia de Orta* 4: 49–55.

Lains e Silva, H. 1956b. *Timor e a cultura do café.* Lisbon: Junta de Investigações do Ultramar.

Lains e Silva, H. 1958. *São Tomé e Príncipe e a cultura do café.* Lisbon: Junta de Investigação do Ultramar.

Landolt, E. 2010. Prof. em. Dr. Hans Ernst Hess, 1920–2009. *Vierteljahrsschr. Naturf. Ges. Zürich* 155: 33–34.

Lange, H. 1874. Die 'Afrikanische Gesellschaft' und die deutsche Expedition der Loangoküste. *Die Gartenlaube* 38: 613–616.

Lanjouw, J. & Stafleu, F.A. 1954. *Index herbariorum*, part 2(1), *Collectors A–D.* Regnum Vegetabile 2. Utrecht: International Bureau for Plant Taxonomy.

Lanjouw, J. & Stafleu, F.A. 1957. *Index herbariorum*, part 2(2), *Collectors E–H.* Regnum Vegetabile 9. Utrecht: International Bureau for Plant Taxonomy.

Lapa e Faro, J.C.P. 1858. Breve notícia sobre o clima de Mossâmedes. *Annaes do Conselho Ultramarino*, Parte não official, sér. 1, Setembro 1858: 501–509.

Launert, E. 1993. Obituary: Arthur Exell. *Independent*, 18 February 1993. https://www.independent.co.uk/news/people/obituary-arthur-exell-1473607.html (accessed 7 December 2021).

Lavranos, J. & Mottram, R. 2017a. The plant gatherings and other vouchers of John J Lavranos: An interpreted checklist from 1954 to 2016. Part 1: List in numerical order. *Cactician* 10: i–xiii, 1–1193. https://www.crassulaceae.ch/de/publications-the-cactician

Lavranos, J. & Mottram, R. 2017b. The plant gatherings and other vouchers of John J Lavranos: An interpreted checklist from 1954 to 2016. Part 2: List in alphabetic order. *Cactician* 11: i–xiii, 1–883. https://www.crassulaceae.ch/de/publications-the-cactician

Lawalrée, A. 2002. Bequaert (Joseph Charles Corneille). *Acad. Roy. Sci. Outre-Mer, Biograph. Belge Outre-Mer* 9: 16–21. http://www.kaowarsom.be/en/notices_BEQUAERT_Joseph_Charles_Corneille (accessed 7 December 2021).

Lawson, G.W. 1985. *Plant life in West Africa*, 2nd ed. Accra: Ghana Universities Press.

Lawson, G.W. & John, D.M. 1987. *The marine algae and coastal environment of Tropical West Africa*, 2nd ed. Nova Hedwigia, Beiheft 93. Stuttgart: J. Cramer in der Gebrüder Bornträger Verlagsbuchhandlung.

Lawson, G.W., John, D.M. & Price, J.H. 1975. The marine algal flora of Angola: Its distribution and affinities. *J. Linn. Soc., Bot.* 70: 307–324. https://doi.org/10.1111/j.1095-8339.1975.tb01652.x

Le Crom, J.-P. 2009. La Croix-Rouge française pendant la seconde guerre mondiale. La neutralité en question. *Vingtieme Siecle, Rev. Hist.* 2009/1(101): 149–162. https://doi.org/10.3917/ving.101.0149

Lebrun, J. 1969. La vegetation psammophile du littoral congolais. *Mém. Acad. Roy. Sci. Outre-Mer, Cl. Sci. Nat. Méd.*, n.s., 18: 1–166.

Leeuwenberg, A.J.M. 1965. Isotypes of which holotypes were destroyed in Berlin. *Webbia* 19: 861–863.

Leitão, L.M.T.M. 2015. *Direito do trabalho em Angola*. Lisbon: Leya.

Lellinger, D.B. 1979. Robert J. Rodin (1922–1978). *Amer. Fern J.* 69: 28.

Leteinturier, B. & Malaisse, F. 2001. Sur les traces des botanistes récolteurs sur gisements cuprifères d'Afrique centro-australe. *Syst. & Geogr. Pl.* 71: 133–163.

Liben, L. 1965. Les récoltes botaniques d'Alfred Dewèvre en Afrique tropicale (1895–1896). *Bull. Jard. Bot. État Bruxelles* 35: 375–388.

Liberato, M.C. 1994. Explorações botânicas nos países africanos de língua oficial portuguesa. *Garcia de Orta, Sér. Bot.* 12: 15–38.

Lima, S. 2012. *Grandes exploradores portugueses*. Lisbon: Leya.

Lindgren, U. 2001. Pogge, Paul. In: *Neue Deutsche Biographie* 20: 578–579 [online version]. https://www.deutsche-biographie.de/pnd117687898.html#ndbcontent (accessed 7 December 2021).

Loeb, E.M., Koch, C. & Loeb, E.-M. K. 1956. Kuanyama Ambo magic. 6. Medicinal, cosmetical, and charm flora and fauna. *J. Amer. Folk.* 69: 147–174.

Lohrer, F.E. 2021. *Leonard John Brass (1900–1971): Botanical collector and explorer*. Venus, FL: Archbold Biological Station. https://www.archbold-station.org/documents/publicationsPDF/LJBrass-CV-Dedication.pdf (accessed 7 December 2021).

Lopes, C.A. 2021. O Batalhão de Marinha Expedicionário a Angola (1914). https://www.momentosdehistoria.com/MH_02_10_Marinha.htm (accessed 7 December 2021).

Louis, H. 1974. Jessen, Otto. In: *Neue Deutsche Biographie* 10: 426–427 [online version]. https://www.deutsche-biographie.de/pnd11712382X.html#ndbcontent (accessed 7 December 2021).

Lovering, J.F. 1983. Gregory, John Walter (1864–1932). In: *Australian dictionary of biography*, vol. 9 [online version]. National Centre of Biography, Australian National University. http://adb.anu.edu.au/biography/gregory-john-walter-6479/text11101 (accessed 7 December 2021).

Lugard, E.J. 1909. Introduction. Pp. 81–89 in: Brown, N.E., The flora of Ngamiland. *Bull. Misc. Inform. Kew* 1909: 81–146. https://doi.org/10.2307/4111525

Lugard, E.J. 1941. *Ceropegia: Ceropegia lugardae*, N.E.Br., *Ceropegia kwebensis*, N.E.Br., *C. floribunda*, N.E.Br. *Cact. Succ. J. (Los Angeles)* 13: 89–92.

Luja, E. 1951. Récit d'un voyage au Mozambique (1900–1902) (avec 2 cartes). *Bull. Soc. Naturalistes Luxemb.* 55: 193–210.

Lux, A.E. 1880. *Von Loanda nach Kimbundu: Ergebnisse der Forschungsreise im äquatorialen West-Afrika (1875–1876).* Wien: Verlag von Eduard Hölzel. https://books.google.at/books?id=s8ANAAAAQAAJ

Macedo, A. 1988. *Na outra margem de Abril: Pequenas histórias de grandes homens.* Lisbon: Publicações Projornal Lda.

Machado, F.G. 1971. *Contribuição para o estudo de plantas ictiotoxicas de Angola.* Nova Lisbon: Instituto de Investigação Agronómica de Angola.

Machado, F.G. 1978. Acerca dum novo reconhecimento ictiológico efectuado na Reserva do Sapal de Castro Marim. *Nat. Paisagem* 4: 29–35.

Mannhardt, I. (transcriber). 2007. *Immigrant ships transcribers guild: SS Alexandra Woermann.* Hamburg Departures Direkt Band 172 (Nov 1905), Volume 373-7 I, VIII A 1 Band 172, Page 2521. https://immigrantships.net/v9/1900v9/alexandra woermann19051130.html (accessed 7 December 2021).

Mansfeld, P.A. 2012. *Hugo Baum: Die Lebensgeschichte eines deutschen Botanikers.* Norderstedt: Books on Demand. http://baum.petermansfeld.de/pdf/hb_2012.pdf (accessed 7 December 2021).

Marques, A.S. 1889. *Expedição portugueza ao Muata-Ianvo. 1884–1888: Os climas e as producções das terras de Malange á Lunda.* Lisbon: Imprensa Nacional.

Martins, D.P. 2005. *Fragmentos de vida: A minha terra.* Aradas: ACAD – Associação Cultural de Aradas.

Martins, J.V. 1993. *Crenças, adivinhação e medicina tradicionais dos tutchokwe do nordeste de Angola.* Lisbon: Instituto de Investigação Científica Tropical.

Martins, J.V. 2001. *Os tutchokwe do nordeste de Angola.* Lisbon: Instituto de Investigação Científica Tropical.

Martins, J.V. 2005. *De Bragança à Lunda: A vida errante de um transmontano aventureiro.* Lisbon: Universitária Editora.

Matias, S. 2002. Cooperativa despede-se de José Trovão. *Póvoa Semanário*, 25 September 2002: 21.

Matos, D., Martins, A.C., Senna-Martinez, J.C., Pinto, I., Coelho, A.G., Ferreira, S.S. & Oosterbeek, L. 2021. Review of archaeological research in Angola. *African Archaeol. Rev.* 38: 319–344. https://doi.org/10.1007/s10437-020-09420-8

Maull, O. 1955. Buchner, Max. In: *Neue Deutsche Biographie* 2: 705 [online version]. https://www.deutsche-biographie.de/pnd116822074.html#ndbcontent (accessed 7 December 2021).

May, L. & Dossenbach, J. 2018. *Eine Botanische Expedition: Die Geschichte des Hess Herbars.* YouTube video, 8:06, posted by ETH-Bibliothek. https://www.youtube.com/watch?v=tQVTVEKkGCE (accessed 7 December 2021).

Mazzocchi-Alemanni, N. 1924. L'Angola e il suo divenire (Impressioni di un viaggio nell'Africa occidentale Portoghese). *Rassegna del Mediterraneo e dell'Espansione Italiana* 4/39: 98–108.

McCracken, D.P. & McCracken, E.M. 1988. *The way to Kirstenbosch.* Annals of Kirstenbosch Botanic Gardens 18. [Cape Town]: National Botanic Gardens.

Mechow, A. von. 1882. Bericht über die von ihm geführte Expedition zur Aufklärung des Kuango-Stromes (1878/81). *Verh. Ges. Erdk. Berlin* 9: 475–489.

Mechow, A. von. 1884. *Karte der Kuango-Expedition. 26 Blätter (1 Uebersichtskarte und 25 Sectionen) Maassstab 1:81200*. Berlin: Verlag von A. Asher. http://nbn-resolving.de/urn/resolver.pl?urn:nbn:de:hebis:30:3-464387

Meeuse, A.D.J. 2005. [Correspondence] From Professor A.D.J. Meeuse FLS. *Linnean* 21(4): 14–15.

Meikle, R.D. 1982. A. A. Bullock, 1906–1980. *Kew Bull.* 36: 657–658. https://www.jstor.org/stable/4117903

Mendes, A.M. 2005. Pescas em Portugal: Ultramar – um apontamento histórico / Fisheries in Portugal: Overseas – an historical note. *Revista Portug. Ci. Veterin.* 100(553–554): 17–32. http://www.fmv.ulisboa.pt/spcv/PDF/pdf3_2005/100_17-32.pdf

Mendes, E.J. 1962. Preliminary report on a botanical journey to the Bié-Cuando-Cubango district, Angola, 1959–60. *Compt. Rend. Réun. Assoc. Pour Étude Taxon. Fl. Afrique Trop.* 4: 333–336.

Mendonça, A.B. & Martins, J.M. 2014. Mateus Martins Moreno Júnior (Parte I). *Almanaque Republicano* , 21 February 2014. http://arepublicano.blogspot.com/2014/02/mateus-martins-moreno-junior-parte-i.html (accessed 7 December 2021).

Mendonça, F.A. 1937. Introdução. Pp. xiii–xvii in: Exell, A.W. & Mendonça, F.A., *Conspectus florae Angolensis*, vol. 1(1). Lisbon: Ministério do Ultramar, Junta de Investigações do Ultramar.

Mendonça, F.A. 1952. John Gossweiler. *Anuário Soc. Brot.* 26: iii–xiii.

Mendonça, F.A. 1962a. Botanical collectors in Angola. *Compt. Rend. Réun. Assoc. Pour Étude Taxon. Fl. Afrique Trop.* 4: 111–121.

Mendonça, F.A. 1962b. Botanical collectors in Mozambique. *Compt. Rend. Réun. Assoc. Pour Étude Taxon. Fl. Afrique Trop.* 4: 145–152.

Meneses, O.J.A. de 1956. Ciperáceas de Angola existentes no herbário do Jardim e Museu Agrícola do Ultramar (LISJC). *Garcia de Orta* 4: 239–264.

Meurer, P.H. 2007. Klaus Stopp (1926–2006). *Imago Mundi* 59: 114–115.

Miller, R.B. 1999. Xylaria at the Forest Products Laboratory: Past, present, and future. *Ann. Mus. Roy. Afrique Centr., Sci. Econ.* 25: 243–254.

Milne-Redhead, E. 1978. Mary Alice Eleanor Richards (1885–1977). *Watsonia* 12: 187–190.

Miriello, L. 2020. Giant sable antelope. On: Carnegie Museum of Natural History website. https://carnegiemnh.org/giant-sable-antelope/ (accessed 7 December 2021).

Misiani, S. 2017. De la colonización interna a la reforma agraria: El itinerario de Nallo Mazzocchi Alemanni. Pp. 221–275 in: Misiani, S. & Gómez Benito, C. (eds.), *Construyendo la nación: Reforma agraria y modernización rural en la Italia del siglo XX.* Zaragoza: Prensas de la Universidad de Zaragoza.

Mission Rohan-Chabot 1928. *Angola et Rhodesia (1912–1914): Mission Rohan-Chabot, sous les auspices du Ministère de l'Instruction Publique et de la Société de Géographie*, tome 4, *Histoire naturelle*, fasc. 3, *Insectes (coléoptères et hyménoptères); Arachnides; Mollusques; Fougères.* Paris: Impr. nationale.

Monard, A. 1930. *Voyage de la mission scientifique suisse en Angola (1928–1929).* Neuchâtel: Imprimerie Paul Attinger.

Monard, A. 1935. Contribution à la mammologie d'Angola et prodrome d'une faune d'Angola. *Arq. Mus. Bocage* 6: 1–314.

Monteiro, J.J. 1875. *Angola and the River Congo*, 2 vols. London: Macmillan.

Monteiro, R. 1891. *Delagoa Bay: Its natives and natural history.* London: George Philip & Son. https://doi.org/10.5962/bhl.title.17539

Monteiro, R.F.R. 1957. Aspectos da exploração florestal no distrito do Moxico. *Garcia de Orta* 5: 129–146.

Monteiro, R.F.R. 1962. Le massif forestier du Mayumbe angolais. *Bois Forêts Trop.* 82: 3–17.

Monteiro, R.F.R. 1965. Correlação entre as florestas do Maiombe e dos Dembos. *Bol. Inst. Invest. Ci. Angola* 1: 257–265.

Monteiro, R.F.R. 1970. *Estudo da flora e da vegetação das florestas abertas do planalto do Bié.* Luanda: Instituto de Investigação Científica de Angola.

Monteiro, R.[F.]R. & Frade, E.C. 1960. *Essências florestais de Angola: Estudo das suas madeiras*, vol. 1, *Região dos Dembos.* Memórias e Trabalhos 1. Luanda: Instituto de Investigação Científica de Angola.

Montenegro, J.P. 1957–1964. Esboço de um vocabulário toponímico de Angola. *Bol. Inst. Angola* 9: 75–84; Ibid. 10: 93–95; Ibid. 11: 175–187; Ibid. 12: 117–129; Ibid. 14: 101–110; Ibid. 15: 135–146; Ibid. 16: 95–110; Ibid. 17: 111–123; Ibid. 18: 77–87; Ibid. 20: 95–102. http://memoria-africa.ua.pt/Library/BIA.aspx

Morais, A.T. 1958. Porque é notável a *Welwitschia mirabilis*. *Bol. Inst. Angola* 11: 53–66. http://memoria-africa.ua.pt/Library/BIA.aspx

Moreira, I., Sampaio Martins, E. & Pinto Basto, M.F. 2009. Breve nota biográfica de Grandvaux Barbosa. Pp. xi–xii in: Barbosa, L.A.G., *Carta fitogeográfica de Angola*, facsimile ed. Coimbra: Associação do Ensino Superior em Ciências Agrárias dos Países de Língua Portuguesa (ASSESCA-PLP) / Associação das Universidades de Língua Portuguesa (AULP).

Mouta, F. & O'Donnell, H.F. 1933. *Carte géologique de l'Angola: Notice explicative.* Lisbon: Ministério das Colónias.

Mullan,G.J. 2010. William Iredale Stanton 1930–2010. *Proc. Univ. Bristol Spelaeol. Soc.* 25(1): 8–9.

Müller, M.A.N. 1983. *Grasse van Suidwes-Afrika / Namibië.* [Windhoek]: Direktoraat Landbou en Bosbou, Department Landbou en Natuurbewaring.

Müller, M.A.N. 1984. *Grasses of South West Africa / Namibia.* [Windhoek]: Directorate of Agriculture and Forestry, Department of Agriculture and Nature Conservation.

Normand, D. 1988. René Letouzey, 1918–1989. *J. Agric. Tradit. Bot. Appl.* 35: 325–326.

Noronha, E. 1936. *Os exploradores Capelo e Ivens.* Lisbon: Agência Geral das Colónias.

Nowell, C.E. 1982. *The Rose-colored map: Portugal's attempt to build an African empire from the Atlantic to the Indian Ocean.* Lisbon: Junta de Investigações Científicas do Ultramar.

Nunes, P. 1959. A pesca em Angola. *Bol. Inst. Angola* 12: 103–115.

Oliver, E.G.H. 1991. Hedley Brian Rycroft (1918–1990). *Bothalia* 21: 109–114. https://doi.org/10.4102/abc.v21i1.870

Olpp, J. 1937. Missionar August Wulfhorst zum Gedächtnis. *Ber. Rheinischen Missionsges.* 94: 54–58.

Paiva, J. 2005. Valor e impacto científico das explorações botânicas a Angola organizadas por L.W. Carrisso. Pp. 37–75 in: Freitas, H., Amaral, P., Ramires, A. & Sales, F. (coords.), *Missão Botânica: Angola 1927–1937.* Coimbra: Imprensa da Universidade e Coimbra.

Palmer, T.S. 1943. Hubert Lynes. *The Auk* 60: 482–483. https://doi.org/10.2307/4079300

Papenburg, G. 2010. Van der Ploeg-symposium geslaagd. *Twirre, Natuur in Fryslân* 21: 47.

Pato, B. 1894. *Memórias: Scenas de infancia e homens de lettras,* vol. 1. Lisbon: Tipographia da Academia Real das Sciencias. https://purl.pt/248/4/ (accessed 7 December 2021).

Patterson, K. 1988. Epidemics, famines, and population in the Cape Verde Islands, 1580–1900. *Int. J. African Hist. Stud.* 21: 291–313. https://doi.org/10.2307/219938

Paulian, R. 2004. *Un naturaliste ordinaire: Souvenirs.* Paris: Société nouvelle des Éditions Boubée.

Pauly, A. 2001. Bibliographie des Hyménoptères de Belgique précédée de notices biographiques (1827–2000); Première partie. *Notes Fauniques de Gembloux* 44: 37–84.

Pavlakis, D. 2016. *British humanitarianism and the Congo Reform Movement, 1896–1913.* New York: Routledge. PDF e-book. https://doi.org/10.4324/9781315570136

Pearson, A. 2016. *Waterwitch*: A warship, its voyage and its crew in the era of anti-slavery. *Atlantic Stud.* 13(1): 99–124. https://doi.org/10.1080/14788810.2015.1109804

Pearson, H.H.W. 1910. The travels of a botanist in South-West Africa. *Geogr. J.* 35: 481–511. https://doi.org/10.2307/1777772

Pearson, H.H.W. 1911. On the collections of dried plants obtained in South-West Africa by the Percy Sladen Memorial Expeditions, 1908–1911. *Ann. S. African Mus. / Ann. S.-Afrikaanse Mus.* 9: 1–19.

Pechuël-Loesche, E. 1887. *Kongoland.* Jena: Hermann. Costenoble. https://catalog.hathitrust.org/Record/008642829

Pechuël-Loesche, E. 1907. *Volkskunde von Loango.* Stuttgart: Strecker & Schröder. https://archive.org/details/volkskundevonlo00loegoog

Pedro, J.G. & Barbosa, L.A.G. 1955. A vegetação. Pp. 67–226 in: *Esboço do reconhecimento ecológico-agrícola de Moçambique*, vol. 2. Lourenço Marques: Centro de Investigação Científica Algodoeira.

Pereira, A.M. 1908a. A cultura regada do arroz. *Bol. Offic. Angola* 37 and 46.

Pereira, A.M. 1908b. Posto Experimental Algodoeiro do Quilombo. *Bol. Offic. Angola* 24.

Pereira, A.M. 1908c. O Arroz de Montanha. *Bol. Offic. Angola* 31.

Pereira, R.M. 2005. Raça, Sangue e Robustez: Os paradigmas da Antropologia Física colonial portuguesa. *Cad. Estud. Africanos* 7/8: 209–241. https://doi.org/10.4000/cea.1363

Pessanha, M.V.T. 1963. *A cultura da cebola.* Luanda: Junta Provincial de Povoamento de Angola.

Phillips, R.C. 1876. Richard Cobden Philipps to Sir Joseph Dalton Hooker, 30 November 1876. Identifier: KADC0922. Directors' Correspondence 184/64 Archives, Royal Botanic Gardens, Kew. https://plants.jstor.org/stable/10.5555/al.ap.visual.kadc0922 (accessed 7 December 2021).

Phillips, R.C. 1877. Richard Cobden Philipps to Sir William Thiselton-Dyer, 4 December 1877. Identifier: KADC0927. Directors' Correspondence 184/69. Archives, Royal Botanic Gardens, Kew. https://plants.jstor.org/stable/10.5555/al.ap.visual.kadc0927 (accessed 7 December 2021).

Phillips, R.C. 1888. The Lower Congo: A sociological study. *J. Roy. Anthropol. Inst. Gr. Brit.* 17: 213–237. https://doi.org/10.2307/2841931

Phillips, W.S. 1963. Resolutions of respect: Dr. Homer L. Shantz, 1876–1958. *Bull. Ecol. Soc. Amer.* 44(2): 59–61.

Pieterse, H. 2019. Smuts the botanist. Pp. 114–126 in: Du Pisani, K., Kriek, D. & De Jager, C. (eds.), *Jan Smuts: Son of the veld, pilgrim of the world.* Pretoria: Protea Book House.

Pinto, C.S. 1937. *A vida breve e ardente de Serpa Pinto.* Lisbon: Agência Geral das Colónias.

Pinto, H.V. 1961. *A exploração florestal em Manica e Sofala.* Gazeta do Agricultor, Serie A, Cientifica e Técnica 10. Lourenço Marques: Serviços de Agricultura e Serviços de Veterinária.

Pinto, O.R. 1984. O algodoeiro como cultura oleaginosa. P. 5 in: *Primeiras Jornadas de Engenharia dos Países de Língua Oficial Portuguesa.* Comunicação 5 – tema 1. Lisbon: O Centro.

Pinto Basto, M.F. 1970. Alargamento da área de distribuição de algumas espécies não frequentes em Angola. *Bol. Soc. Brot.*, sér. 2, 44: 291–293.

Plischke, H. 1953. Bastian, Adolf. In: *Neue Deutsche Biographie* 1: 626–627 [online version]. https://www.deutsche-biographie.de/pnd118653423.html#ndbcontent (accessed 7 December 2021).

Plug, C. 2020. Peter, Prof Gustav Albert (botany). In: S2A3 Biographical Database of Southern African Science. http://www.s2a3.org.za/bio/Biograph_final.php?serial=2182 (accessed 7 December 2021).

Pogge, P. 1880. *Im Reiche des Muata Jamwo: Tagebuch meiner im Auftrage der Deutschen Gesellschaft zur Erforschung Aequatorial-Afrika's in die Lunda-Staaten unternommenen Reise.* Berlin: Verlag von Dietrich Reimer. https://books.google.at/books?id=dD5_M6G6bBUC

Polhill, D. & Polhill, R.[M.] 2015. *East African plant collectors.* Richmond: Royal Botanic Gardens, Kew.

Polhill, R.M. 1980. Helen Faulkner, 1888–1979. *Kew Bull.* 34: 619–620. https://www.jstor.org/stable/4119058

Polhill, R.[M.] 1995. Obituary: Jan Gillett. *Independent,* 23 March 1995. https://www.independent.co.uk/news/people/obituary-jan-gillett-1612376.html (accessed 7 December 2021).

Pombo, R. 1935. Angola medicina indígena. *Diogo Cão* 4: 105–112; Ibid. 5: 149–151.

Poppendieck, H.H. 2004. A new species of *Cochlospermum* (Cochlospermaceae) from Angola with notes on its collector Ilse von Nolde and her botanical illustrations. *Schumannia* 4 = *Biodivers. & Ecol.* 2: 225–235.

Porto, N., Bandeirinha, J.A. & Dias, N. 1999. *Angola a preto e branco: Fotografia e ciência no Museu do Dundo 1940–1970.* Coimbra: Museu Antropológico da Universidade de Coimbra.

Poynton, R.J. [1977?]. *Report to the Southern African Regional Commission for the Conservation and Utilization of the Soil (SARCCUS) on tree planting in southern Africa*, vol. 1, *The pines.* [Pretoria]: Department of Forestry.

Poynton, R.J. 1979. *Report to the Southern African Regional Commission for the Conservation and Utilization of the Soil (SARCCUS) on tree planting in southern Africa*, vol. 2, *The eucalypts.* [Pretoria]: Department of Forestry.

Poynton, R.J. 1984. *Characteristics and uses of selected trees and shrubs cultivated in South Africa*, 4th ed. Bulletin 39. [Pretoria]: Directorate of Forestry, Department of Environmental Affairs.

Poynton, R.J. 2010. *Report to the Southern African Regional Commission for the Conservation and Utilization of the Soil (SARCCUS) on tree planting in southern Africa*, vol. 3, *Other genera.* Pretoria: Department of Forestry.

Prahl, P. 1904. Dr. Friedrich Naumann. *Mitth. Thüring. Bot. Vereins*, n.s., 19: 1–7.

Pusch, J., Barthel, K.-J. & Heinrich, W. 2015. *Die Botaniker Thüringens.* Haussknechtia Beiheft 18. Jena: Thüringische Botanische Gesellschaft.

Quintanilha, A. 1975. Quatro gerações de cientistas na história do Instituto Botânico de Coimbra. *Anuário Soc. Brot.* 41: 27–41.

Raimundo, A.R.F. 1985. Panorâmica da vegetação de Angola. *Garcia de Orta, Sér. Estudos Agron.* 12(1–2): 139–144.
Ranki, M. 2018. Sylvi Esteri Soini 1920–2018. *Aluevisti* 16, 18 April 2018: 14 [online]. http://80.246.156.34/arkisto/2018/16/index.html#1/z (accessed 7 December 2021).
Raper, P.E., Möller, L.A. & Du Plessis, L.T. 2014. *Dictionary of southern African place names.* Johannesburg: Jonathan Ball.
Rattray, J. 1886. Account of a botanical journey to the west African coast, with a list of plants found. *Trans. Bot. Soc. Edinburgh* 16: 472–480. https://doi.org/10.1080/03746608609468304
Rattray, J. 1894. Introductory remarks. *Trans. Linn. Soc. London, Zool.* 6: 2–7.
Rebelo, B. 1881. José Alberto d'Oliveira Anchieta. *O Occidente* 92: 153–155; Ibid. 93: 163; Ibid. 94: 171.
Redinha, J. 1953. *Campanha etnográfica ao Tchiboco (Alto-Tchicapa): Notas de viagem*, vol. 1. Subsídios para a História, Arqueologia e Etnografia dos Povos da Lunda 19. Dundo: Companhia de Diamantes de Angola, Museu do Dundo.
Reed, N.T. 2010. Four generations with a passion for Angola. *Angola Memorial Scholarship Fund Newsletter* 2010, Summer: 2–3.
Regala, F.T. 2014. António de Barros Machado (01/10/1912 – 30/05/2002): Figura ímpar da espeleologia portuguesa. *Trogle* 6: 46–61.
Reynolds, G.W. 1950. *The aloes of South Africa.* Johannesburg: The Trustees, The Aloes of South Africa Book Fund.
Reynolds, G.W. 1960. Hunting aloes in Angola. *African Wild Life* 14: 13–25.
Reynolds, G.W. 1966. *The aloes of tropical Africa and Madagascar.* Mbabane, Swaziland: The Trustees, The Aloes Book Fund.
Richards, C. & Place, J. 1960. *East African explorers.* London: Oxford University Press.
Riley, N.D. 1960. Heinrich Ernst Karl Jordan, 1861–1959. *Biogr. Mem. Fellows Roy. Soc.* 6: 106–133. https://doi.org/10.1098/rsbm.1960.0027
Rocha, V. 1965. Contribuição ao estudo da histoplasmose em Angola. *Bol. Sanitário* 1961: 99–102.
Rodin, R.J. 1985. The ethnobotany of the Kwanyama Ovambos. *Monogr. Syst. Bot. Missouri Bot. Gard.* 9: 1–163. https://doi.org/10.5962/bhl.title.149934
Rodrigues, M.C. (ed.) 1990–1993. *Homenagem a J. R. dos Santos Júnior*, 2 vols. Lisbon: Instituto de Investigação Científica Tropical.
Rohan-Chabot, J. 1914. Explorations dans l'Angola et la Rhodésia (1912–1914). *Géographie* 29: 233–239.
Roivainen, H. 1974. Contribution to the flora of South West Africa. *Ann. Bot. Fenn.* 11: 231–249.
Romariz, C. 1952. Colecções botânicas do Instituto Botânico de Lisboa. *Bol. Soc. Portug. Ci. Nat.*, ser. 2, 4: 57–73.
Romeiras, M.M. 1999. Subsídios para o conhecimento dos colectores botânicos em Angola. *Revista Ci. Agrár.* 22: 73–83.
Ronge, G. 1966. Güßfeldt, Richard *Paul* Wilhelm. In: *Neue Deutsche Biographie* 7: 289–290 [online version]. https://www.deutsche-biographie.de/pnd12870814X.html#ndbcontent (accessed 7 December 2021).
Rosenberg, S. & Weisfelder, R.F. 2013. *Historical dictionary of Lesotho.* Plymouth: Scarecrow Press.
Roux, E. 1944. *S.P. Bunting: A political biography.* Cape Town: Published by the author.
Roux, E. 1948. *Time longer than rope.* London: Victor Gollancz.
Roux, E. & Roux, W. 1970. *Rebel pity: The life of Eddie Roux.* London: Rex Collings.

Rycroft, B. 1975. *Kirstenbosch*. Pride of South Africa 16. Cape Town: Purnell South Africa.

Rycroft, B. 1980. *Kirstenbosch*. Cape Town: Howard Timmins.

Salbany, A. 1947. *Ecological and conservation studies in the Magaliesberg area: A further contribution*. Masters thesis. University of the Witwatersrand, Johannesburg, South Africa.

Salbany, A. 1953. Aspects de la conservation des sols au Mozambique. *Sols Africains* 2: 320–331.

Salbany, A. 1956. Reconhecimento geral preliminar dos tipos de pastos em Angola. *Agron. Angol.* 16: 39–55.

Sampaio Martins, E. 1990. Boraginaceae. Pp. 59–110 in: Launert, E. & Pope, G.V. (eds.), *Flora Zambesiaca*, vol. 7(4). London: Flora Zambesiaca Managing Committee.

Sampaio Martins, E. 1994a. John Gossweiler: Contribuição da sua obra para o conhecimento da flora angolana. *Garcia de Orta, Sér. Bot.* 12: 39–68.

Sampaio Martins, E. 1994b. Homenagem a botânicos com colaboração relevante no Centro de Botânica do IICT. I. Eduardo Mendes. *Garcia de Orta, Sér. Bot.* 12: 11–14.

Sampaio Martins, E. 2009. Aponogetonaceae. Pp. 65–73 in: Timberlake, J.R. & Martins, E.S. (eds.), *Flora Zambesiaca*, vol. 12(2). London: Royal Botanic Gardens, Kew.

Sampaio Martins, E. & Martins, T.G. 2002. Herbários em Angola: Que futuro? *Garcia de Orta, Sér. Bot.* 16: 1–4.

Santos, M.E.M. 1988. *Viagens de exploração terrestre dos portugueses em África*, 2nd ed. Lisbon: Instituto de Investigação Científica Tropical.

Santos, N.B., Lages, M., Castro, S., Palma, M., Sequeira, M. & Lérias, A. 2016. Os irmãos d'Abranches Bizarro: Pioneiros da estatística psiquiátrica e médica em Portugal. Pp. 29–37 in: Pereira, A.L. & Pita, J.R. (eds.), *VI Jornadas Internacionais da História da Psiquiatria e Saúde Mental*. Coimbra: Grupo de História e Sociologia da Ciência e da Tecnologia-CEIS20 e Sociedade de História Interdisciplinar da Saúde. https://estudogeral.sib.uc.pt/bitstream/10316/46242/1/JornHistPsiquiatria6.pdf (accessed 7 December 2021).

Santos, R.M. 1967. *Plantas úteis de Angola: Contribuição iconográfica*. Luanda: Instituto de Investigação Científica de Angola.

Santos, R.M. 1972. *Contribuição para o conhecimento dos nomes vernáculos das plantas do Cuando Cubango*. Luanda: Instituto de Investigação Científica de Angola.

Santos, R.M. 1982. *Itinerários florísticos e carta da vegetação do Cuando Cubango*. Estudos, Ensaios e Documentos 137. Lisbon: Instituto de Investigação Científica Tropical / Junta de Investigação Científica do Ultramar.

Santos, R.M. 1989. *Plantas úteis de Angola: Contribuição iconográfica*, vol. 2. Lisbon: Instituto de Investigação Científica Tropical.

Santos Júnior, J.R. 1950. A alma do indígena através da etnografia de Moçambique. *Bol. Soc. Geogr. Lisboa* 68(7–8): 399–424.

Saraiva, S., Gonçalves, A.E., Conde, P., Sampaio Martins, E., Figueira, R. & Catarino, L. 2012. António Rocha da Torre e a flora de Moçambique. Pp. 1–15 in: *Atas do Gongresso Internacional Saber Tropical em Moçambique: História, Memória e Ciência*, Lisboa, 24–26 outubro de 2012. Lisbon: Instituto de Investigação Científica Tropical. https://2012congressomz.files.wordpress.com/2013/09/susana_t05c04-2.pdf (accessed 7 December 2021).

Saraiva, T. 2014. Mimetismo colonial e reprodução animal: Carneiros caracul no Sudoeste angolano. *Etnográfica* 18: 209–227. https://doi.org/10.4000/etnografica.3403
Satre, L.J. 2005. *Chocolate on trial: Slavery, politics, and the ethics of business.* Athens: Ohio University Press.
Sauer, C.O. 1959. Homer Leroy Shantz. *Geogr. Rev. (New York)* 49: 278–280.
Schinz, H. 1891. *Deutsch-Südwest-Afrika: Forschungsreisen durch die deutschen Schutzgebiete Gross-Nama- und Hereroland, nach dem Kunene, dem Ngami-See und der Kalaxari, 1884–1887.* Oldenburg und Leipzig: Schulzesche Hof-Buchhandlung und Hof-Buchdruckerei.
Schmitz, A. 1963. Les muhulu du Haut-Katanga méridional. *Bull. Jard. Bot. État Bruxelles* 32: 221–299. https://doi.org/10.2307/3667284
Schnee, H. 1920. *Deutsches Kolonial-Lexikon*, vol. 3. Leipzig: Quelle & Meyer. https://archive.org/details/bub_gb_bY4zAQAAMAAJ (accessed 7 December 2021).
Schoeman, C. 2017. *The historical Overberg.* Cape Town, Century City: Zebra Press., an imprint of Penguin Random House South Africa.
Scholtz, J. du P. 1941. *Uit die geskiedenis van die naamgewing aan plante en diere in Afrikaans*. Cape Town: Nasionale Pers.
Schunack, W., Braun, A. & Albrecht, S. 2006. Prof. Dr. rer. nat. Klaus Stopp in memoriam. *Deutsche Apotheker-Zeitung* 146(27): 119–120.
Schütt, O.H. 1881. *Reisen im südwestlichen Becken des Congo.* Berlin: Verlag von Dietrich Reimer. http://digital.slub-dresden.de/id493275525
Semedo, C.M.B. 1982. *O Jardim e Museu Agrícola do Ultramar: Breve apontamento da sua história e da sua actividade*. Lisbon: Junta de Investigações do Ultramar.
Serpa Pinto 1881a. *Como eu atravessei Africa*, 2 vols. London: S. Low, Marston, Searle, & Rivington.
Serpa Pinto 1881b. *How I crossed Africa*, 2 vols. Hartford, CT: R.W. Bliss.
Setshogo, M.P. 2005. *Preliminary checklist of the plants of Botswana.* Southern African Botanical Diversity Network Report 37. Pretoria and Gaborone: SABONET.
Settesoldi, L., Tardelli, M. & Raffaelli, M. 2005. *Esploratori Italiani nell'Africa Orientale fra il 1870 ed il 1930: Missioni scientifiche con raccolte botaniche rilievi geografici er etnografici.* Florence: Centro Studi Erbario Tropicale.
Shantz, H. 1919–1920. *Travel notes on a trip through Africa from the Cape to Cairo.* Typescript published electronically 2009 under: Homer Shantz Smithsonian Expedition of 1919–1920. Tucson: University of Arizona Libraries & the University of Arizona Herbarium. https://uair.library.arizona.edu/item/271810 (accessed 7 December 2021).
Shantz, H.L. & Turner, B.L. 1958. *Photographic documentation of vegetational changes in Africa over a third of a century.* College of Agriculture Report 169. Tucson: University of Arizona.
Shantz, H.L., Marbut, C.F. & Kincer, J.B. 1923. *The vegetation and soils of Africa.* New York: National Research Council and the American Geographical Society.
Sieiro, D.M. 1974a. *Contribuição para o estudo do regime alimentar da palanca preta gigante (Hippotragus niger variani Thomas).* Série Técnica 44. Nova Lisboa: Instituto de Investigação Agronómica de Angola.
Sieiro, D.M. 1974b. *Herbívoros selvagens de Angola*. Série Técnica 45. Nova Lisboa: Instituto de Investigação Agronómica de Angola.
Sillitoe, F.S. 1944. Morley Thomas Dawe, O.B.E., F.L.S., F.R.G.S. *J. Kew Guild* 1943, 5: 301–302.

Silva, J.A. 1964. *Gorongosa: Experiência de um caçador de imagens*. Lourenço Marques: Empresa Moderna.

Silva, J.A. 1965. *Gorongosa: Shooting big game with a camera*. Lourenço Marques: Empresa Moderna.

Silva, J.A. 1972. Contribuição para o estudo bioecológico da palanca real (*Hippotragus niger variani*). Lisbon: Junta de Investigações do Ultramar.

Silva, J.D. 1940. *Francisco Newton: Explorador naturalista. (Apontamentos para uma biografia)*. Colecção Pelo Império 68. Lisbon: Agência Geral das Colónias.

Silva, J.J. s.d. [after 1785]. Riscos De alguns Mammaes, Aves e Vermes do Real Museo de Nossa Senhora d'Ajuda. Ditos De Peixes e Vermes de Angola, com o Prospecto da Embocadura do Rio Dande. Ditos De varios Animaes raros de Moçambique com alguns Prospectos e Retratos. Arquivo Histórico dos Museus da Universidade de Lisboa.

Silva. J.J. 1813a. Extracto da viagem, que fez ao sertão da Benguella no anno de 1785 por ordem do Governador e Capitão General do Reino de Angola, o Bacharel Joaquim José da Silva, enviado á aquelle Reino como Naturalista, e depois Secretario do Governo. De Loanda para Benguella. *O Patriota* 1: 97–100; Ibid. 1(2): 86–98; Ibid. 1(3): 49–60.

Silva, J.J. 1813b. Noticias sobre Cabo Negro, extrahidas dos fragmentos da viagem do Doutor Joaquim José da Silva. *O Patriota* 1(6): 71–77.

Silva, M.S. 1996. *Roberto Ivens: O homem, a vida*. Ponta Delgada: Centro de Apoio Tecnológico à Educação.

Silva, M.S. 2005. Roberto Ivens. In: Açorianos de Cultura. Archived web page at https://web.archive.org/web/20050426210742/http://www.cate.rcts.pt/producao_audiovisual/acorianos_cultura/ivens.htm (accessed 7 December 2021).

Simon, W.J. 1983. *Scientific expeditions in the Portuguese overseas territories (1783–1808) and the role of Lisbon in the intellectual-scientific community of the late eighteenth century*. Lisbon: Instituto de Investigação Científica Tropical.

Slewinski, B. 2008. As minhas pesquisas sobre o meu avô Zé. *Gazeta d'Orey* 17: 4–5.

Smith, G.F. & Correia, R.I. de S. 1988. Notes on the ecesis of *Aloe davyana* (Asphodelaceae: Alooideae) in seed-beds and under natural conditions. *S. African J. Sci.* 84: 873.

Smith, G.F. & Correia, R.I. de S. 1989. Aspects of the ecesis of *Aloe davyana* (Asphodelaceae: Alooideae) under natural conditions. Pp. 115–116 in: *15th Annual Congress of the South African Association of Botanists: Abstracts*. University of Pretoria, Pretoria.

Smith, G F. & Correia, R.I. de S. 1992. Establishment of *Aloe greatheadii* var. *davyana* from seed for use in reclamation trials. *Landscape & Urban Planning* 23(1): 47–54.

Smith, G.F. & Figueiredo, E. 2011. Provenance of the material on which the name *Aloe mendesii* Reynolds (Asphodelaceae), a cliff-dwelling species from Angola, is based. *Bradleya* 29: 61–66. https://doi.org/10.25223/brad.n29.2011.a7

Smith, G.F. & Figueiredo, E. 2019. The nomenclature and correct author citation of the names *Kalanchoe marnieriana* H.Jacobsen ex L.Allorge and *Bryophyllum marnierianum* (H.Jacobsen ex L.Allorge) Govaerts (Crassulaceae), with notes on the nomenclature of some Madagascan *Kalanchoe* taxa. *Bradleya* 37: 212–217. https://doi.org/10.25223/brad.n37.2019.a19

Smith, G.F. & Figueiredo, E. 2021. Taxonomic and nomenclatural notes on *Kalanchoe hauseri* Werderm. (Crassulaceae subfam. Kalanchooideae) from Angola, with notes on Friedrich Welwitsch's collecting activities in the country. *Bradleya* 39: 182–187. https://doi.org/10.25223/brad.n39.2021.a18

Smith, G.F. & Williamson, G. 1997. Leslie Charles Leach (1909–1996). *Taxon* 46: 374–376. https://doi.org/10.1002/j.1996-8175.1997.tb05176.x
Smith, G.F. & Willis, C.K. 1999. *Index herbariorum: Southern African supplement*, 2nd ed. Southern African Botanical Diversity Network Report 8. Pretoria: SABONET.
Smith, G.F., Figueiredo, E. & Catarino, L. 2012. Eduardo José dos Santos Moreira Mendes (1924–2011). *Bothalia* 42: 67–68. https://doi.org/10.4102/abc.v42i1.8
Smith, G.F., Figueiredo, E., Crouch, N.R., Oosthuizen, D. & Klopper, R.R. 2016. *Aloe ×inopinata* Gideon F.Sm., N.R.Crouch & Oosth., (Asphodelaceae) [*Aloe arborescens* Mill. × *Aloe chortolirioides* A.Berger var. *chortolirioides*]: A nothospecies from the Barberton Centre of Endemism, Eastern South Africa. *Haseltonia* 22: 55–63. https://doi.org/10.2985/026.022.0111
Smith, G.F., Figueiredo, E., Loureiro, J. & Crouch, N.R. 2019a. *Kalanchoe ×gunniae* Gideon F.Sm & Figueiredo (Crassulaceae), a new South African nothospecies derived from *Kalanchoe paniculata* Harv. × *Kalanchoe sexangularis* N.E.Br. *Bradleya* 37: 141–150. https://doi.org/10.25223/brad.n37.2019.a9
Smith, G.F., Figueiredo, E. & Bernhard, S. 2019b. Notes on the taxonomy and nomenclature of *Kalanchoe brevisepala* (Humbert) L.Allorge and its basionym, *K. millotii* Raym.-Hamet & H.Perrier subsp. *brevisepala* Humbert, and *Kalanchoe dinklagei* Rauh (Crassulaceae). *Bradleya* 37: 87–96. https://doi.org/10.25223/brad.n37.2019.a31
Smith, G.F., Figueiredo, E. & Silva, V. 2020. The contributions to and controversies introduced into research on the Aphodelaceae subfam. Alooideae and Crassulaceae by Flávio Resende (1907–1967) in the mid-20th century. *Bradleya* 38: 141–157. https://doi.org/10.25223/brad.n38.2020.a16
Smith, G.O. 2011. *An arid Eden: A personal account of conservation in the Kaokoveld.* Johannesburg & Cape Town: Jonathan Ball.
Soares, F.A. 1959. *Madeiras de Angola na zona do caminho de ferro de Benguela.* Lisbon: [s.n.].
Soares, F.A. 1961. *Viveiros e plantação de eucaliptos em Angola.* Lisbon: Missão de Estudos Agronómicos do Ultramar.
Soini, S. 1981. Agriculture in northern Namibia, Owambo and Kawango 1965–1970. *J. Agric. Sci. Finland* 53: 168–209. https://doi.org/10.23986/afsci.72069
Soini, S. 2015. *Sota-ajan ylijäämä.* Kotka: Mikko Ranki.
Sousa, F., Figueiredo, E. & Smith, G.F. 2010. *Cyphostemma mendesii* (Vitaceae), a new species from Angola. *Phytotaxa* 7: 35–39. https://doi.org/10.11646/phytotaxa.7.1.4
Soyaux, H. 1879. *Aus West-Afrika 1873–1876: Erlebnisse und Beobachtungen*, 2 vols. Leipzig: F.A. Brockhaus.
Soyaux, H. 1888. *Deutsche Arbeit in Afrika: Erfahrungen und Betrachtungen.* Leipzig: F.A. Brockhaus.
Spoerndli, J. 1945. The Mbari question. *Anthropos* 37/40(4/6): 891–893.
Spooner, H. 1925. Morley Thomas Dawe F.L.S. F.R.G.S. *J. Kew Guild* 1925, 4: 280–282.
Stafleu, F.A. & Cowan, R.S. 1976. *Taxonomic literature: A selective guide to botanical publications and collections with dates, commentaries and types*, 2nd ed., vol. 1, *A–G*. Utrecht: Bohn, Scheltema & Holkema. https://doi.org/10.5962/bhl.title.48631
Stafleu, F.A. & Cowan, R.S. 1979. *Taxonomic literature: A selective guide to botanical publications and collections with dates, commentaries and types*, 2nd ed., vol. 2, *H–Le*. Utrecht: Bohn, Scheltema & Holkema; The Hague: dr. W. Junk. https://doi.org/10.5962/bhl.title.48631

Stafleu, F.A. & Cowan, R.S. 1981. *Taxonomic literature: A selective guide to botanical publications and collections with dates, commentaries and types*, 2nd ed., vol. 3, *Lh–O*. Utrecht: Bohn, Scheltema & Holkema; The Hague: dr. W. Junk. https://doi.org/10.5962/bhl.title.48631

Stafleu, F.A. & Cowan, R.S. 1983. *Taxonomic literature: A selective guide to botanical publications and collections with dates, commentaries and types*, 2nd ed., vol. 4, *P–Sak*. Utrecht/Antwerp: Bohn, Scheltema & Holkema; The Hague/Boston: dr. W. Junk. https://doi.org/10.5962/bhl.title.48631

Stafleu, F.A. & Cowan, R.S. 1985 *Taxonomic literature: A selective guide to botanical publications and collections with dates, commentaries and types*, 2nd ed., vol. 5, *Sal–Ste*. Utrecht/Antwerp: Bohn, Scheltema & Holkema; The Hague/Boston: dr. W. Junk. https://doi.org/10.5962/bhl.title.48631

Stafleu, F.A. & Cowan, R.S. 1986. *Taxonomic literature: A selective guide to botanical publications and collections with dates, commentaries and types*, 2nd ed., vol. 6, *Sti–Vuy*. Utrecht/Antwerp: Bohn, Scheltema & Holkema; The Hague/Boston: dr. W. Junk. https://doi.org/10.5962/bhl.title.48631

Stafleu, F.A. & Cowan, R.S. 1988. *Taxonomic literature: A selective guide to botanical publications and collections with dates, commentaries and types*, 2nd ed., vol. 7, *W–Z*. Utrecht/Antwerp: Bohn, Scheltema & Holkema; The Hague/Boston: dr. W. Junk. https://doi.org/10.5962/bhl.title.48631

Stafleu, F.A. & Mennega, E.A. 1998. *Taxonomic literature: A selective guide to botanical publications and collections with dates, commentaries and types*, 2nd ed., suppl. 5, *Da–Di*. Königstein: Koeltz Scientific Books. https://doi.org/10.5962/bhl.title.48631

Stafleu, F.A. & Mennega, E.A. 2000. *Taxonomic literature: A selective guide to botanical publications and collections with dates, commentaries and types*, 2nd ed., suppl. 6, *Do–E*. Konigstein: Koeltz Scientific Books. https://doi.org/10.5962/bhl.title.48631

Stanley, H.M. 1878. *Through the dark continent*, 2 vols. New York: Harper & Brothers.

Stassen, N. 2010. Die Dorslandtrekke na Angola en die redes daarvoor (1874–1928). *Historia* 55: 32–54. http://ref.scielo.org/pm7srr (accessed 7 December 2021).

Stopp, K. 1958. Die verbreitungshemmenden Einrichtungen in der südafrikanischen Flora. *Bot. Stud.* 8: 1–103.

Stopp, K. 1964. Die *Ceropegia*-Arten der Umbraticola-Gruppe. *Bot. Jahrb. Syst.* 83: 115–125.

Stopp, K. 1971. Neue *Ceropegia*-Aufsammlungen in Angola. *Bot. Jahrb. Syst.* 91: 469–475.

Swinscow, T.D.V. 1972. Friedrich Welwitsch, 1806–72: A centennial memoir. *Biol. J. Linn. Soc.* 4: 269–289. https://doi.org/10.1111/j.1095-8312.1972.tb00695.x

Tams, G. 1845. *Visit to the Portuguese possessions in south-western Africa*, 2 vols. London: published by T. C. Newby.

Tanghe, M. 1992. Paul Duvigneaud (1913–1991). *Belg. J. Bot.* 125: 3–15.

Taquelim, M. 2008. *Desenhando em viagem: Os cadernos de África de Roberto Ivens*. Masters thesis. Faculdade de Belas-Artes, Universidade de Lisboa, Lisbon, Portugal. https://core.ac.uk/download/pdf/12423224.pdf (accessed 7 December 2021).

Taruffi, D. 1918. O planalto de Benguela e o seu futuro agrícola. *Bol. Soc. Geogr. Lisboa* 7/9: 185–226.

Teixeira, J.B. 1957. A acção do padre Antunes como naturalista botânico. *Portugal em África* 14(79): 38–47.

Teixeira, J.B. 1962. Le naturaliste Joaquim José da Silva et les itinéraires des expéditions qu'il a effectuées en Angola, de 1783 à 1804. *Compt. Rend. Réun. Assoc. Pour Étude Taxon. Fl. Afrique Trop.* 4: 103–109.

Teixeira, J.B. 1968. *Parque Nacional do Bicuar: Carta da vegetação (1.ª aproximação) e memória descritiva.* Nova Lisboa: Instituto de Investigação Agronómica de Angola.

Teixeira, J.B. 1970. Hamamelidaceae. Pp. 29–30 in: Exell, A.W., Fernandes, A. & Mendes, E.J. (eds.), *Conspectus florae Angolensis*, vol. 4. Lisbon: Junta de Investigações do Ultramar/Instituto de Investigação Científica de Angola.

Teixeira, J.B., Cardoso de Matos, G. & Baptista de Sousa, J.N. 1967. *Parque nacional da Quiçama: Carta da vegetação e memória descritiva.* Nova Lisboa: Instituto de Investigação Agronómica de Angola.

Teusz, S. 2018. Julius Eduard Teusz (1845–1912)—biolog, odkrywca, podróżnik. *Afryka* 48: 143–155. https://journals.indexcopernicus.com/search/article?articleId=1963038 (accessed 7 December 2021).

Thiers, B. 2021. Index Herbariorum. New York Botanical Garden's Virtual Herbarium. http://sweetgum.nybg.org/ih/ (accessed 7 December 2021).

Thompson, P.K. & Page, C.R. 1986. *Gazetteer of Angola*, 2nd ed. Washington, D.C.: Defense Mapping Agency.

Thorold, C.A. 1975. *Diseases of cocoa.* Oxford: Oxford University Press.

Tinley, K.L. 1977. *Framework of the Gorongosa ecosystem.* D.Sc. thesis. University of Pretoria, Pretoria, South Africa. https://repository.up.ac.za/handle/2263/24526 (accessed 7 December 2021).

Tisserant, C. 1950. *Catalogue de la flore de l'Oubangui-Chari.* Mémoire de l'Institut d'Études Centrafricaines 2. Brazzaville: ORSTOM.

Tittley, I. 2008. James Henry 'Jim' Price (1932–2007). *Phycologist* 74: 48–49.

Toffelmier, G. 1967. Edwin Meyer Loeb (1894–1966). *Amer. Anthropol.* 69: 200–203. https://doi.org/10.1525/aa.1967.69.2.02a00070

Tönjes, H. 1911. *Ovamboland: Country, people, mission; with Particular reference to the largest tribe, the Kwanyama.* Reprint 1996, Windhoek: Namibia Scientific Society.

Torre, A.R. 1940. As minhas herborizações em Moçambique. *Bol. Geral Colón.* 184: 9–28.

Torres, J.R. (ed.) 1904. *Portugal—Dicionário Histórico, Corográfico, Heráldico, Biográfico, Bibliográfico, Numismático e Artístico*, vol. 1. Lisbon: J. Romano Torres.

Trimen, R. 1887–1889. *South African butterflies: A monograph of the extra-tropical species*, 3 vols. London: Trübner. https://doi.org/10.5962/bhl.title.8997

Troelstra, A.S. 2017. *Bibliography of natural history travel narratives.* Leiden: Brill. https://doi.org/10.1163/9789004343788

Tucker, J.T. 1927. *Drums in the darkness: The story of the mission of the United Church of Canada in Angola, Africa.* Toronto: Committee on Literature, General Publicity and Missionary Education of the United Church of Canada.

Tuckey, J.H. & Smith, C. 1818. *Narrative of an expedition to explore the river Zaire, usually called the Congo, in South Africa, in 1816, under the direction of Captain J.H. Tuckey R.N.* London: John Murray. https://doi.org/10.5962/bhl.title.15375

Turner, B.L. 2016. *Homer LeRoy Shantz, leader of the expedition.* Gruver, TX: Texensis.

Turner, D. 1998. Obituary: J.G. Williams. *Independent*, 15 January 1998. https://www.independent.co.uk/news/obituaries/obituary-j-g-williams-1138759.html (accessed 7 December 2021).

Tylor, E.B. 1905. Professor Adolf Bastian: Born June 26, 1826; Died February 3, 1905. *Man* 5: 138–143.
Urban, I. 1892. Der Königl. Botanische Garten und das Botanische Museum zu Berlin in den Jahren 1878–1891. *Bot. Jahrb. Syst.* 14: 9–64.
Urban, I. 1916. *Geschichte des Königlichen Botanischen Museums zu Berlin-Dahlem (1815–1913) nebst Aufzählung seiner Sammlungen.* Dresden: Heinrich.
Van de Casteele, J. 1952. Vanderyst. Col. 872–875 in: *Biographie Coloniale Belge*, vol. 3. Brussels: Institut Royal Colonial Belge.
Van Jaarsveld, E.[J.] & Nagel, R. 1999. *Tavaresia thompsonii* Van Jaarsveld & Nagel, a new species from S. Angola. *Asklepios* 76: 9–10.
Van Steenis, C.G.G.J. 1976. Dedication to Herman Johannes Lam. Pp. 6–14 in: *Flora Malesiana*, ser. I, vol. 7 (1972–1976). http://portal.cybertaxonomy.org/flora-malesiana/node/230 (accessed 7 December 2021).
Van Steenis-Kruseman, M.J. 1950. *Flora Malesiana*, ser. I, vol. 1(1), *Malaysian plant collectors and collections.* Djakarta: Noordhoff-Kolff.
Vasconcelos, C.C. 1942. *Um soldado de África, Paulo Amado de Melo Ramalho da Cunha e Vasconcelos.* Colecção Pelo Império 79. Lisbon: Agência Geral das Colónias.
Vasse, G. 1909. *Trois années de chasse au Mozambique.* Paris: Librairie Hachette.
Vegter, I.H. 1976. *Index herbariorum*, part 2(4), *Collectors M.* Regnum Vegetabile 93. Utrecht: International Bureau for Plant Taxonomy.
Vegter, I.H. 1983. *Index herbariorum*, part 2(5), *Collectors N–R.* Regnum Vegetabile 109. Utrecht: International Bureau for Plant Taxonomy.
Vegter, I.H. 1986. *Index herbariorum*, part 2(6), *Collectors S.* Regnum Vegetabile 114. Utrecht: International Bureau for Plant Taxonomy.
Vegter, I.H. 1988. *Index herbariorum*, part 2(7), *Collectors T–Z.* Regnum Vegetabile 117. Utrecht: International Bureau for Plant Taxonomy.
Verdcourt, B. 1998. Edgar W. B. H. Milne-Redhead M.B.E, LS.O., T.D. (1906–1996). *Watsonia* 22: 128–137.
Verdoorn, I.C. & Codd, L.E.[W.] 1966. Hedendaagse plantkundiges en plant-versamelaars. 1. Frederick Ziervogel van der Merwe. 2. Leslie Charles Leach. *Bothalia* 8 (Bylaag no. 1): 59–64.
Victor, J.E., Smith, G.F., Ribeiro, S. & Van Wyk, A.E. 2015. Plant taxonomic capacity in South Africa. *Phytotaxa* 238: 149–162. https://doi.org/10.11646/phytotaxa.238.2.3
Vieira, C.C. 2006. *Os Portugueses e a travessia do continente africano: Projectos e viagens (1755–1814).* Masters thesis. Universidade de Lisboa, Lisbon, Portugal.
Vieira, G. 2006. En des lieux et en des temps agités, un missionaire comme beaucoup d'autres: le père Eugène Ehrhart (1865–1949). *Mém. Spiritaine* 23: 80–113.
Vieira, G. 2012–2017. *Le Père Duparquet*, 4 vols. Paris: Karthala.
Vilhunen, T. 1995. *To the east and south – Missionaries as photographers 1890–1930.* Helsinki: The Finnish Evangelical Lutheran Mission.
Villain, F. 1929. Les peuples et les missions du Cunène. *Annales de la Congrégation des Pères du Saint-Esprit*, February 1929: 49–56.
Villiers, J.F. 1989. René Letouzey (1918–1989). *Bull. Mus. Natl. Hist. Nat., B, Adansonia* 11: 325–332.
Visser, R.P.W. 2013. Lam, Herman Johannes (1892–1977). In: Biografisch Woordenboek van Nederland [online version]. http://resources.huygens.knaw.nl/bwn1880-2000/lemmata/bwn4/lam (accessed 7 December 2021).
Vollesen, K. 1987. The native species of *Gossypium* (Malvaceae) in Africa, Arabia and Pakistan. *Kew Bull.* 42: 337–349. https://doi.org/10.2307/4109688

Wagenitz, G. 2009. Die Erforscher der Pflanzenwelt von Berlin und Brandenburg. *Verh. Bot. Vereins Berlin Brandenburg* Beih. 6: 157–556.
Wagner, M. 1992. Note on the Shantz Collection, Tucson, Arizona. *History in Africa* 19: 445–449. https://doi.org/10.2307/3172013
Walker, C.C. 2010. Gilbert Westacott Reynolds: His study of *Aloe* and a bibliography of his work. *Bradleya* 28: 111–124. https://doi.org/10.25223/brad.n28.2010.a13
Walker, J.F. 2004. *A certain curve of horn*. New York: Grove Press.
Wallace, G.B. & Wallace, M.M. 1944. Supplement to the revised list of plant diseases in Tanganyika Territory. *E. Afric. Agric. J. Kenya* 10(1): 47–49. https://doi.org/10.1080/03670074.1944.11664406
Wallenstein, F.P. 1956. Relatório da estação agricola da Humpata. In: *Reunião Técnica Anual* 1(2): 21 pp.
Warburg, O. (ed.) 1903. *Kunene–Sambesi-Expedition H. Baum 1903*. Berlin: Verlag des Kolonial-Wirtschaftlichen Komitees. https://doi.org/10.5962/bhl.title.37083
Watt, G. 1926. *Gossypium. Bull. Misc. Inform. Kew* 1926: 193–210. https://doi.org/10.2307/4111806
Wawra, H. & Peyritsch, J. 1860. *Sertum benguelense. Sitzungsber. Kaiserl. Akad. Wiss. Wien, Math.-Naturwiss. Cl.* 38: 543–586.
Werdermann, E. 1937. I. Neue Sukkulenten aus dem Botanischen Garten Berlin-Dahlem. II. *Repert. Spec. Nov. Regni Veg.* 42: 1–7. https://doi.org/10.1002/fedr.19370420103
West, O. 1965. *Fire in vegetation and its use in pasture management with special reference to tropical and subtropical Africa*. Mimeographed Publication No. 1/1965. Hurley, Berkshire, England: Commonwealth Bureau of Pastures and Field Crops.
West, O. 1971. Fire, man and wildlife as interacting factors limiting the development of climax vegetation in Rhodesia. *Proc. Annual Tall Timbers Fire Ecol. Conf.*, 1971: 121–145.
West, O. 1974. *A field guide to the aloes of Rhodesia*. Salisbury: Longman Rhodesia.
Wiencek, H. 1981. Portugal of the navigators. Pp. 89–168 in: Milton, J. & Wiencek, H. (eds.), *Frontiers of Europe: Russia of the czars, Portugal of the navigators*. London: Cassell.
Wild, H. 1953. *A Southern Rhodesian botanical dictionary of native and English plant names*. Salisbury: Herbarium, Branch of Botany and Plant Pathology, Department of Agriculture.
Wild, H., Biegel, H.M. & Mavi, S. 1972. *A Rhodesian botanical dictionary of African and English plant names*. Salisbury: National Herbarium, Department of Research and Specialist Services, Ministry of Agriculture.
Wildeman, E. de. 1948. Dewèvre (Alfred-Prosper). Col. 307–311 in: *Biographie Coloniale Belge*, vol. 1. Brussels: Institut Royal Colonial Belge.
Willgeroth, G. 1929. Fritzsche, Hermann Richard. Pp. 133–134 in: *Die Mecklenburgischen Ärzte von den ältesten Zeiten bis zur Gegenwart*. Schwerin: Verlag der Landesgeschäftsstelle des Meckl. Aerztevereinsbundes.
Williams, A. 1999. John G. Williams (1913–97). *Ibis* 141: 349.
Williams, I.J.M. 1972. *A revision of the genus Leucadendron (Proteaceae)*. Contributions from the Bolus Herbarium 3. Rondebosch: Bolus Herbarium, University of Cape Town.
Williams, J.G. 1963. *A field guide to the birds of East and Central Africa*. London: Collins.
Williams, L.O. & Gilbert, N.W. 1958. *Report of plant exploration to Belgian Congo, Angola and South West Africa: January – May 1958*. Beltsville, MD: United States Department of Agriculture, Agricultural Research Service, Crops

Research Division, New crops Research Branch. Typescript. https://www.biodiversitylibrary.org/bibliography/158793 (accessed 7 December 2021).

Wissenbach, M.C.C. 2011. As feitorias de urzela e tráfico de escravos: Georg Tams, José Ribeiro dos Santos e os negócios da África centro-ocidental na década de 1840. *Afro-Ásia* 43: 43–90. https://doi.org/10.9771/aa.v0i43.21220

Wissmann, H. [von] 1889. *Unter deutscher Flagge quer durch Afrika von West nach Ost.* Berlin: Verlag von Walther & Apolant.

Wissmann, H. von, Wolf, L., François, C. von & Mueller, H. 1891. *Im Innern Afrikas*, 3rd ed. Leipzig: F.A. Brockhaus. https://doi.org/10.5479/sil.314286.39088000808436

Witherby, W.F. 1943. Obituary: Rear-Admiral Hubert Lynes, R.N., C.B., C.M.G. (1874–1942). *British Birds* 36 (June 1942–May 1943): 156–158.

Wongtschowski, B. 2003. *Between Woodbush and Wolkberg: Googoo Thompson's story,* 3rd ed. Pretoria: Protea Book House.

Wright, C.H. 1913. *Moraea revoluta*, C.H. Wright [Iridaceae–Moraeeae]. Pp. 305–306 in: XLVIII.—Diagnoses Africanae—LV. *Bull. Misc. Inform. Kew* 1913: 299–307. https://doi.org/10.2307/4118441

Wurmser, R. 1971. Notice sur la vie et l'oeuvre de Henri Humbert (1887–1967). *Notices et Discours Acad. Sci. Paris* 5: 662–672. https://www.academie-sciences.fr/pdf/eloges/humbert_cr1968.pdf

Young, R.N.G. 1933. An account of a botanical excursion through northern Angola. *S. African Biol. Soc. Pam.* 6: 25–32.